解州关帝庙组群正立面图
Front elevation of the groups of the Haizhou Guandi Temple

解州关帝庙雉门前院横剖面图
Cross-section of the front yard of the Zhi Gate of the Haizhou Guandi Temple

解州关帝庙午门前院横剖面图

Cross-section of the front yard of the Meridian Gate of the Haizhou Guandi Temple

解州关帝庙御书楼前院横剖面图
Cross-section of the front yard of the Yushu Storied Building of the Haizhou Guandi Temple

解州关帝庙崇宁殿前院横剖面图

Cross-section of the front yard of the Chongning Hall of the Haizhou Guandi Temple

0 1 2   4   5m

解州关帝庙春秋楼前院横剖面图
Cross-section of the front yard of the Chunqiu Storied Building of the Haizhou Guandi Temple

解州关帝庙组群纵剖面图（一）
Longitudinal section I of the groups of the Haizhou Guandi Temple

解州关帝庙组群纵剖面图（二）
Longitudinal section II of the groups of the Haizhou Guandi Temple

解州关帝庙组群侧立面图
Side elevation of the groups of the Haizhou Guandi Temple

结义园组群正立面图

Front elevation of the groups of the Jieyi Garden

结义园组群纵剖面图
Longitudinal section of the groups of the Jieyi Garden

结义园组群侧立面图
Side elevation of the groups of the Jieyi Garden

常平关帝庙山门前院横剖面图
Cross-section of the front yard of the Changping Guandi Temple's Mountain Gate

比例尺: 0 0.5 1 2m

常平关帝庙组群纵剖面图
Longitudinal section of the groups of the Changping Guandi Temple

常平关帝庙组群侧立面图
Side elevation of the groups of the Changping Guandi Temple

Traditional Chinese Architecture Surveying and
Mapping Series:
Shrines and Temples Architecture

HAIZHOU GUANDI TEMPLE

Compiled by School of Architecture, Tianjin University &
Guandi Temple Cultural Relics Preservation Bureau in Haizhou, Shanxi
Chief Edited by WANG Qiheng
Edited by GUO Huazhan, WANG Qiheng

China Architecture & Building Press

国家出版基金项目

「十二五」国家重点图书出版规划项目

中国古建筑测绘大系·祠庙建筑

解州关帝庙

天津大学建筑学院 山西解州关帝庙文物保管所 合作编写

王其亨 主编

郭华瞻 王其亨 编著

中国建筑工业出版社

Contents

目　录

Project Information

Location: Wuyi Road, Haizhou Town, Yanhu District, Yuncheng City, Shanxi Province

Construction Time: the Song Dynasty

Area: 11.5 hectares

Responsible Unit: Heritage Management Institute of Haizhou Guandi Temple, Shanxi

Surveying and Mapping Unit: School of Architecture, Tianjin University

Surveying and Mapping Time: 2008

项目信息

地　　址　山西省运城市盐湖区解州镇五一路

始建年代　宋代

占地面积　11.5 公顷

主管单位　山西解州关帝庙文物保管所

测绘单位　天津大学建筑学院

测绘时间　2008 年

Preface

Haizhou Guandi[①] Temple is the temple in memory of the famous general GUAN Yu of Shu Kingdom during Three Kingdoms Period. It is located outside the west gate of Ming and Qing dynasties' Haizhou city site, which now belongs to Haizhou town, Yanhu District, Yuncheng City, Shanxi Province. The temple faces Zhongtiao Mountain on the south and is backed by the salt ponds and saltpeter ponds of Haizhou. After the first construction, Haizhou Guandi Temple had gone through a progressive development in the Song and Yuan dynasties and had finally become the so-called Ancestral Temple of Guandi (*Guandi Zu Miao*) in the Ming and Qing dynasties. The whole temple complex covers an area of 11.5 hectares and contains two separate complexes, namely Guandi Temple and Jieyi[②] Garden. Guandi Temple was designed to a temple-palace style that imitated the imperial palaces to match GUAN Yu's deity name—Guandi (GUAN Yu the God). While Jieyi Garden, to the south of Guandi Temple, focused mainly on the story of Oath of the Peach Garden, when GUAN Yu was not as illustrious and influential as later. The theme of the whole Haizhou Guandi Temple is presenting GUAN Yu's life achievements and enduring cultural values. Changping Guandi Temple, on the other hand, is located to the south of the salt ponds in Changping village. This temple covers an area of 1.3 hectares. It is said that GUAN Yu was born in Changping village and the temple was built on the site of GUAN Yu's old residence—the Ancestral Pagoda (*Zuzhai Ta*) in Changping Guandi Temple is believed to be built on the well where GUAN Yu's parents drowned themselves. As the temple is also used to offer sacrifice to GUAN Yu's ancestor GUAN Longpang, people also call it the Family Temple of Guandi (*Guandi Jia Miao*).

The grand Haizhou Guandi Temple is a materialization of the belief in GUAN Yu, which had gradually flourished since the Northern Song Dynasty and reached its peak in the Ming and Qing dynasties. At the same time, it is also the epitomization of the performance, production and reproduction of righteousness (*Yi*), the core value of traditional Chinese society; and the cultural relic that core political ideas, such as great unification, respecting authority and despising hegemony, which have continued from ancient times to today in China, are embedded in. After the development during the Song and Yuan dynasties and the management of the Qing Dynasty, Haizhou Guandi Temple has became the temple with the largest scale, the highest level, the longest history, the most complete layout and the richest cultural connotations among all temples dedicated to GUAN Yu

导　言

解州关帝庙是纪念三国时期蜀汉名将关羽的本庙，位于山西省运城市盐湖区解州镇（明清解州城址西门外）。解州关帝庙南面中条山，背靠解州盐池和硝池，肇建以后，历经宋元时期的踵事增华，至明清时期发展成为规模宏大甲天下的关帝祖庙。整个建筑组群占地 11.5 公顷，由关帝庙和结义园两个组群组成。其中关帝庙采用庙城形制，模拟帝王宫室以副关羽帝号之尊，结义园位于关帝庙南侧，着重表现关羽身为布衣时与刘备、张飞桃园结义之情状。整个组群以表现关羽生平功业和恒久文化价值为主旨。常平关帝庙位于盐池南侧的常平村，整个建筑组群占地 1.3 公顷。故老相传，该村为关羽出生之地，该庙基址原为关羽祖宅，现存的祖宅塔下为关羽父母投井葬身之地，且庙内奉祀关羽始祖关龙逢，故也被称为「关帝家庙」。

规模宏大的解州关帝庙是北宋以来逐渐兴盛、至明清时期臻于极盛的关羽信仰的物化表现：同时也是「义」这一中国传统社会核心价值观的表现、生产和再生产的最集中

all over the world. It is a masterpiece of Chinese ancient temples. In 1988, the Sate Council listed Haizhou Guandi Temple in the third batch of Major National-level Protected Historical and Cultural Sites. In 2012, building complexes of Guansheng (GUAN Yu the sage) were on the tentative list of China's World Cultural Heritages.

1. Guandi as a god

GUAN Yu got the posthumous title *Zhuangmiu*[③] in 260, the third year of Shu Kingdom's Jingyao Period, Three Kingdoms Period. In the Tang Dynasty (618-907), GUAN Yu gradually gained attention from the government, but it was in the Northern Song Dynasty (960-1127) that he was finally deified by the government. The Northern Song Dynasty is a period when Taoism flourished. Taoist priests of that time said that GUAN Yu had defeated Chiyou, an ancient tribe leader who died and became a monster in salt ponds of Haizhou, and rebuilt the salt ponds to normalize the salt tax. This saying coincided with the words in *Book of Rites: Sacrificial Rites* (*Li Ji*, one of the Confucian classics): "the wise king of ancient times stipulated that sacrificial ceremonies should be given to those who ruled people by law, who died from overwork on royal affairs, who brought peace and stability to the country, who prevented disasters and who saved people from tragedies…These are all people who have made contributions to the public (therefore are worth sacrificing)." It is obvious that what GUAN Yu had done about salt ponds was beneficial to people, so he was deified and awarded as *Chongning Zhenjun*[④]. It is said that the belief in GUAN Yu had started to expand since he got this title and as a consequence, Haizhou Guandi Temple got the chance to develop. Gradually, this temple has become one of the most important traditional Chinese culture relics.

In most cases, gods worshipped in temples would be forgotten by people or even be replaced by a new god as time went by. However, in the case of GUAN Yu, the truth is that not only was he not forgotten, but also the posthumous titles of him became more and more respectable over time. As mentioned above, the tradition of conferring titles on GUAN Yu started in the Song Dynasty. In 1102, the first year of Chongning period, and 1108, the second year of Daguan period in the Northern Song Dynasty, GUAN Yu was entitled *Zhonghui Gong*[⑤] and *Wuan Wang*[⑥] respectively by Emperor Huizong; while in 1187, the fourteenth year of Chunxi period in the Southern Song Dynasty, Emperor Xiaozong put all GUAN Yu's titles together and made it *Zhuangmiu Yiyong Wuan Yingji Wang*[⑦]. During the Yuan Dynasty, this title was slightly changed: in 1328, the first year of Tianli period, Emperor Wenzong changed GUAN Yu's title into *Xianling Yiyong Wuan Yingji Wang*[⑧]. The Ming Dynasty witnessed a great change in GUAN Yu's posthumous titles. In 1368, the first year of Hongwu reign, Hongwu Emperor resumed GUAN Yu's old title *Hanqianjiangjun Shoutinghou*[⑨]. While in 1531, the tenth year of Jiajing reign, this title was revised by Emperor Jiajing to *Hanqianjiangjun*

曾被列入中国世界文化遗产预备名单。

中国古代祠庙建筑的杰作，1988年被国务院列入第三批『全国重点文物保护单位』，2012年，『关圣文化建筑群』

尤其是清代的经营，成为海内外规模最宏大、等级最高、历史最久、格局最完整、文化内涵最丰富的关庙，是

代表；更是中国延续至今的『大一统』和『尊王贱霸』等核心政治理念的文化堡垒。历经宋元以来历代的发展、

一、关帝其神

继三国蜀汉后主景耀三年（260年）关羽获谥号『壮缪』之后，关羽之神从唐代开始进入国家视野，至北宋正式成为获得国家褒封的正神。在特别崇奉道教的北宋，道士所宣扬的『关羽破蚩尤，复盐池，从而使盐税恢复』的神话恰与儒家经典《礼记⊙祭法》中所规定的祭祀原则相吻合：『夫圣王之制祭祀也，法施于民则祀之，以勤死事则祀之，能御大灾则祀之，能捍大患则祀之……皆有功烈于民者也。』显然，关羽因盐池生产而于国有功，并获得了『崇宁真君』的封号。相传得自皇帝的这一封号使得关羽信仰开始发展起来，解州关帝庙也因此获得了发展，逐渐成为中国传统文化的重要圣地。

一般情况下，祠庙之神会因年远代湮而逐渐被淡忘，甚至被新的神取而代之。但是关羽之神的封号却随着时代发展不断加崇：北宋徽宗崇宁元年（1102年），封『忠惠公』；大观二年（1108年），封『武安王』，至南宋孝宗淳熙十四年（1187年），累加封号至『壮缪义勇武安英济王』；元文宗天历元年（1328年），封为『显

Hanshoutinghou [10]. In 1590, the eighteenth year of Wanli reign, the title of GUAN Yu was completely changed by Emperor Wanli into *Xietian Huguo Zhongyi Di* [11]. This is the first time that GUAN Yu was honored as Di, which means gods in Chinese. In 1614, the forty-second year of Wanli reign, Emperor Wanli granted GUAN Yu once more a longer title *Sanjiefumodadi shenweiyuanzhentianzun Guanshengdijun* [12], which proved that the belief in GUAN Yu had flourished at that time. In the Qing Dynasty, GUAN Yu enjoyed a more widespread reputation. In 1652, the ninth year of Shunzhi reign, the first emperor of the Qing Dynasty, Emperor Shunzhi, named GUAN Yu *Zhongyishenwu guanshengdadi* [13]. In 1703, the forty-second year of Kangxi reign, Emperor Kangxi rewarded Haizhou Guandi Temple a horizontal board inscribed with four Chinese characters "义炳乾坤", which means the righteousness famous all over the world, written by himself. In 1725, 1726 and 1732, that is, the third, forth and tenth years of Yongzheng reign respectively, Emperor Yongzheng bestowed the hereditary official positions, *Wujingboshi* [14], on GUAN Yu's descendants in Luoyang, Haizhou and Dangyang. In 1760, the twenty-fifth year of Qianlong reign, Emperor Qianlong revised GUAN Yu's posthumous title *Zhuangmiu* into *Shenyong* [15]. In 1853, the third reign year of Emperor Xianfeng, the sacrifices to GUAN Yu were promoted to middle-level sacrifices (*Zhongsi*), same as the sacrifices to Confucius. Four years later, in 1857, Emperor Xianfeng granted Haizhou Guandi Temple a horizontal board inscribed with "万世人极" written by himself, which means a person more outstanding than people of a generation. In 1879, the fifth reign year of Emperor Guangxu, another horizontal inscribed board with "佐天育物" written the emperor, which means assisting the emperor and caring the country, was given to Haizhou Guandi Temple. The same year, GUAN Yu got his highest-ranking posthumous title *Xuandeyizan Jingchengsuijing Baominhuguo Weixianrenyong Zhongyishenwulingyou Guanshengdadi* [16], which contains twenty-six Chinese characters in total.

Apart from the respect from the ruling class, GUAN Yu also owned his reputation in religious areas. In the Sui and Tang dynasties, GUAN Yu was regarded as samghārāma (*Qielan*), the guardian god of Buddhism. No later than the Yuan Dynasty, he was also deemed as the guardian god of Taoism. In the Ming and Qing dynasties, with the development of the trade in Shanxi Province, GUAN Yu was further regarded as the God of Wealth (*Caishen*). It can be seen that GUAN Yu's roles as god were enriched with time went by, and his functions as god also increased. In the Qing Dynasty, with the development of the society and the enrichment of the belief in him, GUAN Yu was honored as sage (*Shengren*), the same status as Confucius, and was entitled the Confucius of Shanxi Province (*Shanxi Fuzi*) and the Sage of Valor (*Wusheng*) for his extraordinary achievements. During the Republic of China era, GUAN Yu, along with YUE Fei, a famous general of the Song Dynasty, became the main worshiped gods of the temple of military—the Guanyue Temple (Fig.1, Fig.2).

The names of the temples dedicated to GUAN Yu also changed along with the change of his godhood.

灵义勇武安英济王』；明洪武元年（1368年），复其故封，为『汉前将军汉寿亭侯』；嘉靖十年（1531年），订正其封号为『汉前将军汉寿亭侯』；至万历十八年（1590年），为『协天护国忠义帝』，此为关羽帝号之始；万历四十二年（1614年），加封为『三界伏魔大帝神威远震天尊关圣帝君』，这一封号也证明了关羽信仰在此时已相当兴盛；清顺治九年（1652年），封关羽为『忠义神武关圣大帝』；康熙四十二年（1703年），御书『义炳乾坤』匾额；雍正三年（1725年）、四年（1726年）、十年（1732年）分别将关帝洛阳、解州及当阳后裔授为五经博士，世袭罔替；乾隆二十五年（1760年），乾隆皇帝专门改其谥号『壮缪』为『神勇』；咸丰三年（1853年），关帝春秋祭祀正式升入中祀，得与孔子并列；咸丰七年（1857年），为解州关帝庙御书『万世人极』匾额；光绪五年（1879年），颁解州关帝庙匾额曰『佐天育物』，同年，关帝封号达到极致，为『宣德翊赞精诚绥靖保民护国威显仁勇武灵佑关圣大帝』，共26字。

除了受到国家的不断加封尊崇外，关羽还在隋唐时期即成为佛教护法伽蓝，至迟至元代又成为道教护法神。明清时期又随着晋商的发展而成为财神，神的职能进一步丰富。清代，随着社会的发展和关羽信仰内涵的不断丰富，关羽进一步被尊为『圣人』，成为与孔子并列的『万世人极』的『山西夫子』『武圣』；至民国时期，关羽还与岳飞一起成为武庙『关岳庙』的主神（图一、图二）。

关庙的名称亦随着关羽神格的变化而变化：唐代，已有『壮缪侯庙』[22]之称；宋元时期兴建的多称关王庙，明后期才开始出现『关帝庙』这一称呼，清代所建的则普遍称关帝庙。因解州是关羽的故乡，解州庙是关公信仰的发源地之一，更是关公信仰的中心，因此，解州关庙被称作解州关帝庙，简称为解庙；又因其与其余关帝庙之间的分香关系而被尊称为关帝祖庙。

图一 解州关帝庙万代瞻仰坊石刻封号

图二 解州关帝庙崇宁殿『万世人极』匾

Fig.1 Inscriptions of GUAN Yu's titles on Wandai Zhanyang Memorial Archway⑰, Haizhou Guandi Temple
Fig.2 Horizontal board inscribed with "Wanshi Renji" hung in the Chongning Hall

In the Tang Dynasty, the name Zhuangmiuhou Temple[⑧] already existed. The temples dedicated to GUAN Yu built in the Song and Yuan dynasties were mainly called Guanwang Temple (Temple of GUAN Yu the king). The name Guandi Temple (Temple of GUAN Yu the god) appeared in the late Ming Dynasty and was popularized in the Qing Dynasty. The temple of GUAN Yu in Haizhou got the name Haizhou Guandi Temple (sometimes abbreviated as Hai Temple), because Haizhou is the hometown of GUAN Yu and one of the origins and the center of belief in GUAN Yu. Considering its relationship with other Guandi Temples in terms of offerings, people sometimes also address this temple the Ancestral Temple of Guandi (*Guandi Zu Miao*).

2. Historical Evolution

(1) The Development in the Song, Jin and Yuan Dynasties

It is said that Haizhou Guandi Temple already existed in the Chen (557-589) and Sui dynasties (581-618). By then, as the government had not recognized GUAN Yu as a god, detailed information of the temple remained unknown. According to the records, the earliest known large-scale construction of Haizhou Guandi Temple took place in 1014, the seventh year of Dazhongxiangfu period in the Northern Song Dynasty. During the construction, a new temple was rebuilt based on the former temple sites. Thus, it can be inferred that Haizhou Guandi Temple had existed at least before the year 1014. In 1092, the seventh year of Yuanyou period in the Song Dynasty, Haizhou's local government renovated the temple again. The reconstruction and renovation above were both carried out under edicts given by emperors of the time, indicating the degree to which GUAN Yu was valued as a god by the government. A few years after the renovation, Emperor Huizong of the Northern Song Dynasty gave GUAN Yu the title *Zhonghuigong* in 1102, the first year of Chongning period, and the title *Wuanwang* in 1108, the second year of Daguan period. Since GUAN Yu became a senior god in the sacrificial system of the country, Haizhou Guandi Temple had become a significant sacrificial temple in China.

In the Jin and Yuan dynasties, several renovation projects took place in Haizhou Guandi Temple. During the Jin Dynasty, there were renovations both in 1163, the third year of Dading period, and 1204, the forth year of Taihe period. During the Yuan Dynasty, in 1266, the third year of Zhiyuan period, JIANG Shanxin, the head Taoist priest of Haizhou Guandi Temple, built a residence for the priests, which is later renamed the Chongning Palace[⑨], in the east of Haizhou Guandi Temple (the site of which still remains now). In 1303, the seventh year of Dade period, a strong earthquake struck Haizhou Guandi Temple and cause great damages. It is recorded that a man named ZHANG Zhian renewed the temple after the earthquake. Soon after, a large-scale renovation project was held again from 1312, the first year of Huangqing period, to 1323, the third year of Zhizhi period in the Yuan

二、历史沿革

（一）宋金元时期的发展

相传，解州自陈、隋（581—618年）时期即有关庙，但此时关羽之神尚未受到国家的正式承认，该庙具体形制亦不详。据可考的历史，可见，该庙已知最早的较大规模重建发生于北宋真宗大中祥符七年（1014年），是在前代庙宇基础上重修而成，可见，解州关庙的历史要早于这一时期；宋元祐七年（1092年），地方官重修。这两次重修均为奉诏敕修重修，充分体现了关羽之神获得国家重视的程度。至北宋徽宗崇宁元年（1102年），关羽获封『忠惠公』；大观二年（1108年），晋封『武安王』。自关羽正式成为国家祠祀系统中的高级神之后，解州关庙也成为重要的祠庙建筑。

金元时期，解州关庙屡次修缮。金大定三年（1163年）、泰和四年（1204年），均曾重修；元至元三年（1266年），住持道士姜善信于庙东侧创建了道士居住的道院，即后来的崇宁宫；大德七年（1303年），解州关庙受到大地震的波及而损坏，『提点崇宁宫张志安』撤而新之，皇庆元年（1312年）至至治三年（1323年），解州关庙再次较大规模重修；泰定元年（1324年），重修道院崇宁宫；泰定二年（1325年）至元统三年（1335年），解州关庙由蔡荣重修。

Dynasty. In 1324, the first year of Taiding period, the Taoist monastery Chongning Palace was rebuilt. Later, from 1325, the second year of Taiding period, to 1335, the third year of Yuantong period, Haizhou Guandi Temple was renovated again under CAI Rong's charge.

(2) The Apex in the Ming Dynasty

The development of GUAN Yu's image as a god and Haizhou Guandi Temple went to its peak in the Ming Dynasty. In 1425, the first year of Hongxi period, a Bell Tower was built in Haizhou Guandi Temple and the peripheral columns of the temple's main hall were changed . In 1478, the fourteenth year of Chenghua period, the eight peripheral columns in the front of the Chongning Hall were changed into stone columns and two ornamental columns (*Biao*) were erected on both sides of the main gate. In 1490, the third year of Hongzhi period, PU Zhao, a student of the Impeiral Academy (*Guozijian*), proposed to hold official sacrificial ceremonies in Haizhou Guandi Temple in springs and autumns every year. Since then, the sacrificial ceremonies of Haizhou Guandi Temple were among the official sacrificial ceremonies list. During the same period, a U-shape stage named Kaiyan Pavilion was built, the south part of which was demolished later in the Zhengde period (1506-1521); and the east and west wings were desolated. From 1510 to 1514, the fifth to the ninth year of Zhengde period, the residential hall of Haizhou Guandi Temple was renovated. From the above records, it is obvious that until Zhengde period there was a main gate, two ornamental columns and a stage in the front of Haizhou Guandi temple, a main hall in the center, and a residential hall in the rear. In other words, a typical Chinese sacrificial-temple layout, consisting of an Outer Court (*Qianchao*, which usually consists of the main hall of the complex) with practical functions and an Inner Court (*Houqin*, which usually contains the residential hall of the complex) with recreational functions, was already formed at that time. While the stage building located at the central axis showed strong regional characteristics.

In 1523, the second year of Jiajing period, the complex of Haizhou Guandi Temple was further improved, resulting in a complex consisting of gate walls, archways, an offering hall, a main hall, a residential hall and covered corridors. Meanwhile, as the entrance of the temple was fairly close to the avenue leading to the west gate of Haizhou City, on which many people would walk, a partition wall with two brick gates on the east and west sides was set up in the front of the temple to keep irrelevant people out and to extend the frontal part of the temple. In 1524, the third year of Jiajing period, the head Taoist priest of that time, YANG Yancheng, renovated the Chongning Palace. In 1530, the ninth year of Jiajing period, another reconstruction project was carried out, in which change of archways (*Paifang*) and the screen wall (*Yingbi*) were mentioned. The existence of archways, the screen wall and the brick gates proved the development of the frontal part of the temple in Jiajing period. Twenty-five years later, in the thirty-fourth year of Jiajing period, Haizhou Guandi Temple was destroyed

（二）明代的高峰

明代是关羽神格发展的高峰时期，也是解州关帝庙格局发展的高峰时期。洪熙元年（1425 年），创建钟楼台砌，更换正殿檐柱；成化十四年（1478 年），改崇宁殿前檐柱为 8 根石柱，并树表于门之左右两侧；弘治三年（1490 年），郡人监生蒲昭奏准春秋祭祀，解州关庙正式进入地方祀典；弘治时期，解州关庙内建戏楼「开颜楼」，该戏楼的南楼于正德（1506—1521 年）时期撤掉，东西两楼亦年久失修；正德五年（1510 年）至九年（1514 年），重修寝宫。至此，可以知道，正德年间，解州关庙前有大门及表、戏楼，中有大殿，后有寝宫，已经形成祠庙建筑典型的前朝后寝格局，并且中轴线上的戏楼较为突出，表现出了较强的地方色彩。

嘉靖二年（1523 年），形制进一步发展，组群已有门墙、坊牌、正殿、寝宫、行廊等建筑。同时，因该庙门前紧临出解州城西门的大道，往来杂沓，遂于庙前东西两侧各建砖门一座，「使非有事于庙者，则往来于屏墙之外」，发展了庙的前部格局。嘉靖三年（1524 年），时任道正杨衍澄增修崇宁宫；嘉靖九年（1530 年），重修该庙时已提及「坊壁台砌」，说明除东西砖门外，还有牌坊及影壁，庙制进一步发展。嘉靖三十四年（1555 年），河东大地震，解州关庙摧圮殆尽，但关羽坐像却奇迹般地完整保存下来；嘉靖三十五年至三十七年（1556—1558 年）重修午门、正殿、寝殿、乐楼等，仍采用旧有的规模制度；嘉靖四十四年（1565 年），重修开颜楼之东西二楼，并在原南楼的位置建牌坊一座。此时，解州关庙规制已有较大发展，南部中轴线上有影壁、屏墙，左右两侧是砖门；中部有大门、午门、牌坊，牌坊北侧是东西二戏楼及献殿，再往北是正殿崇宁殿，崇宁殿后为寝宫；两侧为行廊。

completely in the Hedong earthquake, leaving only the statue of GUAN Yu. From 1556 to 1558, namely the thirty-fifth year to the thirty-seventh year of Jiajing period, the Meridian Gate (*Wumen*), the main hall, the residential hall and the stage building were rebuilt consecutively, following the former buildings' scales and appearances. In 1565, the forty-fourth year of Jiajing period, the west and east wings of Kaiyan Pavilion, the stage building, were renovated again, while an archway was added in where the former south part stood. Till then, Haizhou Guandi Temple has basically formed the layout that remained up to now. On the southern part of the central axis, there was the partition wall, on both ends of which stood brick gates, and the screen wall; on the middle part, there was the Meridian Gate, archways, the stage building with wings and the offering hall; on the northern part, there was the Chongning Hall, the residential hall; and there were covered corridors on the eastern and western sides of the central axis.

The development of Haizhou Guandi Temple went to its peak in Wangli period, the Ming Dynasty. In 1573, the first year of Wangli period, Haizhou Guandi Temple was awarded a horizontal inscribed board with "英烈", which means heroic, on it. Meanwhile, it is said in the archives that the Linjing Pavilion (also called Chunqiu Storied Building[20]), which had twenty-eight columns and was nine *zhang* in height, was built during this period, along with the two wing buildings and the seventy-four-bay covered corridor. Although the two wing buildings had not been named as the Dao Storied Building and the Yin Storied Building[21] yet, the layout of the residential Inner Court, in which the Chunqiu Storied Building located in the center, the Dao and Yin Storied Buildings located to the east and west, while the covered corridor encircled the others as a barrier, had formed. Soon afterwards, the Bell Tower and the Drum Tower were built in the east and west of the temple's front, perfecting the layout of the southern part. It is also recorded in the archives that the lowland in the northern part of the temple was made even to be in accord with the ancient Chinese architecture planning tradition that the northern (rear) part of the complex should be the same as or higher than the southern (front) part. In general, the current layout of Haizhou Guandi Temple had already formed in Wanli period, just as people of that time said: "the scale and layout of Haizhou Guandi Temple now has achieved its peak, and looks virtually like an imperial palace". After the Wanli period, there were still small-scale improvements in the temple, but the overall layout remained basically unchanged. For example, from 1592, the twentieth year of Wanli period, to 1593, the twenty-first year of Wanli period, the exterior space of the temple was ornamented with iron statues, incense burners and lion statues; from 1620, the forty-eighth year of Wanli period, to 1621, the first year of Tianqi period, the Wandai Zhanyang Memorial Archway and other facilities were built (Fig.3).

Beside the development of layout, another monumental event for Haizhou Guandi Temple in Wanli period was the establishment of the Jieyi Garden. From 1620, the forty-eighth year of Wanli period,

万历时期，解州关帝庙的发展进入高峰。万历元年（1573年），解州关庙获赐额『英烈』，与此相应，『建

麟经阁二十八楹，高九丈，翼以二楼，廊七十四。』此时虽尚无刀楼，印楼居中，印楼、刀楼

左右护翼的格局已经形成，且通过周围行廊发展了寝宫后部的格局；不久之后，增建庙前部东西门钟鼓楼，使

庙前部格局发展完善；又因『嫌北势渐下，因覆土为山负宸，长三百，高称之。』于春秋楼后覆土为山，弥补

了解州关帝庙所在地势南高北低这一自然缺陷，也更符合中国传统建筑空间布局的一般特点。综观此时庙制，

确如时人所说：『规制大成，俨然王居矣』，充分奠定了今日所见解州关庙格局的基础。此后，虽重修不断，

但多属于恢复旧制或踵事增华，对关庙总体格局影响不大。如万历二十年至二十一年（1592—1593年），增置

铁像、香炉、狮子等室外陈设，万历四十八年（1620年）至天启元年（1621年）增建万代瞻仰坊等（图三）。

除庙制方面的发展外，万历时期的另一项发展是结义园的创建。万历四十八年至天启元年（1620—1621年），

于庙南侧创建莲池，即后来的结义园，建『山雄水阔』坊、莲亭、老子庙并官厅，道院，形成解州关庙增色的

小型园林。

图四 解州八景之「汉宫桧柏」图（自康熙十二年《解州志》）

图三 解州关帝庙春秋楼关羽坐像

Fig.3　Statue of GUAN Yu in the Chunqiu Storied Building, Haizhou Guandi Temple

Fig.4　One of the eight views of Haizhou: Cypresses Dated Back to the Han Dynasty (From *Haizhou Local Records*, published in the twelfth year of Kangxi period)

to 1621, the first year of Tianqi period, a lotus garden was built to the south of Haizhou Guandi Temple, which was turned into the Jieyi Garden later. In this garden, there was the Shanxiong Shuikuo Memorial Archway[22], the Lotus Pavilion, the Laozi Temple[23], the official hall and Taoism monasteries. The garden was as a credit to Haizhou Guandi Temple.

(3) The Improvement in the Qing Dynasty

In the Qing Dynasty, as the government concentrated more and more on the worship for GUAN Yu (Fig.4), the layout and the form of Haizhou Guandi Temple further improved. From 1683, the twenty-second year of Kangxi reign, to 1684, the twenty-third year of Kangxi reign, the Qisheng Shrine[24] of the temple was reconstructed. In 1702, the forty-first year of Kangxi reign, a big fire caused serious damages to the temple. The next year, Emperor Kangxi visited Haizhou Guandi Temple during his journey to the western areas and awarded the temple a horizontal board inscribed with "义炳乾坤", which was now hung inside the temple, and subsidized the reconstruction of the temple with his own properties. The kindness of Emperor Kangxi and his belief in GUAN Yu made this a big event at that time.

In 1725, the third year of Yongzheng period, Emperor Yongzheng conferred the title Duke on GUAN Yu's father, grandfather and great-grandfather, and issued the edict to build the Chongsheng Shrine[25]. After the shrine was completed, Prince Guo[26] came and visited Haizhou Guandi Temple, painted a portrait of GUAN Yu by himself and composed a poem. From 1755 to 1758, namely the twentieth year of Qianlong Period to the twenty-third year of Qianlong period, the stage was moved to the inside of the Zhi Gate[27], which led to the increase of the height of the Zhi Gate; two archways, the Dayi Cantian Memorial Archway and the Jingzhong Guanri Memorial Archway[28], were built on either side of the Meridian Gate; the pond in front of the temple was dredged and the Junzi Pavilion[29] was added in the garden as well. In 1762, the twenty-seventh year of Qianlong reign, the former Bagua Pavilion was changed into the Yushu Storied Building[30]; the buildings in the front of Linjing Pavilion (that is, the Chunqiu Storied Building) were renamed the Dao Storied Building and the Yin Storied Building; the form of the garden in front of the temple was further improved and the garden got its official name—the Jieyi Garden (Fig.5, Fig.6). According to archives, figures of ZHOU Cang and LIAO Hua[31] were set up on both sides of the Meridian Gate; figure of YANG Yi was added in the Bujiang Shrine[32] to the east of the Zhi Gate, while figure of Chitu the horse was located in the Zhuifengbo Shrine[33] to the west of the Zhi Gate during this period. From 1773 to 1777, the thirty-eighth year to the forty-second year of Qianlong period, the Jiaozhong Hall[34], the Long Corridor, the Zhou Pavilion[35] and many other constructions were built in the Jieyi Garden, which improved the layout of the garden. In 1809, the fourteenth year of Jiaqing period, a Bell Pavilion was erected to the west of the main hall, the Chongning Hall; accessorial buildings of about thirty bays were built; and decorated walls

（二）清代的完善

清代，随着官方对关羽崇奉的不断加强，解州关庙的格局也进一步发展完善（图四）。康熙二十二年（1683—1684年），重修启圣祠；四十一年（1702年）四月，关帝庙遭火灾而毁坏严重；四十二年（1703年），康熙帝西巡途中谒庙，御书『义炳乾坤』匾额，悬于殿内并发内帑修复旧制，成为一时盛事。

雍正三年（1725年），敕封关公三代为公爵，和硕果亲王诣庙，亲画圣像并题诗；乾隆二十年至二十三年（1755—1758年），移建乐楼至雉门内侧，雉门随之增高，于午门两侧增建牌坊，即『大义参天』坊和『精忠贯日』坊；并疏浚庙前池塘，建君子亭；二十七年（1762年），改八卦楼为御书楼，又改麟经阁前楼为刀、印两楼，同时，进一步完善庙前园林格局，并正式命名其为『结义园』（图五、图六）；『周仓廖化』配立午门两旁，祠内增入功曹杨仪，仍为部将祠，西有追风伯祠，即亦兔马。乾隆三十八年至四十二年（1773—1777年），于结义园内新建教忠堂、长廊及舟亭等建筑，结义园格局发展完善。嘉庆十四年（1809年），增建解州关帝庙正殿西钟亭、大门外廊房三十余间并照壁两旁花墙，至此，解州关帝庙格局发展完善。

此后，该庙屡遭地震、火灾及战乱等天灾人祸。嘉庆十九年（1814年），解州关帝庙遭地震毁坏，不久后开始重修；咸丰五年（1855年），敕封关帝三代王爵，经咸丰时期的努力，至同治八年（1867年）始复旧观。不过，光绪三年（1877年）至二十三年（1897年），『义壮乾坤』坊、大门、乐楼、钟鼓楼皆遭火灾损毁，『午门、大门、乐楼、东西角门、东华、西华二门、钟楼及牌坊并庙内外廊房百余间、部将祠、追风伯祠、官厅、崇圣祠大门概成灰烬』。光绪二十三年（1897年），改结义园为第一模范国民学校。光绪三十三年（1907年）至宣统元年（1909年），将结义园改建为关岳庙，并增建官厅（图七）。

關聖廟圖

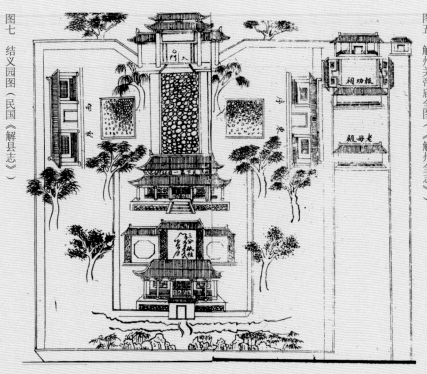

图七　结义园图（民国《解县志》）

图五　解州关帝庙全图（《解州全志》）

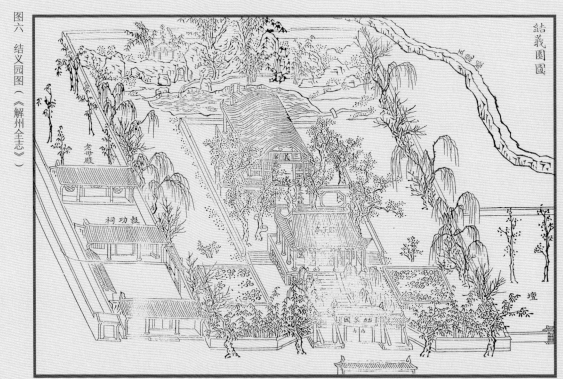

图六　结义园图（《解州全志》）

結義園圖

Fig.5　Panorama of Haizhou Guandi Temple (From *Overall Local Records of Haizhou*)
Fig.6　Panorama of the Jieyi Garden (From *Overall Local Records of Haizhou*)
Fig.7　Panorama of the Jieyi Garden (From *Haixian Local Records*, published in the Republic of China era)

were added on either side of the screen wall. Till then, the development of Haizhou Guandi Temple's layout and form had come to an end.

Afterwards Haizhou Guandi Temple was attacked by earthquakes, conflagrations and wars for many times. In 1814, the nineteenth year of Jiangqing period, an earthquake struck Haizhou Guandi Temple and caused great damages. Reconstruction started soon after the disaster. In 1855, the fifth year of Xianfeng period, the emperor bestowed the title King on GUAN Yu's father, grandfather and great-grandfather. Through the hard working in Xianfeng period, Haizhou Guandi Temple was almost restored as it had been in 1867, the eighth year of Tongzhi period. However, from 1877, the third year of Guangxu reign, to 1897, the twenty-third year of Guangxu reign, the Yizhuang Qiankun Archway[①], the main gate, the stage building, the Bell Tower and the Drum Tower were all damaged in conflagration. Records of the time said that the Meridian Gate, the main gate, the stage building, the east and west corner gates, the Donghua Gate and Xihua Gate[②], the Bell Tower, archways, accessorial buildings of hundreds of bays, the Bujiang Shrine, the Zhuifengbo Shrine, the official hall and the gate of the Chongsheng Shrine were all burnt to ashes. The whole Haizhou Guandi Temple was terribly damaged. In 1897, the twenty-third year of Guangxu period, the Jieyi Garden was turned into the First Exemplary People's School. From 1907, the thirty-third year of Guangxu reign, to 1909, the first year of Xuantong reign, it was further changed into the Guanyue Temple and an official hall was built in it (Fig.7).

(4) The Setback in the Republic of China era

The buildings to the south of Yushu Storied Building damaged in the Guangxu-period fire were gradually restored from 1915 to 1920. However, between 1923 and 1925, the Chongning Hall, the Chunqiu Storied Building, the Dao Storied Building, the Yin Storied Building, accessorial buildings of hundreds of bays and the east and west official halls were damaged again in the war, while the Shengmu Hall and the Sisheng Hall[③] were ruined. The restoring work continued from 1939 to 1940. However, in 1939 the official halls and storerooms were blown up again and restored immediately; and in 1947, the newly built Shengmu Hall and the Sisheng Hall were blasted again in war and consequently were removed. Since then, no more change had been carried out on buildings of Haizhou Guandi Temple, and the layout and the form were finally fixed. Almost all buildings were preserved, except for the Yizhuang Qiankun Archway, the Shengmu Hall, the Guanping Hall and the Guanxing Hall[④] (Fig.8).

3. Characteristics of the Layout and the Form of Haizhou Guandi Temple

Haizhou Guandi Temple is a sacrificial temple. Most sacrificial ceremonies hold in the temple aim to

（四）民国的多舛

中华民国 4 年至 9 年（1915—1920 年），清末火灾中损毁的御书楼以南的诸建筑获得恢复；民国 12 年至 14 年（1923—1925 年），崇宁殿、春秋楼、刀楼、印楼并廊房百余间，东西官厅等处再度于战事中受损，圣母殿、嗣圣殿损毁，民国 28 年至 29 年（1939—1940 年）陆续补修；民国 28 年（1939 年），官厅、官库再被炸坏，旋被修复。1947 年，圣母殿、嗣圣殿于战事中被炸毁拆除。至此，解州关帝庙于历史中定格下来，其中，除义壮乾坤坊、圣母殿及关平殿、关兴殿损毁后一直未能恢复外，其余部分都较完整地保存了下来（图八）。

三、形制特点

解州关帝庙属于祠庙建筑，祭祀活动以崇德报功为主题，并通过神道设教的方法生产和传播主神所代表的核心价值观。解州关帝庙即用来祭祀关羽，并进一步『表忠树义，扬烈报功』，其建筑形制堪称中国古代祠庙建筑的典范。

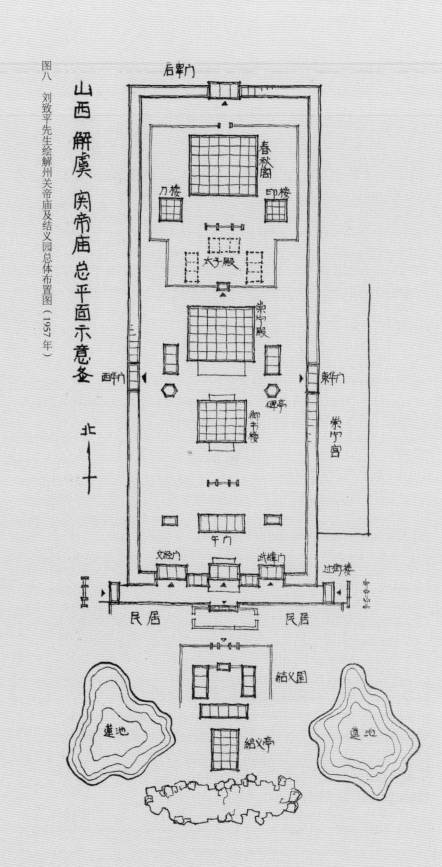

山西 解虞 关帝庙 总平面示意图

图八 刘致平先生绘解州关帝庙及结义园总体布置图（1957年）

Fig.8 Site plan of Haizhou Guandi Temple and the Jieyi Garden by LIU Zhiping (1957)

show respect to GUAN Yu and his virtues, and to found religions to help the production and spread of the core values of GUAN Yu, which are the loyalty and righteousness. With these considerations, the layout and the form of Haizhou Guandi Temple are exemplary among the traditional Chinese sacrificial temples.

(1) Architectural Layout of "Five Gates and Three Courts" and "Outer Court in the front and Inner Court in the Rear" [40]

Buildings of Haizhou Guandi Temple are arranged along the central axis from the south to the north. To the south of the main hall, the Chongning Hall, on the central axis, there stand three layers of gates: the Duan Gate to the south, the Zhi Gate, the Wenjing Gate and the Wuwei Gate in the middle and the Meridian Gate to the north [41]. In front of the Chongning Hall, there is the great platform (*Yuetai*) and ritual implements, such as bronze tripods (*Ding*) and bronze cranes. With the above buildings forming the Outer Court, there is an intact courtyard at the rear of the Chongning Hall, serving as the residential Inner Court: along the central axis, from the south to the north, there is the Niangniang Hall (ruined), the Guanping Hall (ruined) and the Guanxing Hall (ruined); the Qisu Qianqiu Memorial Archway [42], the Dao Stories Building and the Yin Storied Building; and the Chunqiu Storied Building, which is the highlight of the whole courtyard. Further north along the central axis, there is the Houzai Gate [43], which serves as the ending of the whole complex. The Zhi Gate in the front of the temple, the Houzai Gate in the rear of the temple, and many other gates, halls and storied buildings are connected by encircling covered corridors, forming a courtyard of the highest form in ancient China. In general, the overall layout of Haizhou Guandi Temple is an imitation of the typical layout of imperial palaces, which usually have "five gates and three courts", "outer court in the front and inner court in the rear" and "literary functions in the east and martial functions in the west", so as to match GUAN Yu's status as a great god.

(2) Layout and Form as a Temple-palace

More similarities between Haizhou Guandi Temple and imperial palaces could be found in their layouts. Same as imperial palaces, Haizhou Guandi Temple also starts with the Duan Gate and walls attached to it. The frontal area of the Zhi Gate, which was originally an outdoor space, was included by the temple and ornamented with the partition wall, brick gates, the screen wall and archways, to imitate the frontal plaza of the imperial city, and to emphasize the central courtyard in the north of the Zhi Gate. Moreover, the Donghua Gate and the Xihua Gate are located among the eastern and western accessorial buildings that encircle the main courtyard; while the Houzai Gate is located on the north end of the central axis, same as the gates of the imperial palaces.

（一）整体布局采用三朝五门、前朝后寝的宫室制度

解州关帝庙取南向为正方向，中轴线上正殿之南自南而北依次布置端门、雉门及文经门、武纬门、午门等三重、五座门殿，正殿崇宁殿前施高大的月台，陈列铜鼎、铜鹤等礼器；崇宁殿后单独形成一个完整的后寝院：沿中轴线从前至后依次布置娘娘殿（已毁）及关平殿（已毁）、关兴殿（已毁）、气肃千秋坊、刀楼、印楼、春秋楼则形成中轴线上后寝部分的高潮；再往北，后寝院外的中轴线上布置厚载门，形成整个建筑组群的北端收束。通过四周的回廊将自前部雉门起，至北端厚载门止的诸多门、殿、楼、阁联络形成一整个大院落，从而形成了中国古代最高形制的院落回廊形式，并充分模拟了帝王三朝五门、前朝后寝、文东武西的宫室制度，以充分与关帝的神格相应。

（二）采用庙城的规格形式

进一步看，解州关帝庙还整体上模拟了『宫城』的布局形式。首先，通过端门两侧的宫墙将宫城的形象展示出来；其次，雉门前部区域，原本是庙外的部分，经过明嘉靖时期增建屏墙、砖门、影壁和牌坊等一系列发展，不但成为庙制中重要的组成部分，而且进一步将雉门以北的核心回廊院落凸显出来，增加了空间层次；再次，核心院落的东西两廊上，当崇宁殿、御书楼之间分别设东华门、西华门，北侧轴线上设厚载门，这也是直接模仿宫城形制的典型表现。

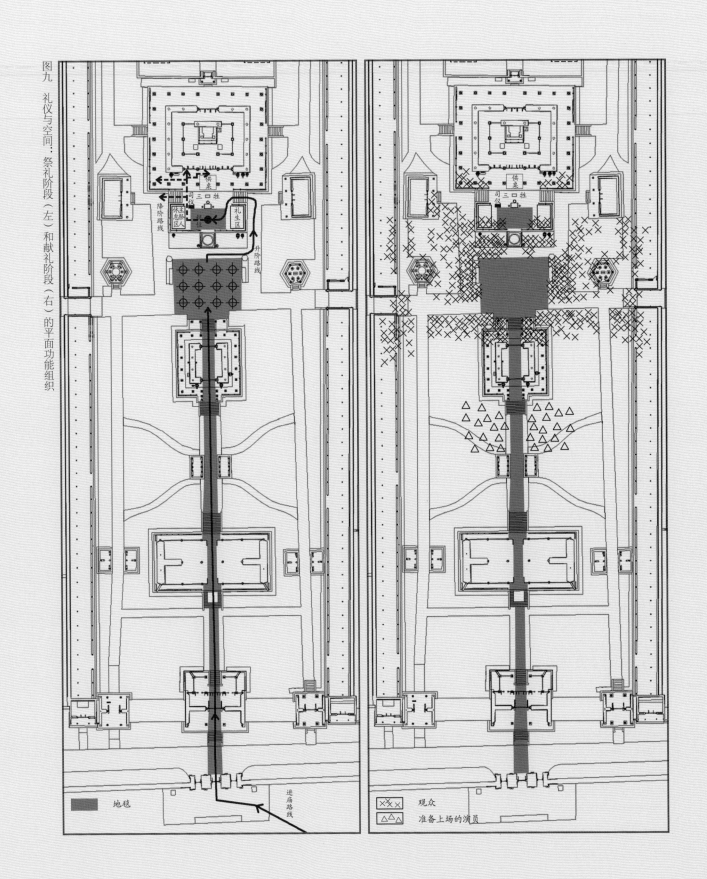

图九 礼仪与空间：祭礼阶段（左）和献礼阶段（右）的平面功能组织

地毯

进庙路线

观众

准备上场的演员

Fig.9 Etiquette and Space: Organization of the Space for Sacrificial Activities (Left) and for Offering Activities (Right)

(3) Central Courtyard Arranged Centered on the Needs of Sacrificial Ceremonies

Since 1490, the third year of Hongzhi period in the Ming Dynasty, sacrifices in Haizhou Guandi Temple had been included as part of the local official sacrificial ceremonies. Later, the level of sacrificial ceremonies in Haizhou Guandi Temple has become higher and higher with the rising importance of GUAN Yu's godhood. In 1853, the third year of Xianfeng period, the sacrificial level was finally promoted to middle-level sacrifices. Sacrificial ceremonies in Haizhou Guandi Temple are ritual activities held by the government that aim to enlighten people and inherit culture through retelling GUAN Yu's stories and sharing core values of these stories. The aims of the ceremonies determine that space in Haizhou Guandi Temple should be spacious and open for people, just like the main courtyard of Haizhou Guandi Temple is now. The main courtyard of Haizhou Guandi Temple consists of the Chongning Hall, the Yushu Storied Building and accessorial buildings both on east and west sides. In front of the Chongning Hall, a platform extends as the space for presenting offerings. In front of the platform, to give audiences enough space to attend the ceremonies, courtyard space between the platform and the Yushu Storied Building is left empty. This space can also be used to perform large-scale singing and dancing activities during *Sanxianli* (a special traditional sacrificial activity). To the north of the Yushu Storied Building on the central axis, there exists a temporary stage of regional features, which can be set up by adding wood panels when needed and torn down quickly. Apart from the above buildings, the east official hall and west official storeroom in the courtyard are preserved as accessorial buildings for the rest of officials and the storage of sacrificial vessels during ceremonies. What's more, to emphasize the main buildings on the central axis and the encircled core space, the Stele Pavilion, the Bell Pavilion and some other buildings are located symmetrically (Fig.9, Fig.10).

(4) Building Names Origins from the Ancient Chinese Core Virtues that GUAN Yu has—Loyalty, Righteousness, Benevolence and Courage

On the east-west axis in the front of the Zhi Gate stand the Yizhuang Qiankun Memorial Archway (ruined) in the east end and the Weizhen Huaxia Memorial Archway[49] in the west end (Fig.11, Fig.12). The abutments of the Bell Tower and the Drum Tower are both inscribed with four Chinese characters, " 关圣义起 ", which means the great righteousness of GUAN Yu the sage. The eastern and western bays of Duan Gate, as well as the east and west archways on either side of the Meridian Gate, are inscribed with " 大义参天 " and " 精忠贯日 "[49]. The horizontal board hung in the Chongning Hall was inscribed with Emperor Qianlong's handwriting " 神 勇 ", which means extremely brave (Fig.13). All these names suggest the traditional Chinese core virtues that GUAN Yu has, namely loyalty, righteousness, benevolence and courage.

（三）核心院落的空间组织围绕祭祀礼仪活动的需要进行

自明弘治三年（1490 年）解州关帝庙正式进入地方祀典以来，随着关羽神格的不断提高，其祭祀的级别也越来越高，清咸丰三年（1853 年），正式升入中祀。祭祀礼仪活动是一项官方举行的，参加的人数众多，并且是公开的文化活动，解州关帝庙的核心院落正是紧密围绕这一需要而展开的。该院落由崇宁殿、御书楼及东西两廊围合而成，轴线上于正殿崇宁殿前设大月台以展礼，月台南侧、御书楼北侧留出庭院空间，在祭祀时刻供各处前来参谒祖庙的团体驻足观礼，并可在『三献』礼环节供规模较大的歌舞活动使用；御书楼北侧轴线上附有一处搭板戏台，这是保留了原有地方色彩的表现。该院落还附有东官厅、西官库，是为参加礼仪活动的官员临时休息和存放重要祭器准备的，属于礼仪活动的附属建筑；还对称设有碑亭、钟亭等建筑，则起到进一步烘托主体建筑和围合核心空间的作用（图九、图十）。

（四）建筑题名紧密围绕关羽所代表的忠、义、仁、勇等中国古代核心价值观展开

雉门前的东西轴线上，东侧当路设置『义壮乾坤』坊（已毁），西侧当路设置『威震华夏』坊（图十一、图十二）；钟楼、鼓楼台砌题额为『关圣义起』；端门东、西次间门额及午门东西牌坊均分别题写『精忠贯日』；正殿崇宁殿前檐下挂乾隆御笔亲题的『神勇』匾（图十三）。所有这些文字符号均意在表现关羽所代表的忠、义、仁、勇等中国古代核心价值观。

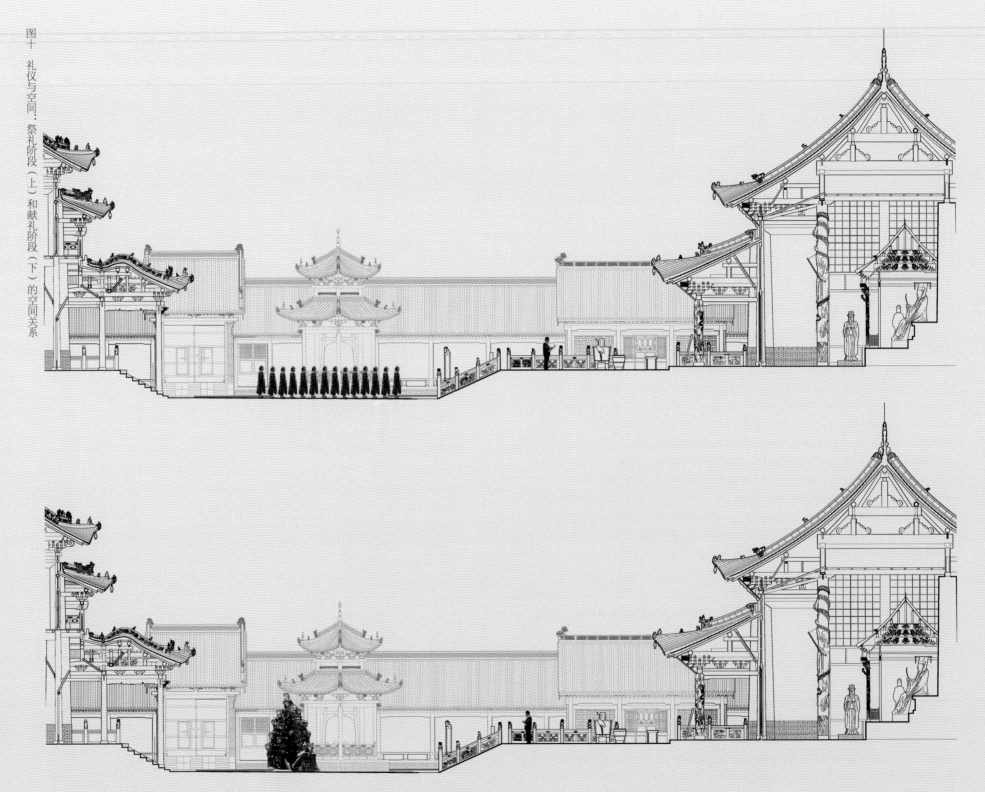

Fig.10 Etiquette and Space: the Spatial Relationship between the Sacrificial Activities (up) and for Offering Activities (down)

图十二　康熙御书『义炳乾坤』匾

图十三　乾隆钦定『神勇』匾

图十一　『威震华夏』坊

Fig.11　the Weizhen Huaxia Memorial Archway
Fig.12　Horizontal inscribed board with "义炳乾坤" written by Emperor Kangxi
Fig.13　Horizontal inscribed board with "神勇" written by Emperor Qianlong

图十四　鲍希曼 1907 年所拍的关羽坐像照片

Fig.14　Photo of the statue of GUAN Yu taken by Ernst Boerschmann in 1907

4. Conservation and Research in Modern Times

The administrative departments and scholars have paid great attention to Haizhou Guandi Temple for a long time. In 1907, the German Architect, Ernst Boerschmann, visited Haizhou Guandi Temple on his architecture-investigation journey in China and took the earliest-known photo of GUAN Yu's statue in the temple (Fig.14). When Japanese invaded and occupied Haizhou in 1941, photos of the Yushu Storied Building, the Xiuyu Tiaoshan Memorial Archway® and other buildings were presented in their publications. As soon as Haizhou was liberated in 1948, Haizhou Guandi Temple gained specialized preservation immediately. In 1952, the Heritage Management Institute of Haizhou Guandi Temple was established. In 1957, the People's Committee of Shanxi Province declared Haizhou Guandi Temple the Major Province-level Protected Historical and Cultural Sites. In the same year, LIU Zhiping, a research fellow of Institute of Architectural Theory and History of China Academy of Building Research and a member of Society for the Study of Chinese Architecture (*Yingzao Xueshe*), carried out a field investigation on Haizhou Guandi Temple and made the temple known comprehensively and systematically to the academic circle. His investigation results, *the Investigation into the Traditional Chinese Architecture in Inner Mongolia and Shanxi Province*, were mimeographed and later published on the first and second volumes of China Academy of Building Research's internal journal (No.8284, No.8285) successively. In the 1970s, multiple preservation projects were carried out under the supervision of the Heritage Management Institute, which mainly focused on remedial repairs and maintenance. In July 2008, Professor WANG Qiheng at Tianjin University, along with 115 university faculties and students, completed the surveying and mapping of Haizhou Guandi Temple, the Jieyi Garden and Changping Guandi Temple thoroughly. During the surveying and mapping, 3D laser scanning and photogrammetric technologies were used, and more than 700 digitalized drawings were finished. From 2009 to 2012, under the guidance of Professor WANG Qiheng, GUO Huazhan, ZHU Yang, WANG Ruijia, LIU Tingting, YE Qing, JIANG Bei and others completed *the Preservation Planning of Haizhou Guandi Temple* and *the Preservation Planning of Changping Guandi Temple* in cooperation with staffs from the Heritage Management Institute of Haizhou Guandi Temple. These two planning were approved by National Cultural Heritage Administration and put into practice in 2012. In recent years, under the auspices of the Heritage Management Institute of Haizhou Guandi Temple, traditional sacrificial ceremonies and cultural activities have been held regularly, resulting in an increase in communication with believers of GUAN Yu from Fujian and Taiwan provinces. It is not too much to say that the belief in GUAN Yu has played an important role in the cross-strait cultural exchange, and that Haizhou Guandi Temple has explored actively on the rational use of culture relics.

四、近现代保护和研究

解州关帝庙很早就引起了管理部门和学术界的重视。早在1907年，德国建筑师恩斯特·鲍希曼（Ernst Boerschmann）在考察中国建筑时就曾到解州关帝庙，拍摄了已知最早的关帝坐像照片（图十四）。1941年，日本侵略者侵占解州期间，在其出版的刊物上刊登了解州关帝庙御书楼及『秀毓条山』坊等照片。1948年，解州解放以后，立即获得了专门保护：1952年，关帝庙文物管理所成立；1957年，山西省人民委员会即将关帝庙公布为山西省重点文物保护单位；同年，中国建筑科学研究院建筑理论及历史研究所研究员、营造学社成员刘致平先生实地调查，使解州关帝庙较系统、全面地为学界所知，调查成果《内蒙古、山西等处古建筑调查纪略》曾以油印的方式印行，并分成上、下两部分发表于中国建筑科学研究院建筑情报研究所的内部印刷品《建筑工程情报资料》第8284号《建筑历史研究》（第一辑）和第8285号《建筑历史研究》（第二辑）卷首。1970年代，在文管所的主持下，解州关帝庙内修缮保护工程不断，结义园及常平关帝庙进行了全面测绘，采用先进的三维激光扫描技术和摄影测量技术，共完成数字化测绘图纸700余张。2009—2012年，王其亨教授所指导郭华瞻、朱阳、王蕊佳、刘婷婷、叶青、江蓓等人配合解州关帝庙文管所共同编制完成了《解州关帝庙保护规划》和《常平关帝庙保护规划》。两部规划均于2012年获得了国家文物局的批复并正式实施。近年来，在解州关帝庙文管所的主持下，通过定期举办传统祭祀文化活动，加强了与福建、台湾等地民间关帝信众及民间社团组织的交流，使关帝信仰成为海峡两岸民间文化交流中的一项重要内容，也使得解州关帝庙在遗产的合理利用方面做出了积极的探索。

注释
Notes

① Guandi is the deity name of GUAN Yu. It means GUAN Yu the God. All footnotes in this book are added by the translator.

② Jieyi means becoming sworn brothers or sisters in Chinese. GUAN Yu, LIU Bei and ZHANG Fei, leading group of the Shu Kingdom, are the best-known Jieyi brothers in China. Their story of becoming sworn brothers is known as the Oath of the Peach Garden (Taoyuan Jieyi) in China.

③ Sometimes it is also called *Zhuangmiu Hou*. According to the posthumous naming rules in the *Yi Zhou Shu*, *zhuang* is meant for brave and strong people that die in wars, *miu* is meant for people who failed to live up to his reputation, and *hou* means the marquis.

④ *Chong* means sublime, *ning* means compassionate. *Zhenjun* is a title for immortals used by Taoist priests.

⑤ *Zhonghui Gong* means the Duke of Loyalty and Kindness. *Zhong* refers to loyalty, *hui* refers to kindness, while *gong* refers to the duke.

⑥ *Wuan Wang* means the King of Military and Peace. *Wu* refers to military, *an* refers to peace, while *wang* refers to the king.

⑦ The new added word *yiyong* means righteousness and courage, *yingji* means valor and self-devoted.

⑧ The word *xianling* that replaces *zhuangmiu* means theophany.

⑨ *Han Qianjiangjun Shoutinghou* is the former title of GUAN Yu. *Han* is the dynasty that GUAN Yu belongs to. *Qianjiangjun* is GUAN Yu's military official position, which refers to general. *Shoutinghou* is the title of GUAN Yu during the Three Kingdoms Period, which means the Marquis of Shouting.

⑩ *Hanshoutinghou* has the same meaning as *Shoutinghou*.

⑪ *Xietian* means obedience to heaven, *huguo* means protection to the country, *zhongyi* means loyal and righteous. *Di* means god.

⑫ *Sanjiefumodadi* means the god subduing devils of the three divisions of the universe. *Shenweiyuanzhentianzun* means god of martial prowess with great fame. *Guanshengdijun* means GUAN Yu the sage and the god.

⑬ *Zhongyishenwu guanshengdadi* means the loyal, righteous, divine and valiant GUAN Yu the sage and the god.

⑭ *Wujingboshi* is the officials who teach the Five Classics.

⑮ It is believed that *zhuangmiu* is a bad posthumous title. So Emperor Qianlong changed it into shenyong, a god posthumous title that means extreme valiant.

⑯ *Xuande* means the great morality. *Yizan* means assisting. *Jingcheng* means absolute sincerity. *Suijing* means safe and stability. *Baominhuguo* means protecting the people and the country. *Weixianrenyong* means imposing, prominent, benevolent and valiant. *Zhongyishenwulingyou* means loyal, righteous, extreme valiant and powerful.

⑰ *Wandai Zhanyang* means respected by thousands of generations.

⑱ 唐代诗人郎君青有诗作"壮缪侯庙别友人":将军秉天姿,义勇冠古今。走黄造战场,一剑万人敌。谁为感恩者,竟是思归客。流落荆巫间,徘徊故乡隔。离筵对祠宇,泪洒暮霞雪。去去勿复言,衔悲向陈迹。见全唐诗。
There is a poem of LANG Junzhou, the poet of the Tang Dynasty, recorded in the *Complete Collection of Tang Poems (Quantangshi)*, named *Farewell to A Friend in Zhuangmiuhou Temple*, which says "(GUAN Yu) the general imposing like a god, famous for his righteousness and great love. Came to the battlefield with illness, still conquered all enemies. Who is now worshipping him? Homesick people whose future is dim. Wandering in the Jing and Wu lands, far away from homeland. Leave the seat to the temple of Zhuangmiuhou, my tear fall as snow. Please leave without a word friend, my sorrows will be buried in the relics and never sent."

⑲ The palace was named after GUAN Yu's title *Chongning Zhenjun*.

⑳ Chunqiu is one of the Confucius Classics, also called Linjing. The building is named after Chunqiu/Linjing because the GUAN Yu status in it is reading Chunqiu.

㉑ The names of the two wing buildings refer to Building of the Sword and Building of the Seal.

㉒ The archway's name means that landscape is magnificent.

㉓ Temple dedicated to Laozi, the founder of Taoism.

㉔ *Qisheng* means enlightment.

㉕ *Chongsheng* means respecting the sage.

㉖ Prince Guo is the son of Emperor Kangxi, brother of Emperor Yongzheng.

㉗ The name originated from *Book of Rites*. Zhi Gate is a kind of gate for emperors.

㉘ The name of the former archway means the great righteousness, the latter one means the great loyalty.

㉙ *Junzi* refers to men of noble characters.

㉚ The former name originates from *Bagua*, or the Eight Diagrams, a traditional Chinese philosophy, while the latter name *Yushu*, which means writings of the emperor, originates from the story that Emperor Kangxi has awarded the temple a board with inscriptions written by him.

㉛ ZHOU Cang and LIAO Hua are GUAN Yu's subordinates in *the Romance of the Three Kingdoms*.

㉜ Bujiang Shrine is the shrine dedicateds to GUAN Yu's subordinates. Now, figures of YANG Yi, WANG Fu and ZHAO Lei were worshipped in the shrine.

㉝ Bujiang Shrine is the shrine dedicateds to GUAN Yu's houre, Chitu. Chitu was entitled the Count of Chasing Wind in the Ming Dynasty.

㉞ *Jiaozhong* means to the loyalty education.

㉟ The Zhou Pavilion refers to a boat-like pavilion.

㊱ *Yizhuang Qiankun* means the righteousness famous all over the world.

㊲ The Donghua Gate and the Xihua Gate are the eastern and western entrance gates of the temple. Names of the gates mean the Gate of the East Prosperity and the Gate of the West Prosperity.

㊳ *Shengmu* in the former hall's name means the goddess, while *Sisheng* in the latter name means the descendants of the sage.

㊴ The Guanping Hall and the Guanxing Hall are halls dedicated to GUAN Ping and GUAN Xing, sons of GUAN Yu.

㊵ The tradition of having "Five Gates and Three Courts" and "Outer Court in the front and Inner Court in the Rear" in the imperial palace origins from the Zhou Dynasty. It is said that palace of emperors should have a series of five gates on the central axis, an outer court for government affairs in the front of the palace and an inner court for residence in the rear to show their legitimization.

㊶ The names of the Duan Gate, the Zhi Gate and the Meridian Gate coincide with the names of gates in the Forbidden City, proving that the gates of Haizhou Guandi Temple are designed to imitate those of the imperial palace. The names of the Wenjing Gate and the Wuwei Gate origins from the word *Wenjingwuwei*, which means to be proficient both in literary and military knowledge.

㊷ *Qisu Qianqiu* means the spirit lasting forever.

㊸ The name origins from *I Ching* and means the social commitment.

㊹ *Weizhen Huaxia* means the prestige known all over China.

㊺ For "大义参天" and "精忠贯日", see footnote 26.

㊻ *Xiuyu Tiaoshan* means the beautiful Zhongtiao Mountain.

图

版

Drawings

解州关帝庙

Haizhou Guandi
Temple

中国古建筑测绘大系·祠庙建筑——解州关帝庙

024

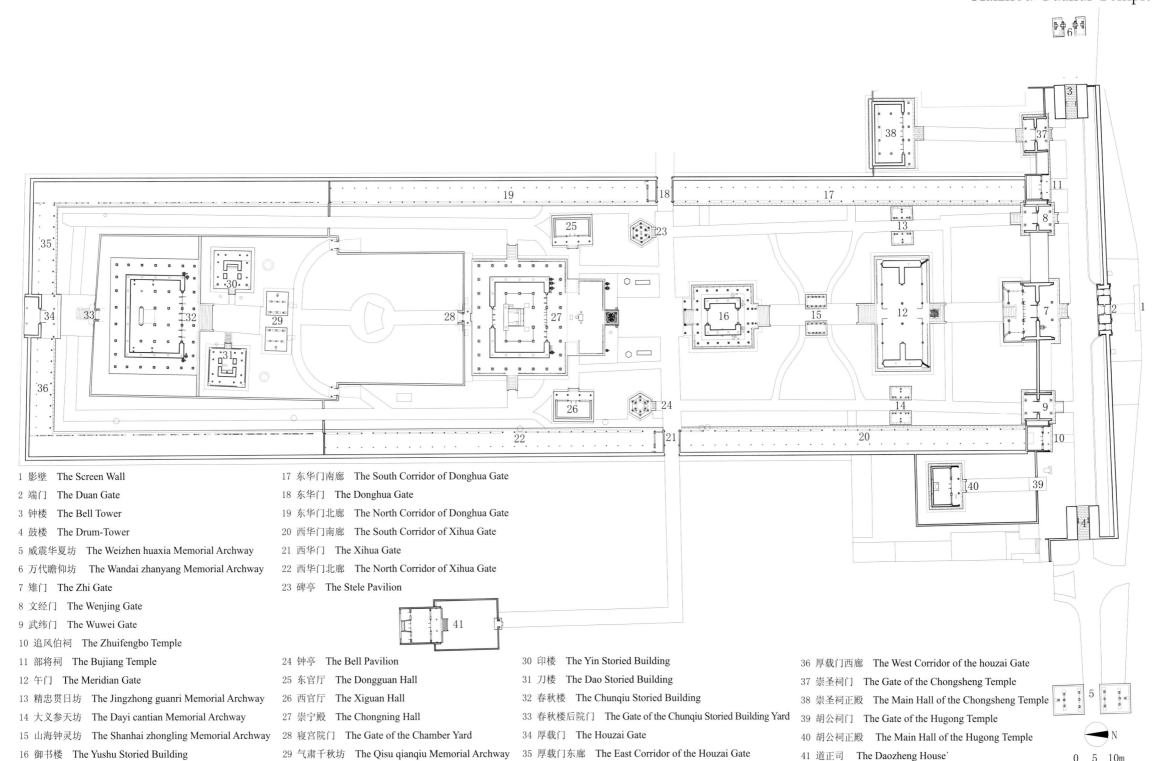

1 影壁   The Screen Wall
2 端门   The Duan Gate
3 钟楼   The Bell Tower
4 鼓楼   The Drum-Tower
5 威震华夏坊   The Weizhen huaxia Memorial Archway
6 万代瞻仰坊   The Wandai zhanyang Memorial Archway
7 雉门   The Zhi Gate
8 文经门   The Wenjing Gate
9 武纬门   The Wuwei Gate
10 追风伯祠   The Zhuifengbo Temple
11 部将祠   The Bujiang Temple
12 午门   The Meridian Gate
13 精忠贯日坊   The Jingzhong guanri Memorial Archway
14 大义参天坊   The Dayi cantian Memorial Archway
15 山海钟灵坊   The Shanhai zhongling Memorial Archway
16 御书楼   The Yushu Storied Building

17 东华门南廊   The South Corridor of Donghua Gate
18 东华门   The Donghua Gate
19 东华门北廊   The North Corridor of Donghua Gate
20 西华门南廊   The South Corridor of Xihua Gate
21 西华门   The Xihua Gate
22 西华门北廊   The North Corridor of Xihua Gate
23 碑亭   The Stele Pavilion
24 钟亭   The Bell Pavilion
25 东官厅   The Dongguan Hall
26 西官厅   The Xiguan Hall
27 崇宁殿   The Chongning Hall
28 寝宫院门   The Gate of the Chamber Yard
29 气肃千秋坊   The Qisu qianqiu Memorial Archway

30 印楼   The Yin Storied Building
31 刀楼   The Dao Storied Building
32 春秋楼   The Chunqiu Storied Building
33 春秋楼后院门   The Gate of the Chunqiu Storied Building Yard
34 厚载门   The Houzai Gate
35 厚载门东廊   The East Corridor of the Houzai Gate

36 厚载门西廊   The West Corridor of the houzai Gate
37 崇圣祠门   The Gate of the Chongsheng Temple
38 崇圣祠正殿   The Main Hall of the Chongsheng Temple
39 胡公祠门   The Gate of the Hugong Temple
40 胡公祠正殿   The Main Hall of the Hugong Temple
41 道正司   The Daozheng House`

N

0 5 10m

解州关帝庙组群平面图
Plan of the groups of the Haizhou Guandi Temple

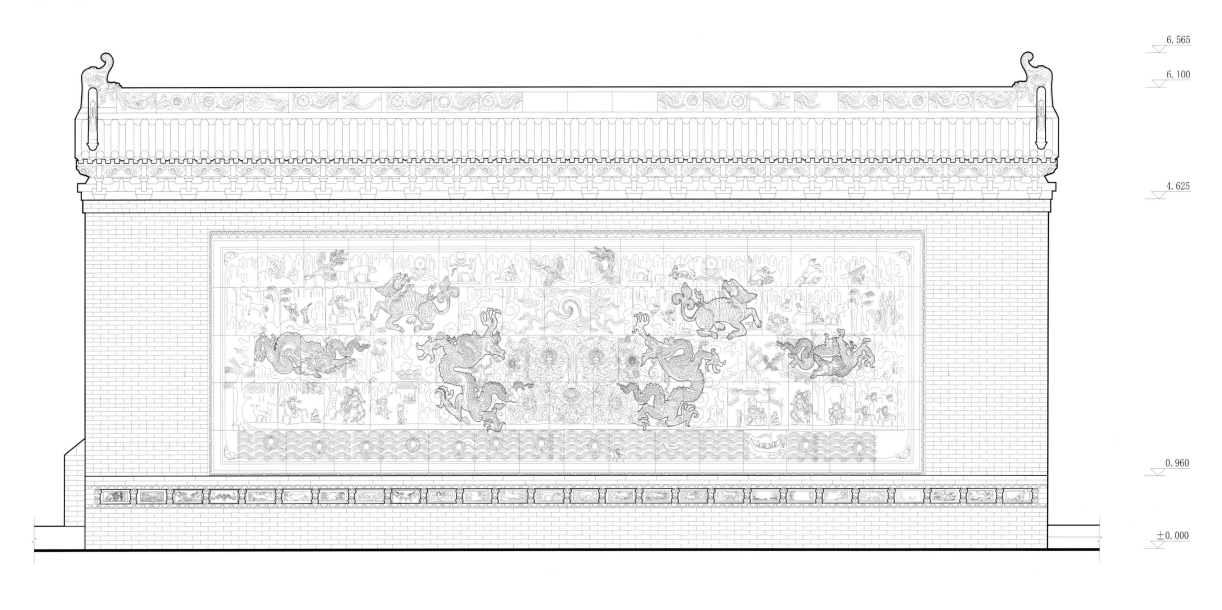

中国古建筑测绘大系·祠庙建筑 —— 解州关帝庙

025

6.565

6.100

4.625

0.960

±0.000

1685   9770   1685

13140

0   0.5   1m

影壁正立面图
Front elevation of the Screen Wall

6.565

6.100

5.121

4.625

0.960

±0.000

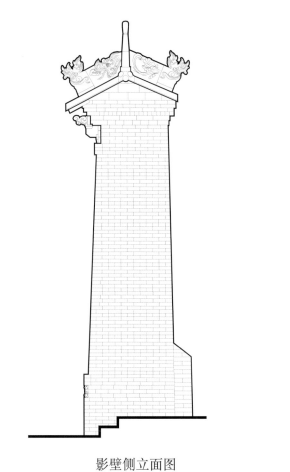

影壁侧立面图
Side elevation of the Screen Wall

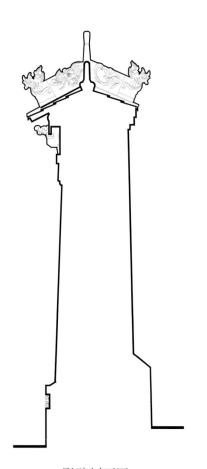

影壁剖面图
Cross-section of the Screen Wall

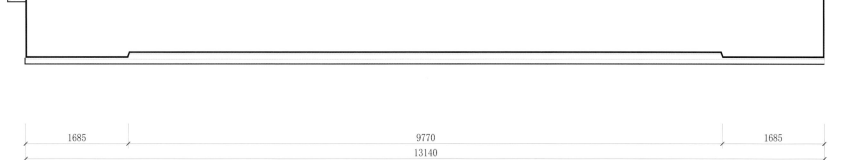

1370

| 1685 | 9770 | 1685 |

13140

影壁平面图
Plan of the Screen Wall

N

0   0.5   1m

中国古建筑测绘大系·祠庙建筑——解州关帝庙

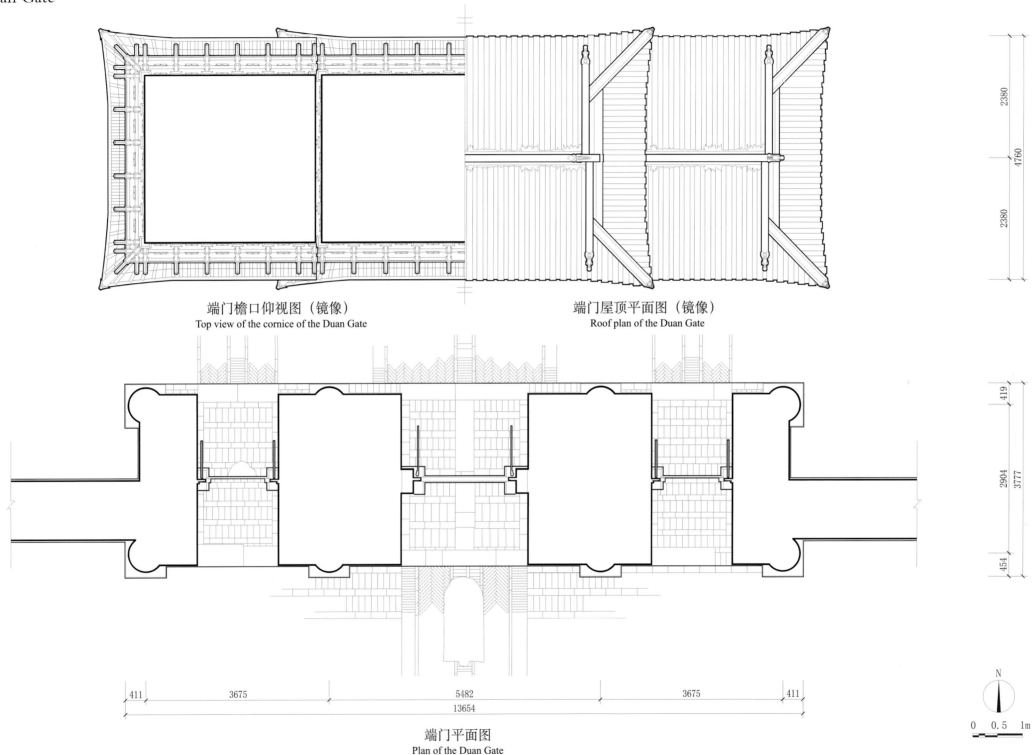

端门檐口仰视图（镜像）
Top view of the cornice of the Duan Gate

端门屋顶平面图（镜像）
Roof plan of the Duan Gate

端门平面图
Plan of the Duan Gate

411 3675 5482 3675 411
13654

2380 4760 2380

419 2904 3777 454

N

0 0.5 1m

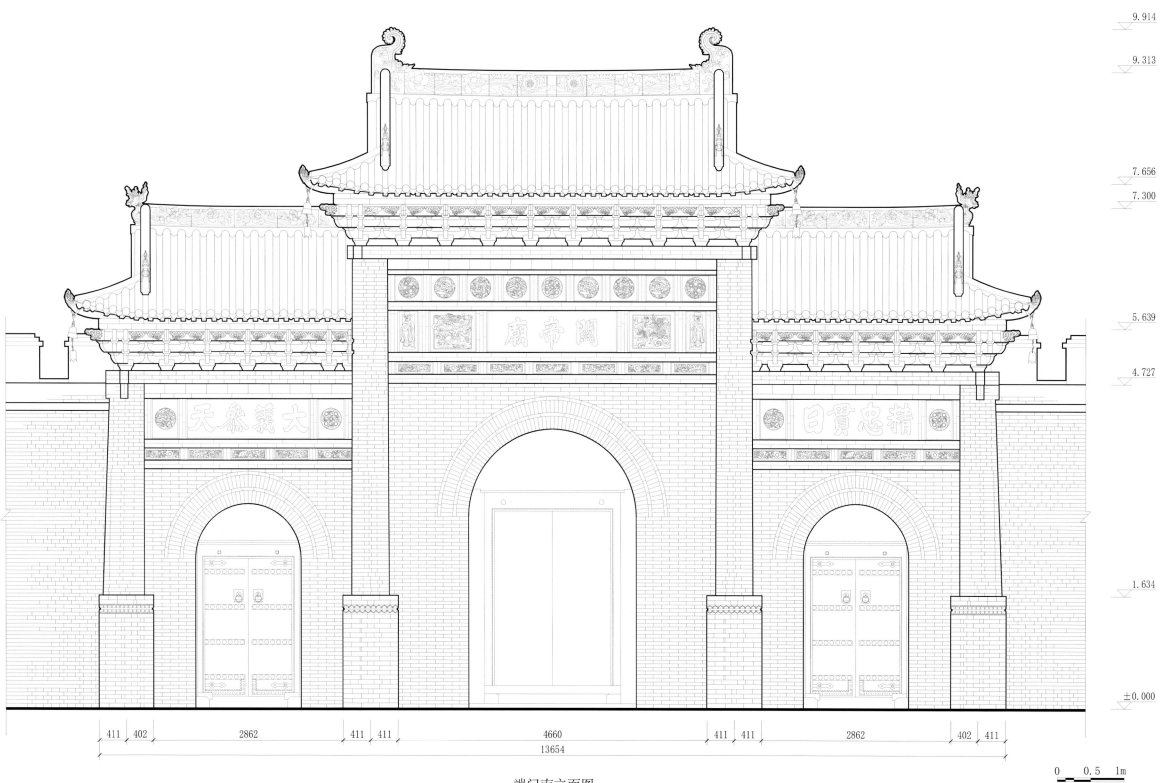

關帝廟

大義參天

精忠貫日

| 411 | 402 | 2862 | 411 | 411 | 4660 | 411 | 411 | 2862 | 402 | 411 |

13654

9.914
9.313
7.656
7.300
5.639
4.727
1.634
±0.000

0   0.5   1m

端门南立面图
South elevation of the Duan Gate

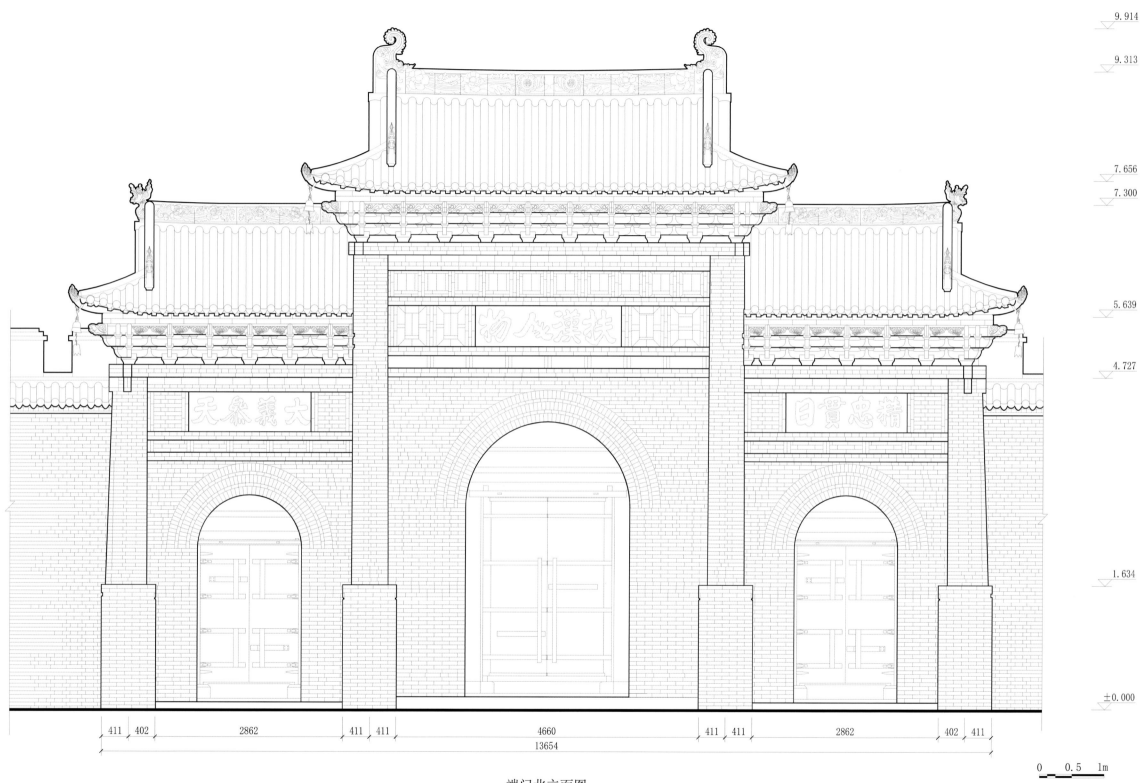

9.914

9.313

7.656
7.300

5.639

4.727

1.634

±0.000

411 402 2862 411 411 4660 411 411 2862 402 411

13654

0　0.5　1m

端门北立面图
North elevation of the Duan Gate

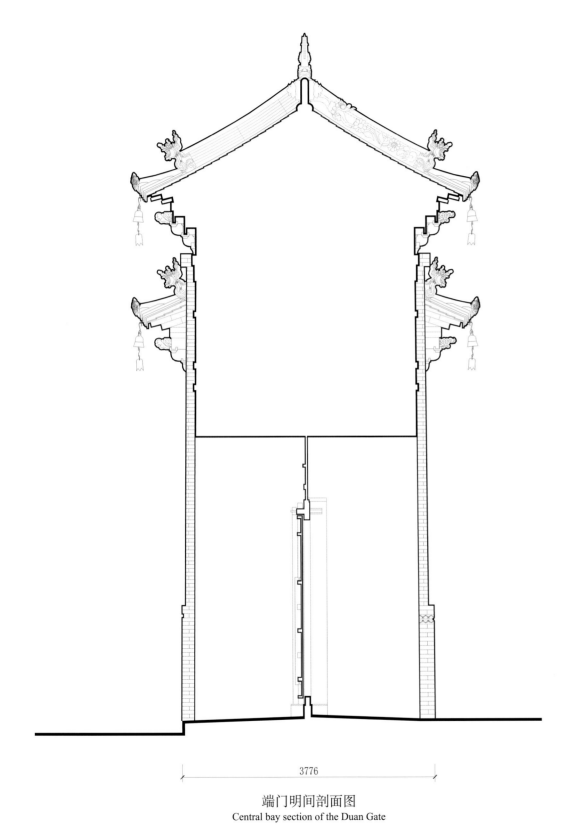

9.914

7.656
7.300

6.282

5.639

4.727

1.634

±0.000

3776

端门明间剖面图
Central bay section of the Duan Gate

3776

端门西立面图
West elevation of the Duan Gate

0   0.5   1m

钟楼
The Bell Tower

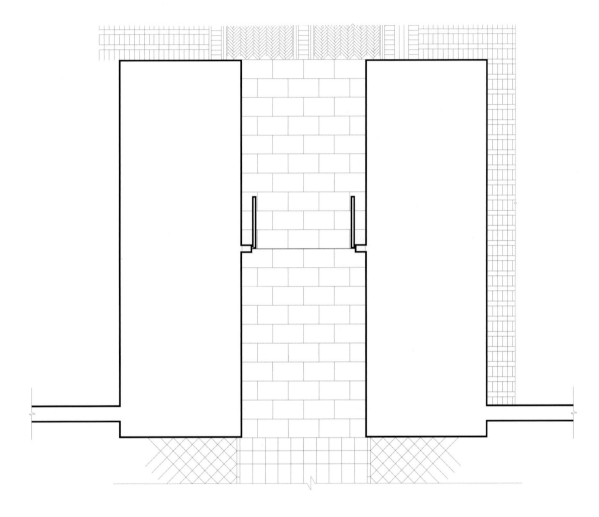

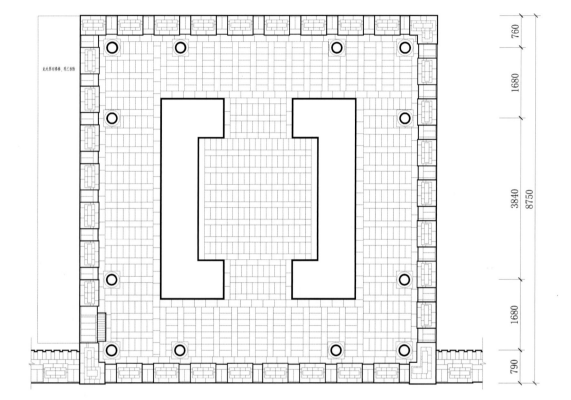

钟楼一层平面图
The First Floor Plan of the Bell Tower

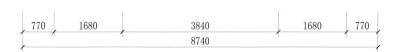

钟楼二层平面图
The Second Floor Plan of the Bell Tower

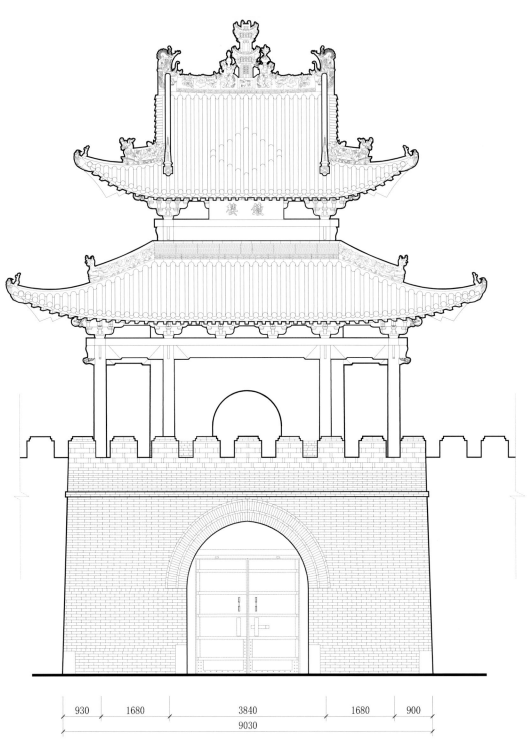

15. 447

14. 864

11. 396

10. 559

8. 406

7. 737

4. 260

±0. 000

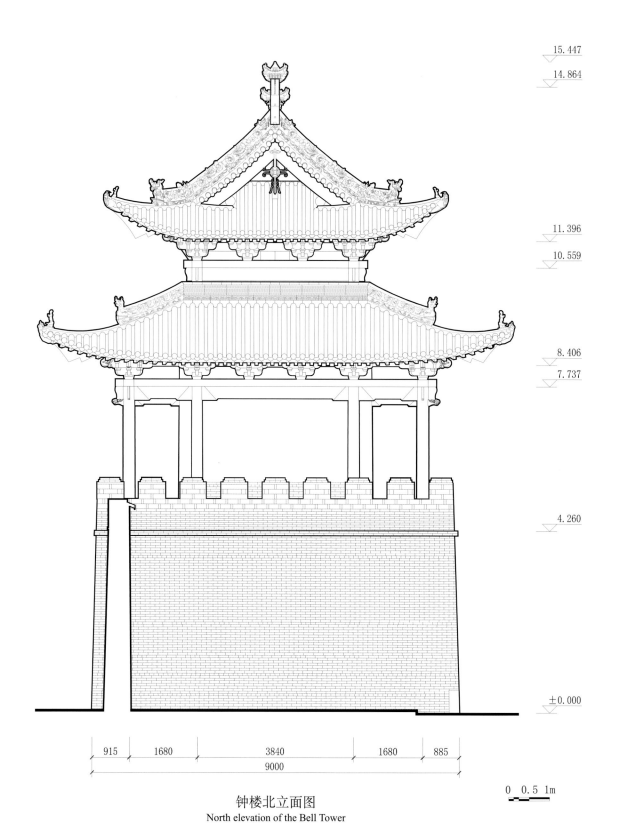

| 930 | 1680 | 3840 | 1680 | 900 |

9030

钟楼西立面图
West elevation of the Bell Tower

| 915 | 1680 | 3840 | 1680 | 885 |

9000

0   0.5 1m

钟楼北立面图
North elevation of the Bell Tower

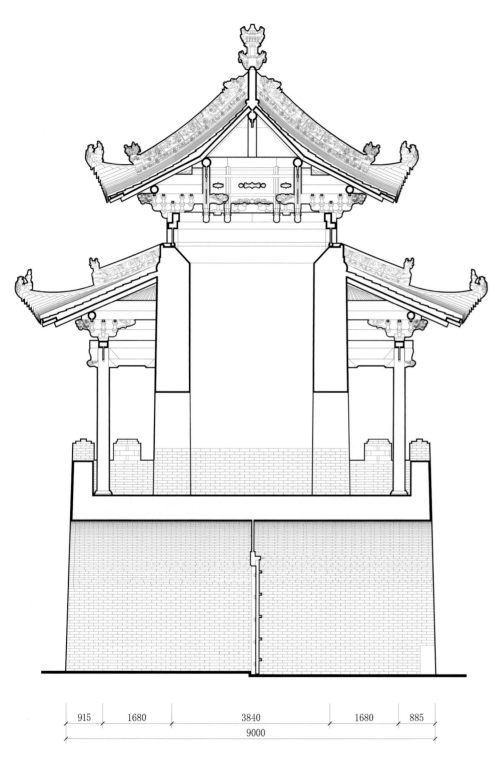

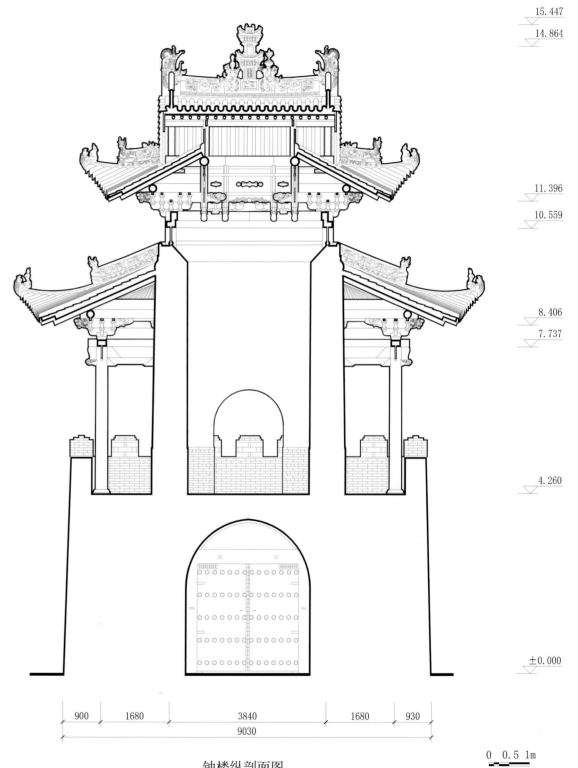

15.447
14.864

11.396
10.559

8.406
7.737

4.260

±0.000

915 1680 3840 1680 885
9000

钟楼横剖面图
Cross-section of the Bell Tower

900 1680 3840 1680 930
9030

0 0.5 1m

钟楼纵剖面图
Longitudinal section of the Bell Tower

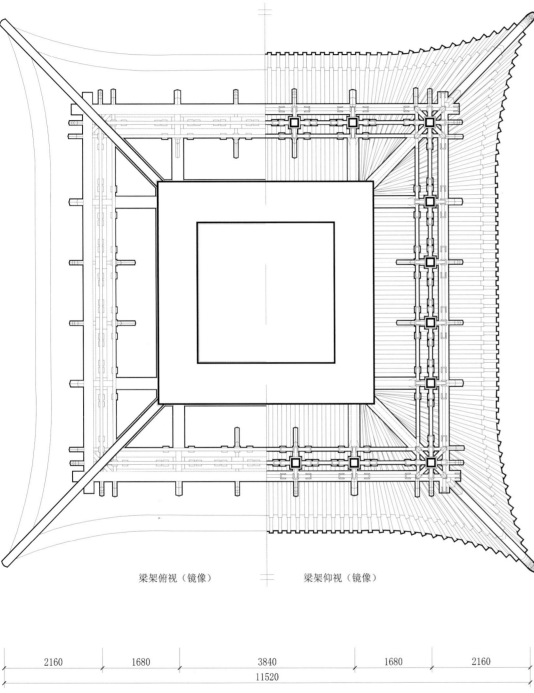

梁架俯视（镜像） 梁架仰视（镜像）

2160 1680 3840 1680 2160

11520

钟楼一层梁架仰俯视图
Top and bottom views of the beam frame of the first-floor of the Bell Tower

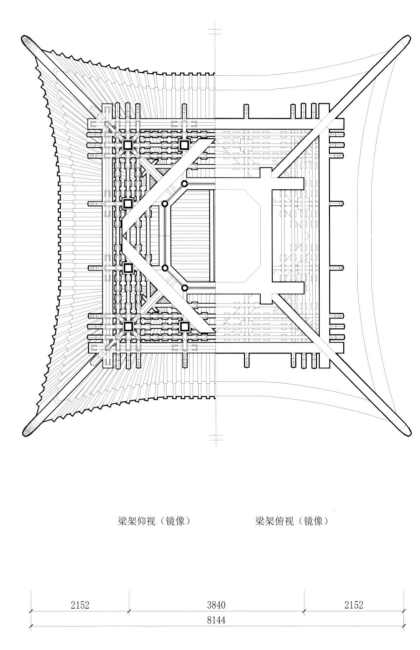

梁架仰视（镜像） 梁架俯视（镜像）

2152 3840 2152

8144

0 0.5 1m

钟楼二层梁架仰俯视图
Top and bottom views of the beam frame of the second-floor of the Bell Tower

鼓楼
The Drum-Tower

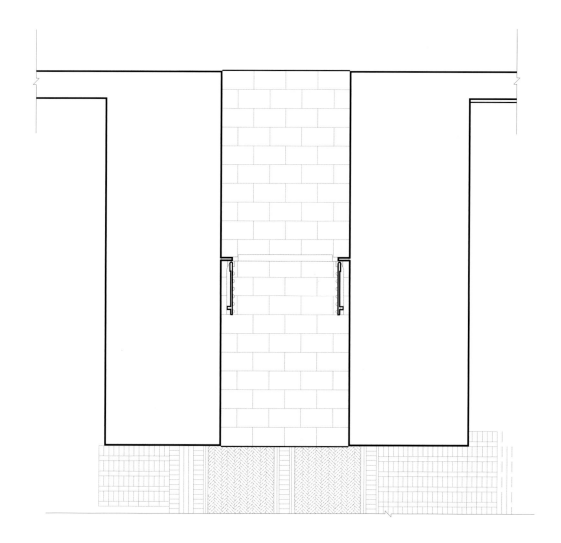

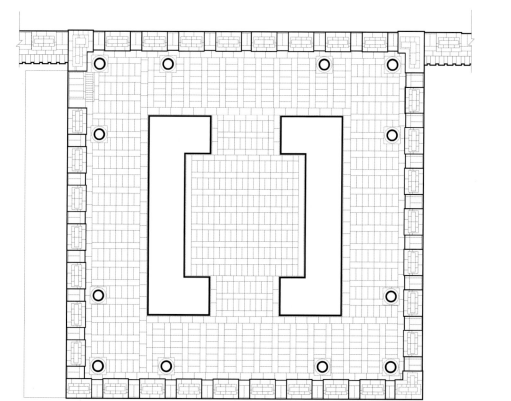

2836　3116　2984

8936

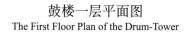

鼓楼一层平面图
The First Floor Plan of the Drum-Tower

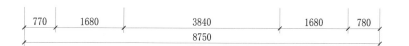

790　1680　3840　1680　8750　760

770　1680　3840　1680　780

8750

N

0 0.5 1m

鼓楼二层平面图
The Second Floor Plan of the Drum-Tower

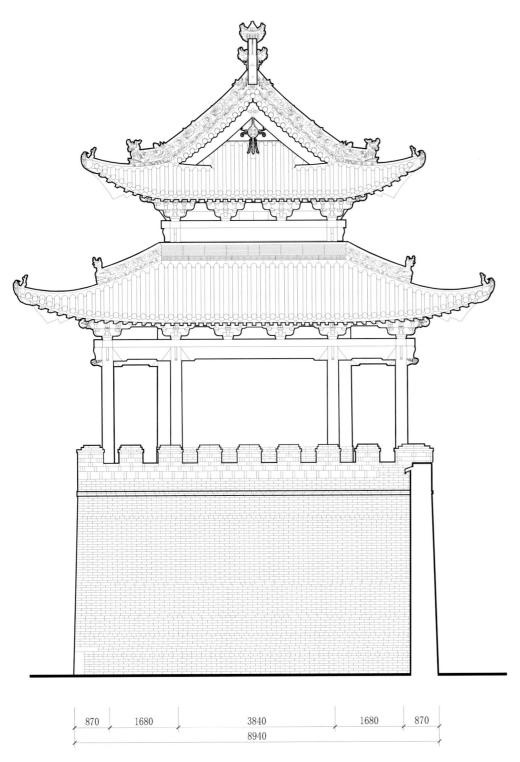

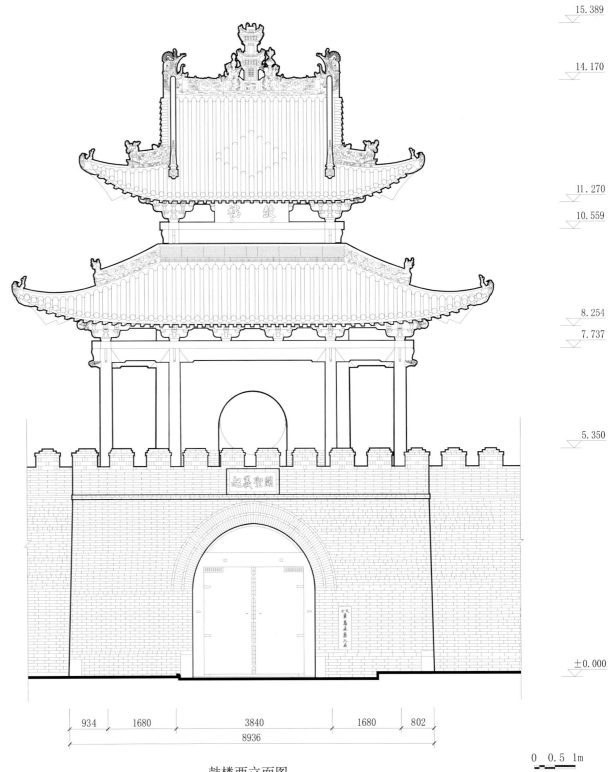

15. 389

14. 170

11. 270

10. 559

8. 254

7. 737

5. 350

±0. 000

036

| 870 | 1680 | 3840 | 1680 | 870 |

8940

鼓楼北立面图
North elevation of the Drum-Tower

| 934 | 1680 | 3840 | 1680 | 802 |

8936

鼓楼西立面图
West elevation of the Drum-Tower

0 0.5 1m

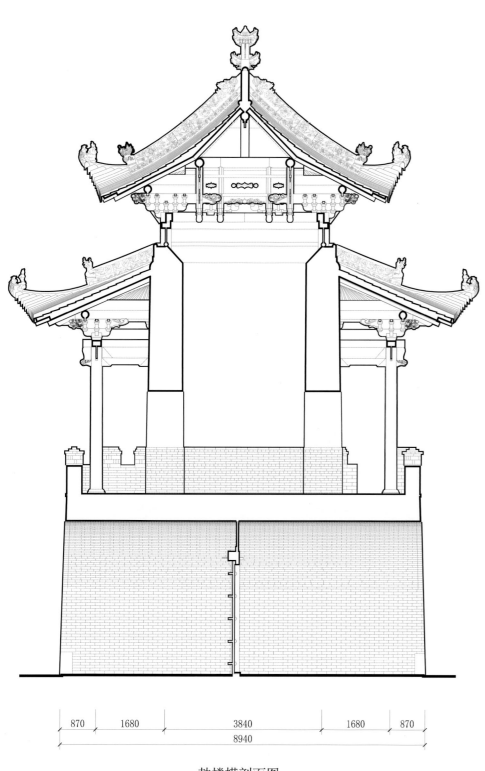

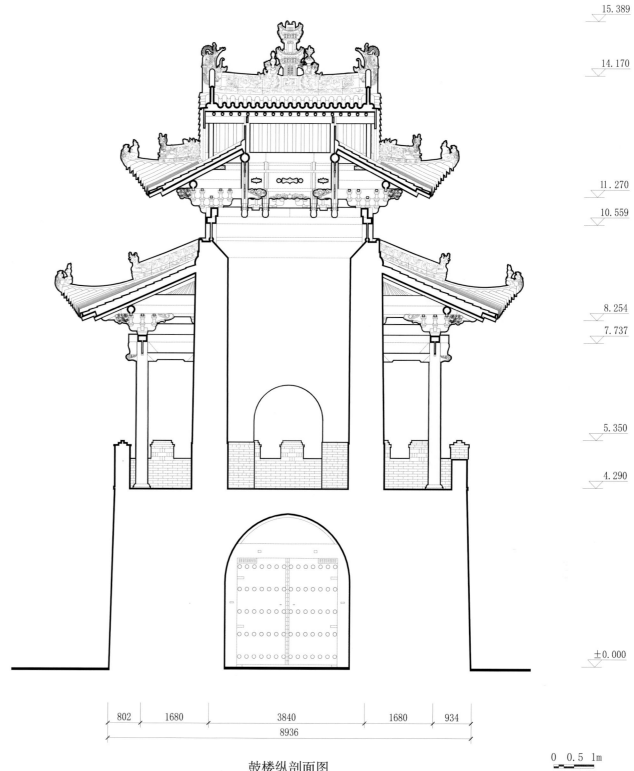

15.389

14.170

11.270

10.559

8.254

7.737

5.350

4.290

±0.000

| 870 | 1680 | 3840 | 1680 | 870 |

8940

| 802 | 1680 | 3840 | 1680 | 934 |

8936

鼓楼横剖面图
Cross-section of the Drum-Tower

鼓楼纵剖面图
Longitudinal section of the Drum-Tower

0 0.5 1m

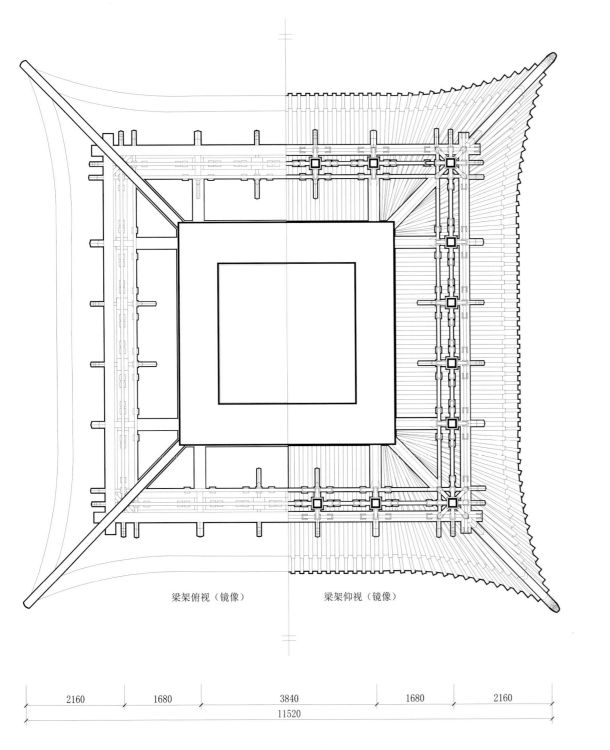

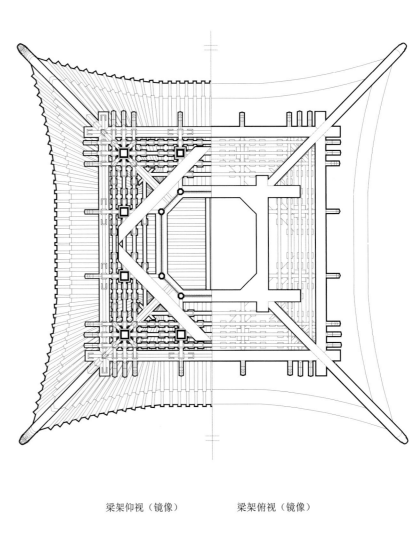

梁架俯视（镜像）　　梁架仰视（镜像）

梁架仰视（镜像）　　梁架俯视（镜像）

| 2160 | 1680 | 3840 | 1680 | 2160 |

11520

| 2167 | 3840 | 2167 |

8174

0　0.5　1m

鼓楼一层梁架仰俯视图

Top and bottom views of the beam frame of the first-floor of the Drum-Tower

鼓楼二层梁架仰俯视图

Top and bottom views of the beam frame of the second-floor of the Drum-Tower

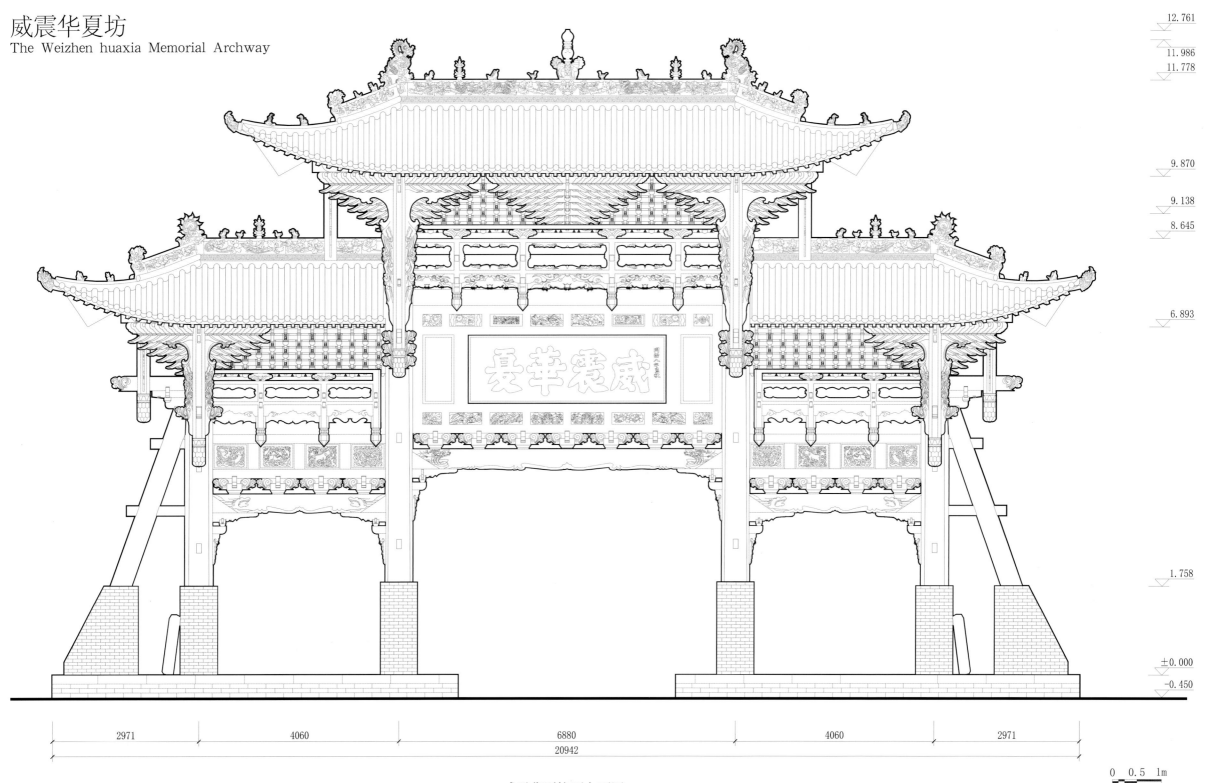

12.761
11.986
11.778
9.870
9.138
8.645
6.893
1.758
±0.000
-0.450

2971　4060　6880　4060　2971
20942

0　0.5　1m

威震华夏坊西立面图
West elevation of the Weizhen huaxia Memorial Archway

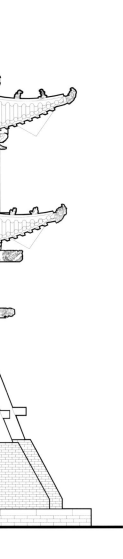

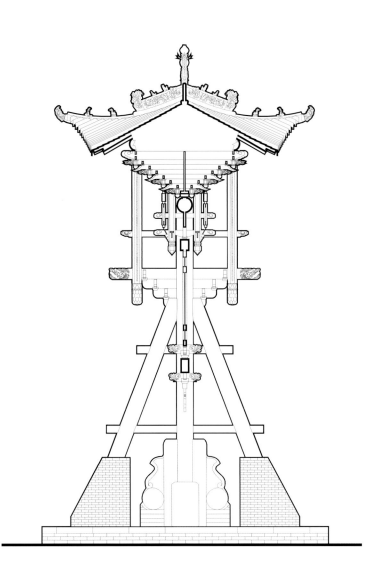

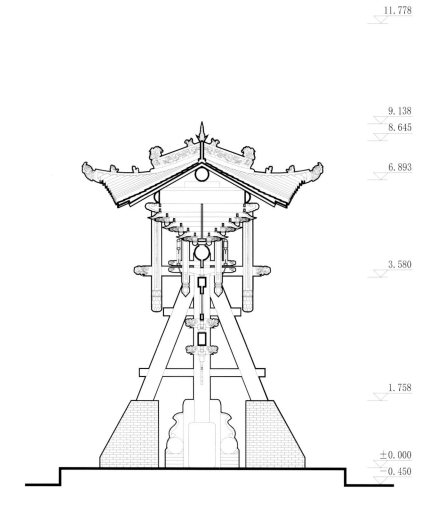

12.761

11.778

9.138

8.645

6.893

3.580

040

1.758

±0.000
-0.450

3910　　　　3910

7820

3910　　　　3910

7820

3910　　　　3910

7820

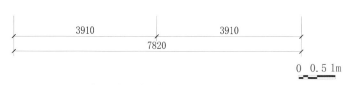

0　0.5　1m

威震华夏坊侧立面图
Side elevation of the Weizhen huaxia Memorial Archway

威震华夏坊明间横剖面图
Central bay section of the Weizhen huaxia Memorial Archway

威震华夏坊次间横剖面图
Side bay section of the Weizhen huaxia Memorial Archway

万代瞻仰坊
The Wandai zhanyang Memorial Archway

8.860

8.500

7.346

7.050

5.916

5.416

4.945

4.486

3.575

±0.000

-0.180

2920   2030

4950

700   1900   3540   1900   700

8740

0   0.5   1m

万代瞻仰坊北立面图

North elevation of the Wandai zhanyang Memorial Archway

万代瞻仰坊东立面图

East elevation of the Wandai zhanyang Memorial Archway

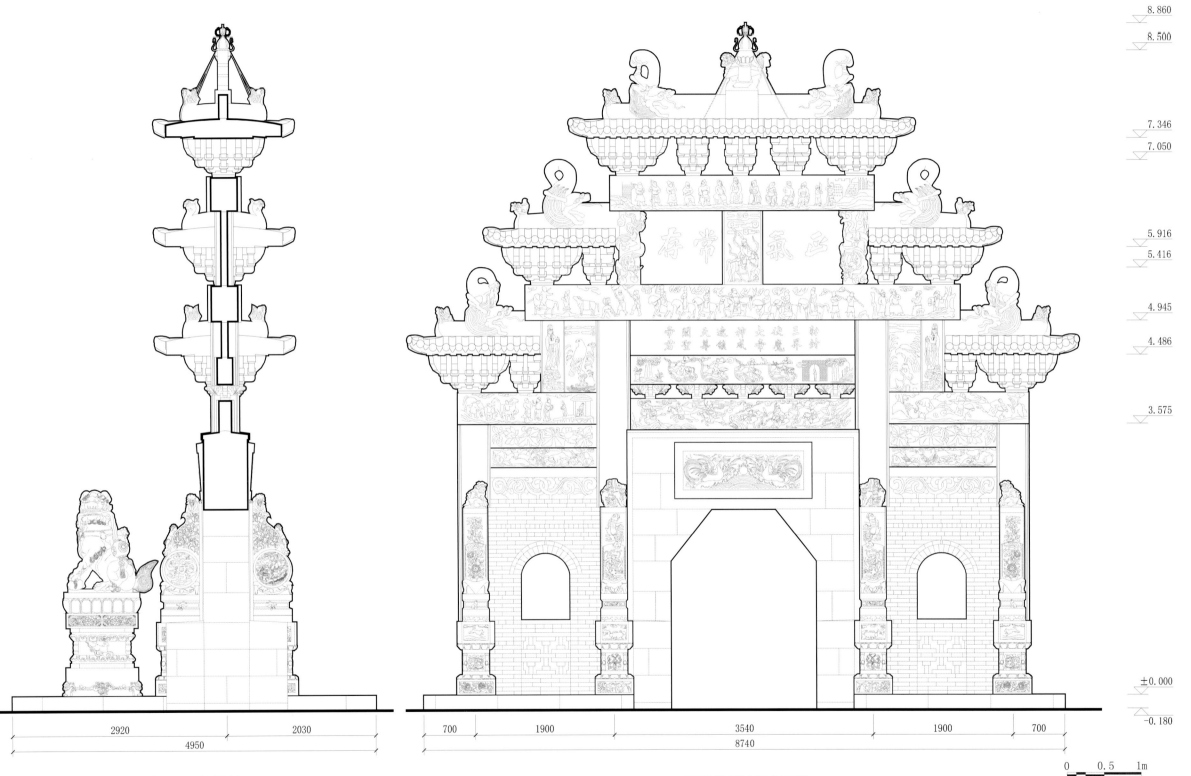

万代瞻仰坊明间横剖面图
Central bay section of the Wandai zhanyang Memorial Archway

万代瞻仰坊西立面图
West elevation of the Wandai zhanyang Memorial Archway

8.860
8.500
7.346
7.050
5.916
5.416
4.945
4.486
3.575
±0.000
-0.180

2920　2030
4950

700　1900　3540　1900　700
8740

0　0.5　1m

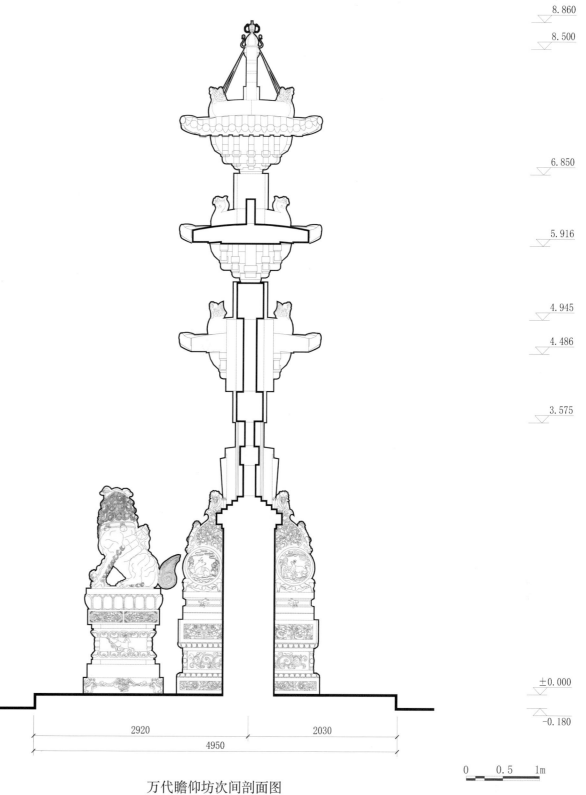

8.860

8.500

6.850

5.916

4.945

4.486

3.575

±0.000

-0.180

0   0.5   1m

2920   2030

4950

万代瞻仰坊梢间剖面图
Side bay section of the Wandai zhanyang Memorial Archway

2920   2030

4950

万代瞻仰坊次间剖面图
Second bay section of the Wandai zhanyang Memorial Archway

中国古建筑测绘大系·祠庙建筑——解州关帝庙

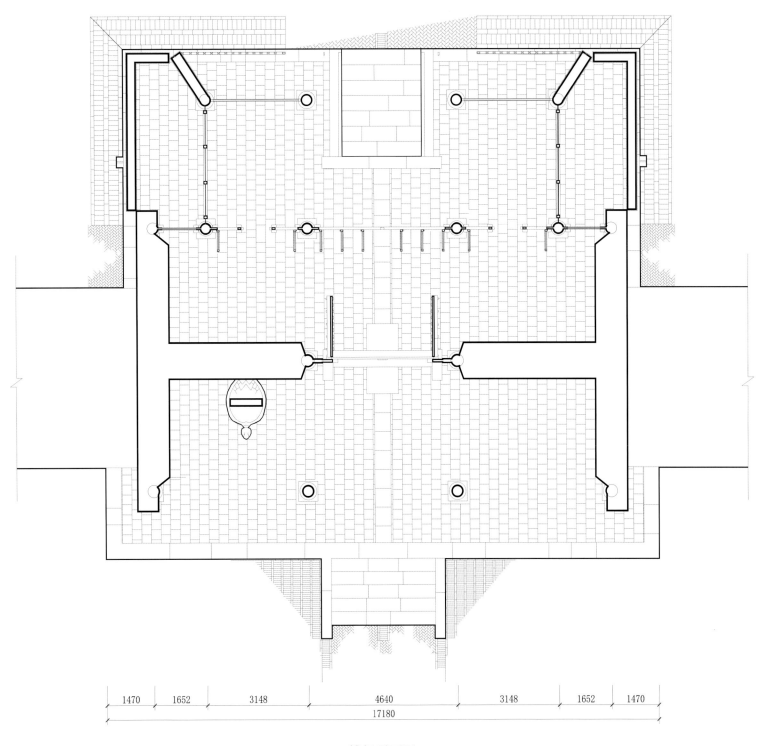

1560
3920
17142
3920
1550
2272

1470　1652　3148　4640　3148　1652　1470
17180

雉门平面图
Plan of the Zhi Gate

N

0　1　2m

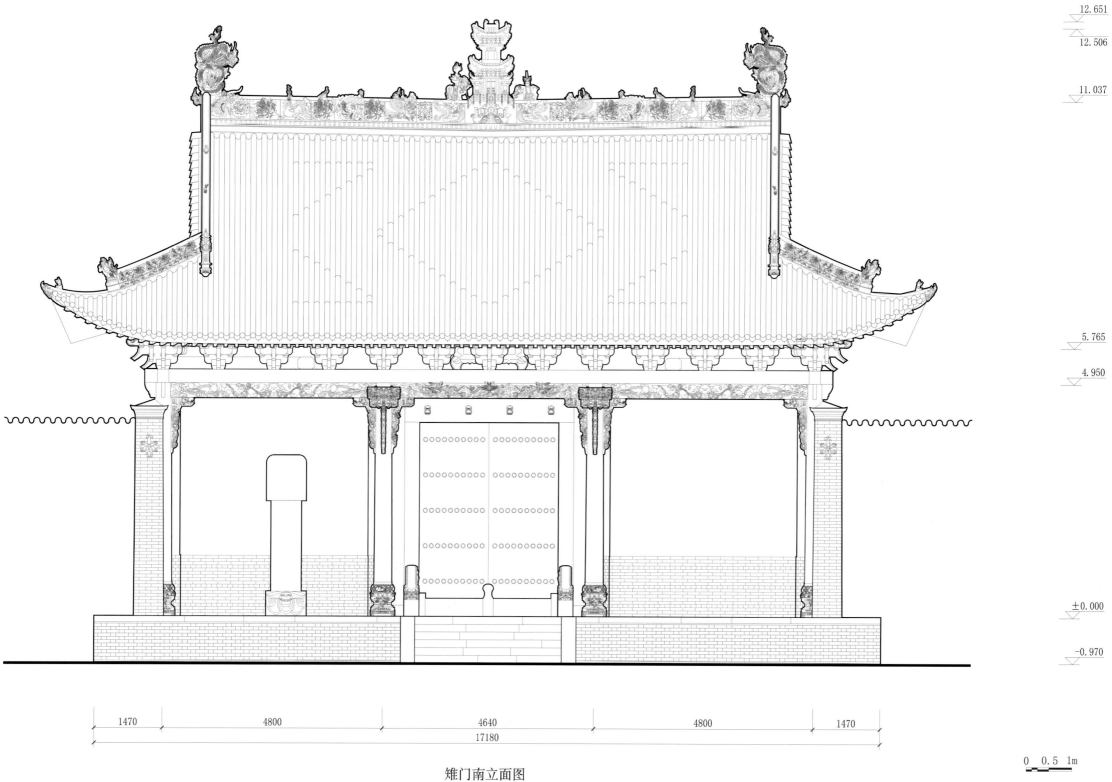

12.651

12.506

11.037

5.765

4.950

±0.000

-0.970

| 1470 | 4800 | 4640 | 4800 | 1470 |

17180

0 0.5 1m

雉门南立面图
South elevation of the Zhi Gate

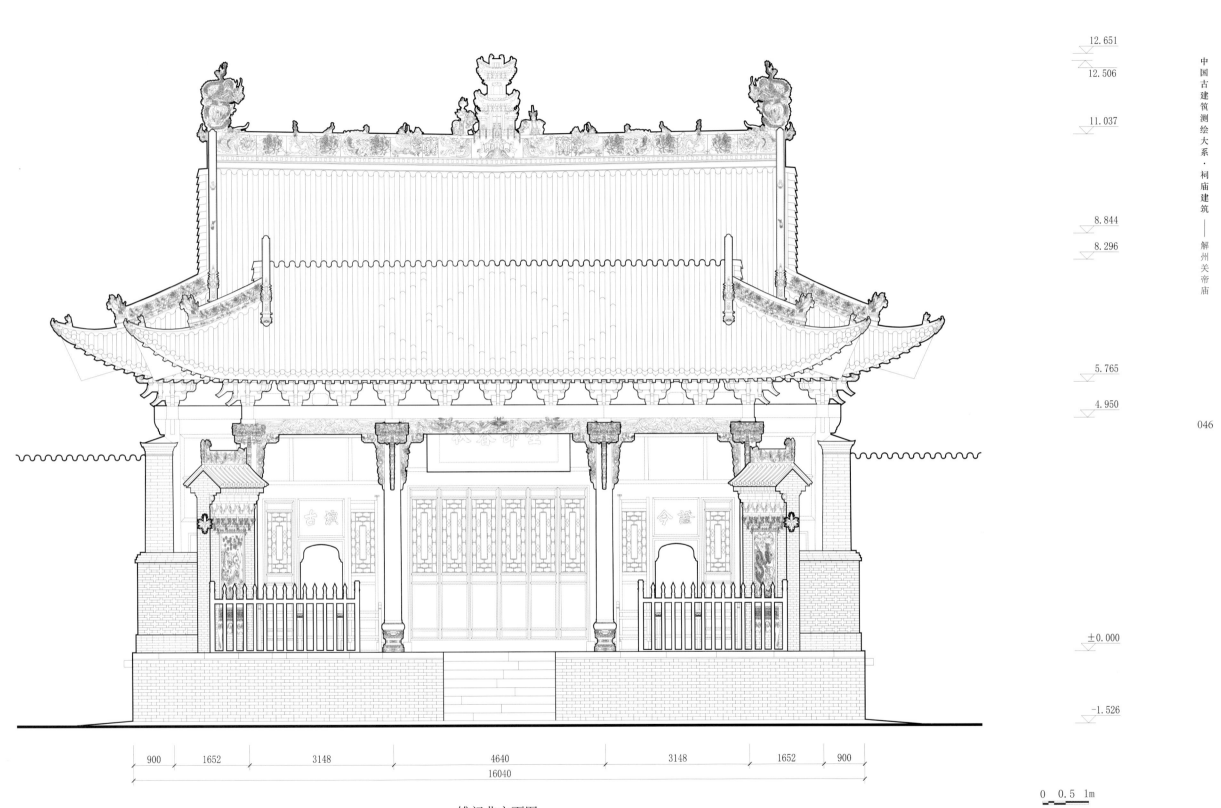

12.651
12.506

11.037

8.844
8.296

5.765

4.950

±0.000

-1.526

| 900 | 1652 | 3148 | 4640 | 3148 | 1652 | 900 |

16040

0   0.5   1m

雉门北立面图
North elevation of the Zhi Gate

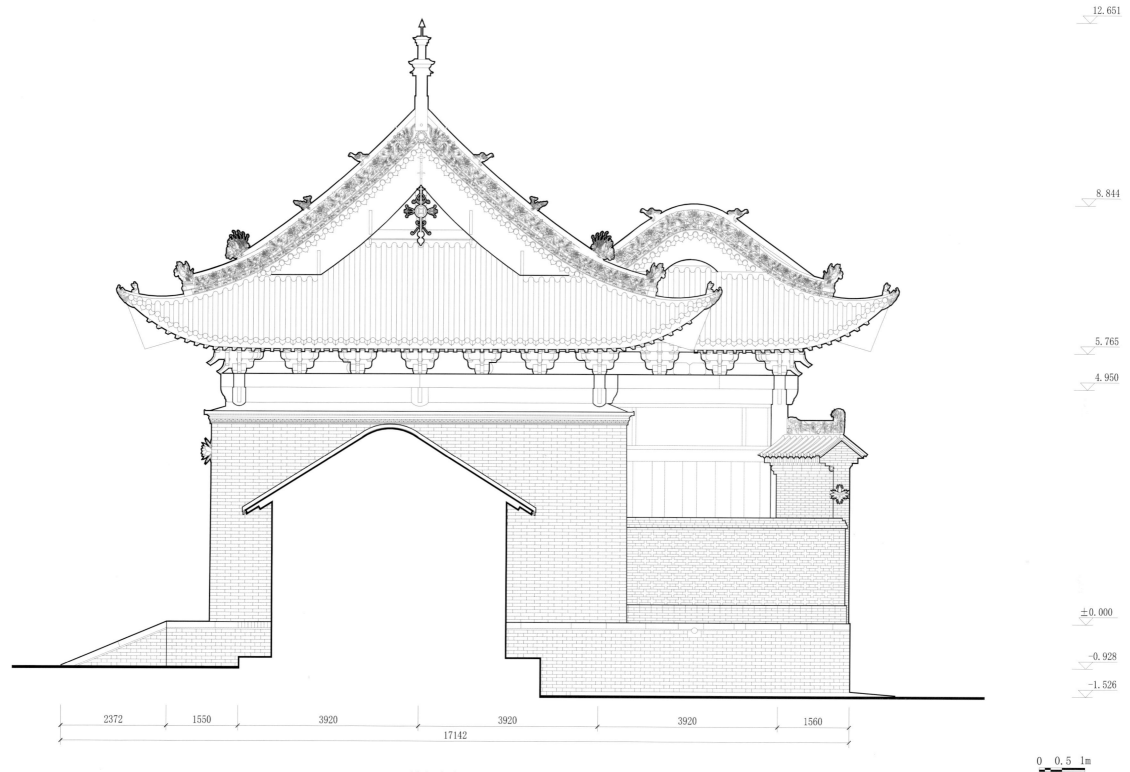

12.651

8.844

5.765

4.950

±0.000

-0.928

-1.526

2372　　1550　　　　3920　　　　　　3920　　　　　　3920　　　　1560

17142

0　0.5　1m

雉门东立面图
East elevation of the Zhi Gate

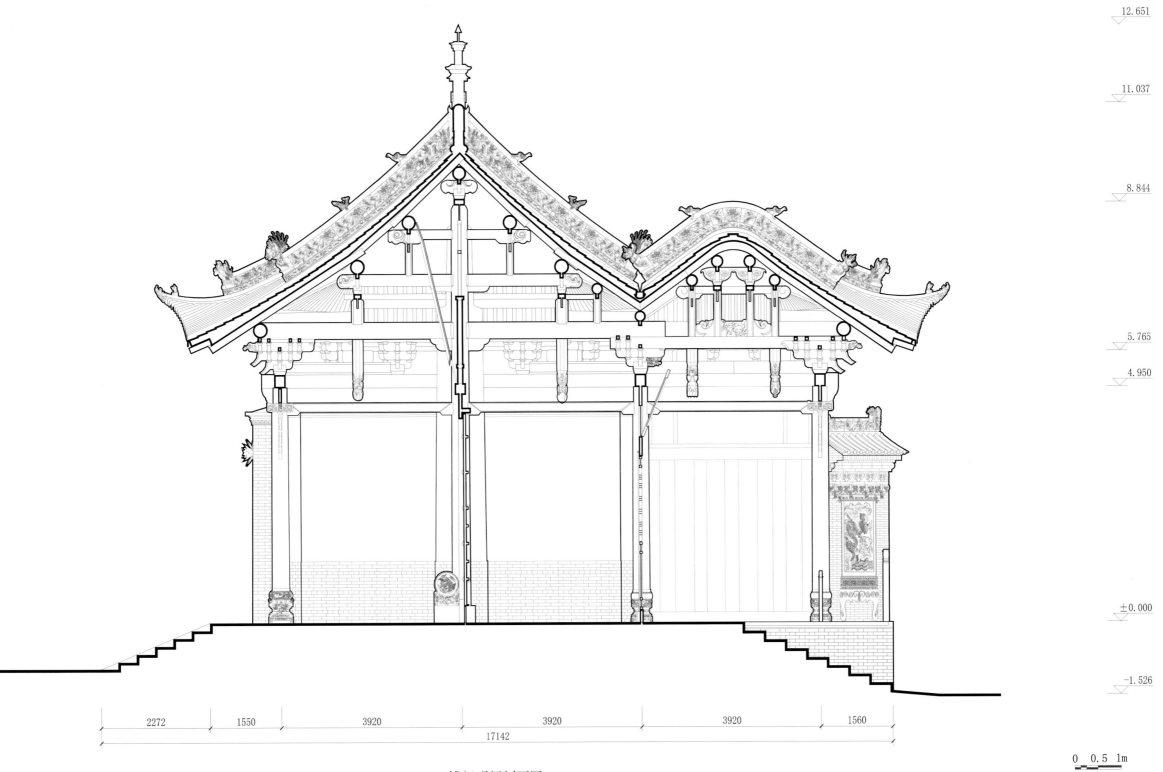

12.651

11.037

8.844

5.765

4.950

±0.000

-1.526

2272　1550　3920　3920　3920　1560

17142

0　0.5　1m

雉门明间剖面图
Central bay section of the Zhi Gate

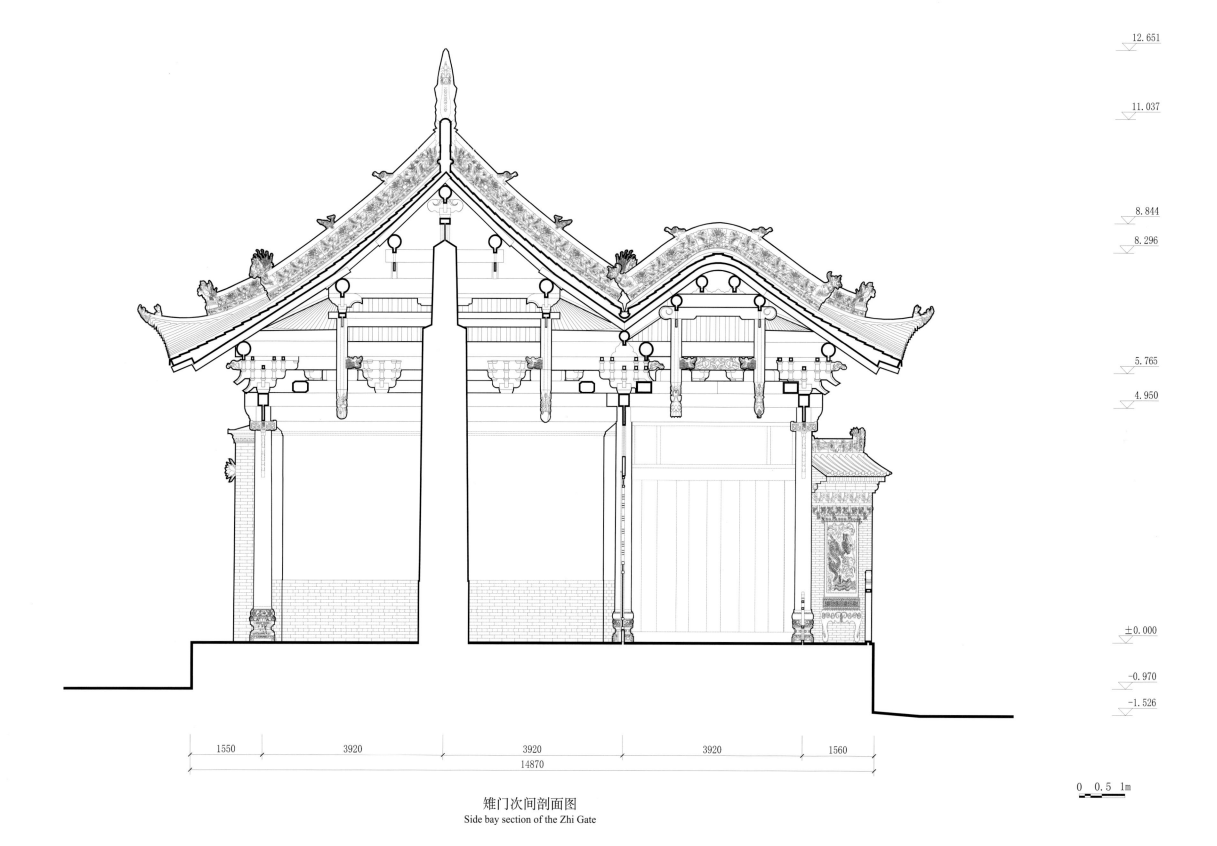

12.651

11.037

8.844

8.296

5.765

4.950

±0.000

-0.970

-1.526

1550　　3920　　3920　　3920　　1560

14870

0　0.5　1m

雉门次间剖面图
Side bay section of the Zhi Gate

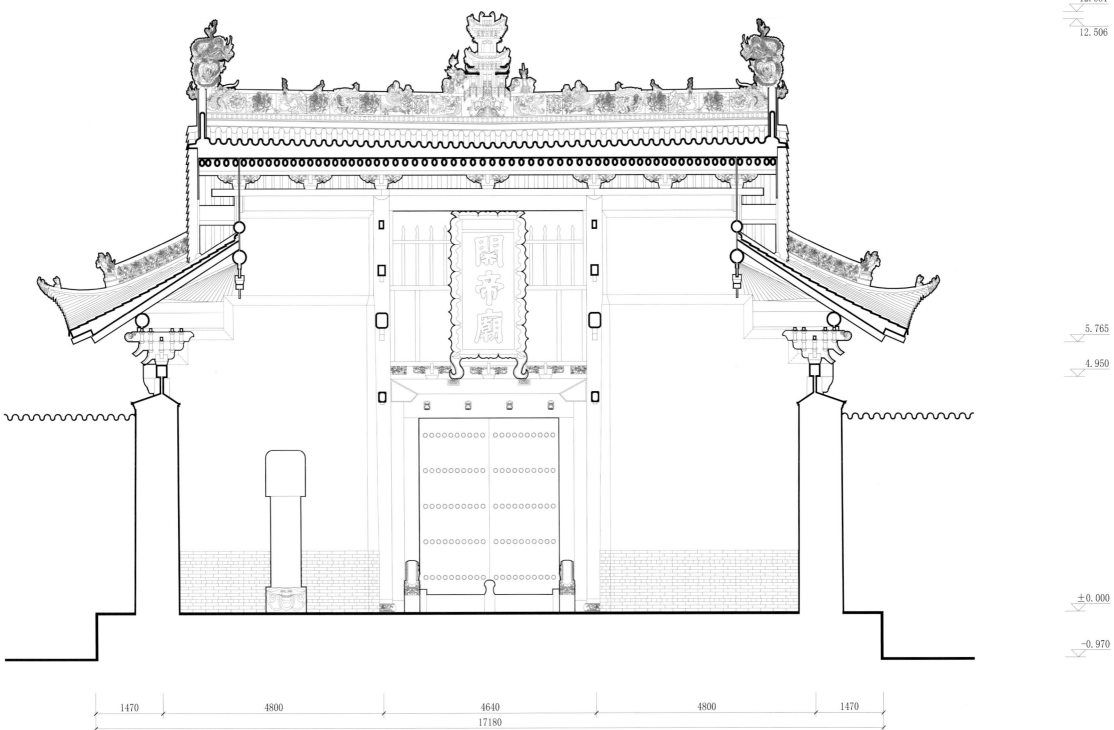

12.651
12.506

5.765
4.950

±0.000

-0.970

| 1470 | 4800 | 4640 | 4800 | 1470 |

17180

0　0.5　1m

雉门前厅纵剖面图
Longitudinal section of the front hallway of the Zhi Gate

关帝庙

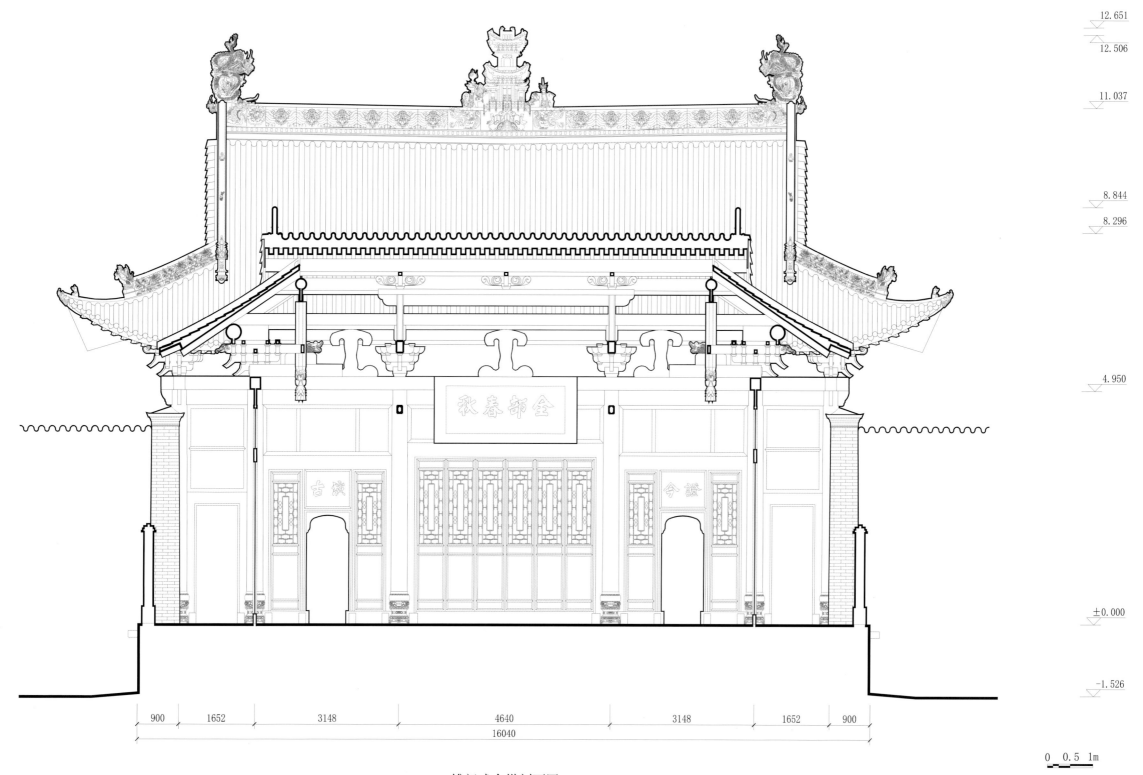

12.651
12.506
11.037
8.844
8.296
4.950
±0.000
-1.526

900 1652 3148 4640 3148 1652 900
16040

0 0.5 1m

雉门戏台纵剖面图
Longitudinal section of the back portico of the Zhi Gate

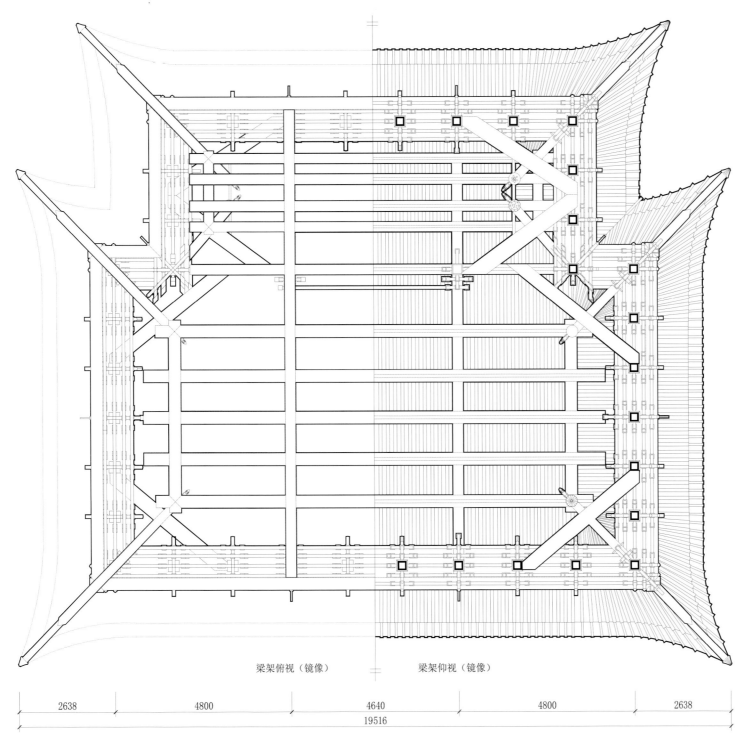

2638

1282

2638

3920

17036

3920

2638

2638   4800   4640   4800   2638

19516

梁架俯视（镜像）   梁架仰视（镜像）

0   1   2m

雉门梁架仰俯视图
Top and bottom views of the beam frame of the Zhi Gate

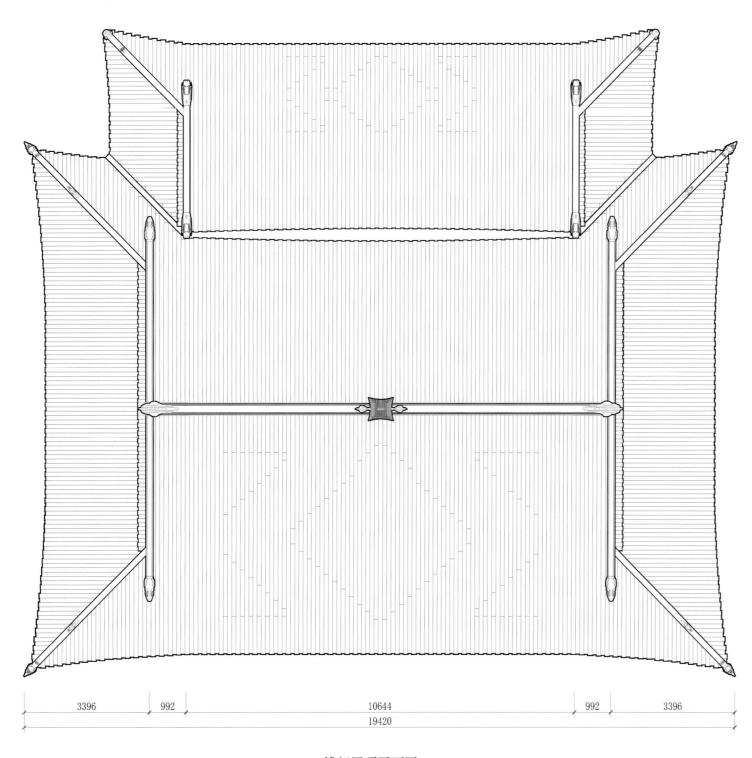

2938

17068

14130

3396 992 10644 992 3396

19420

雉门屋顶平面图
Roof plan of the Zhi Gate

0 1 2m

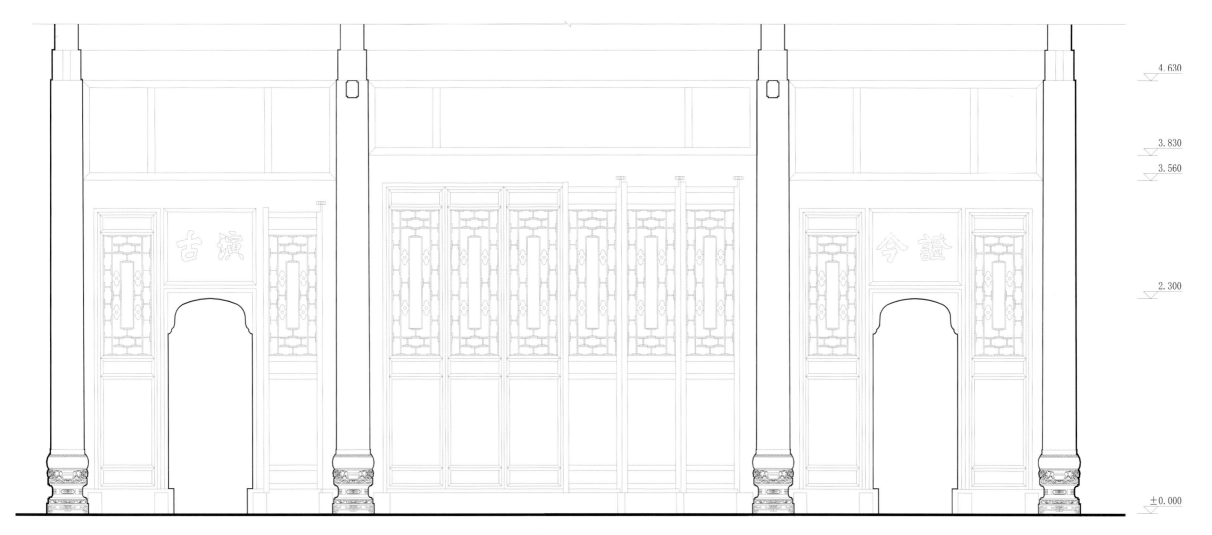

雉门格扇门立面图
Elevation of the partition door of the Zhi Gate

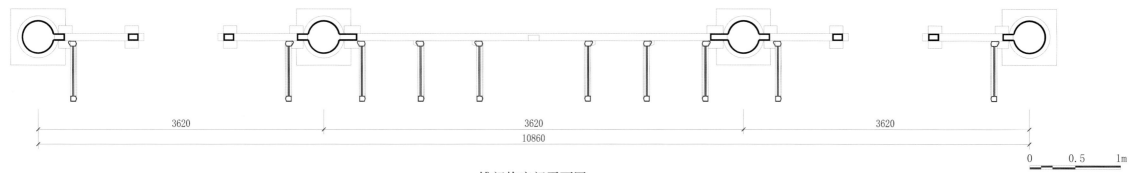

雉门格扇门平面图
Plan of the partition door of the Zhi Gate

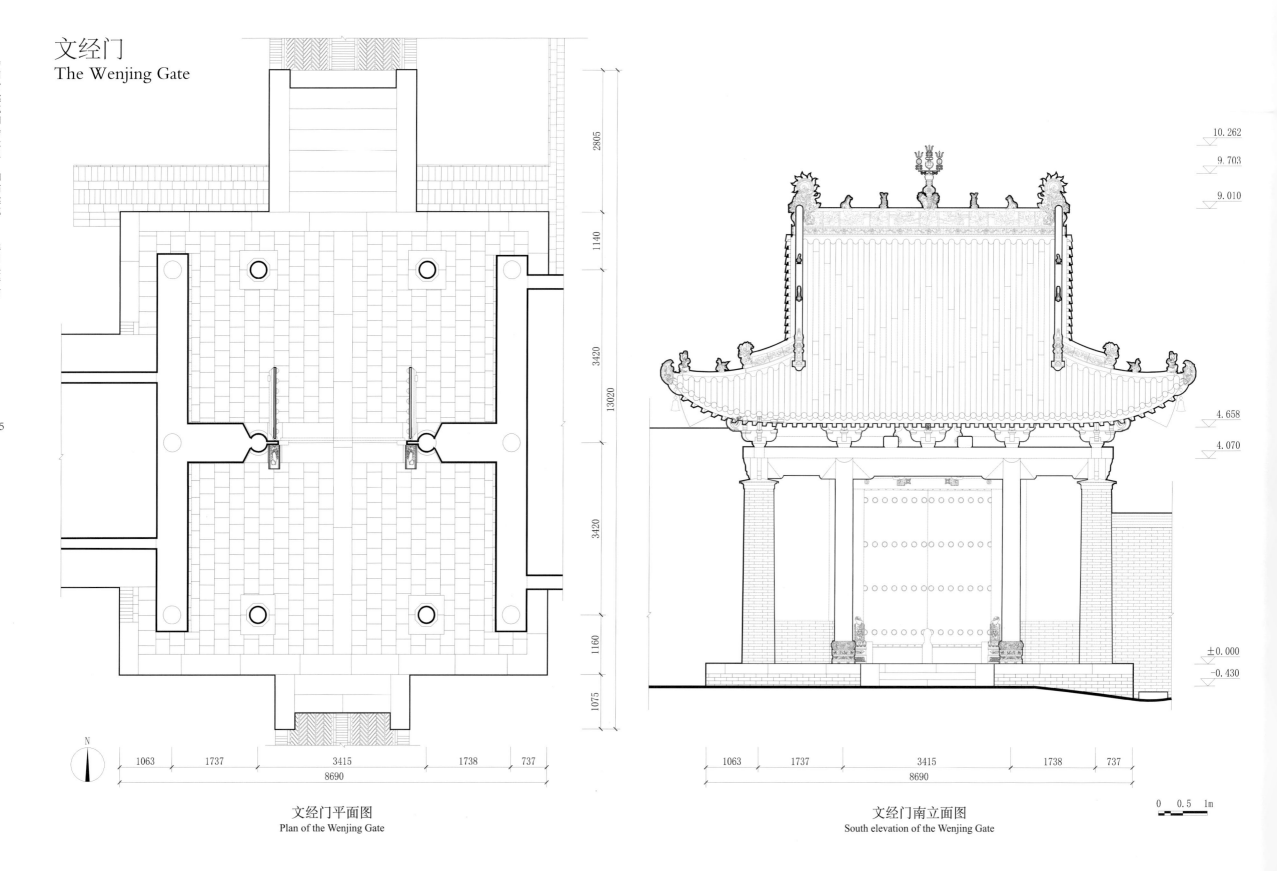

中
国
古
建
筑
测
绘
大
系
·
祠
庙
建
筑
——
解
州
关
帝
庙

文经门平面图
Plan of the Wenjing Gate

文经门南立面图
South elevation of the Wenjing Gate

N

2805
1140
3420
13020
3420
1160
1075

1063 1737 3415 1738 737
8690

10.262
9.703
9.010
4.658
4.070
±0.000
-0.430

1063 1737 3415 1738 737
8690

0 0.5 1m

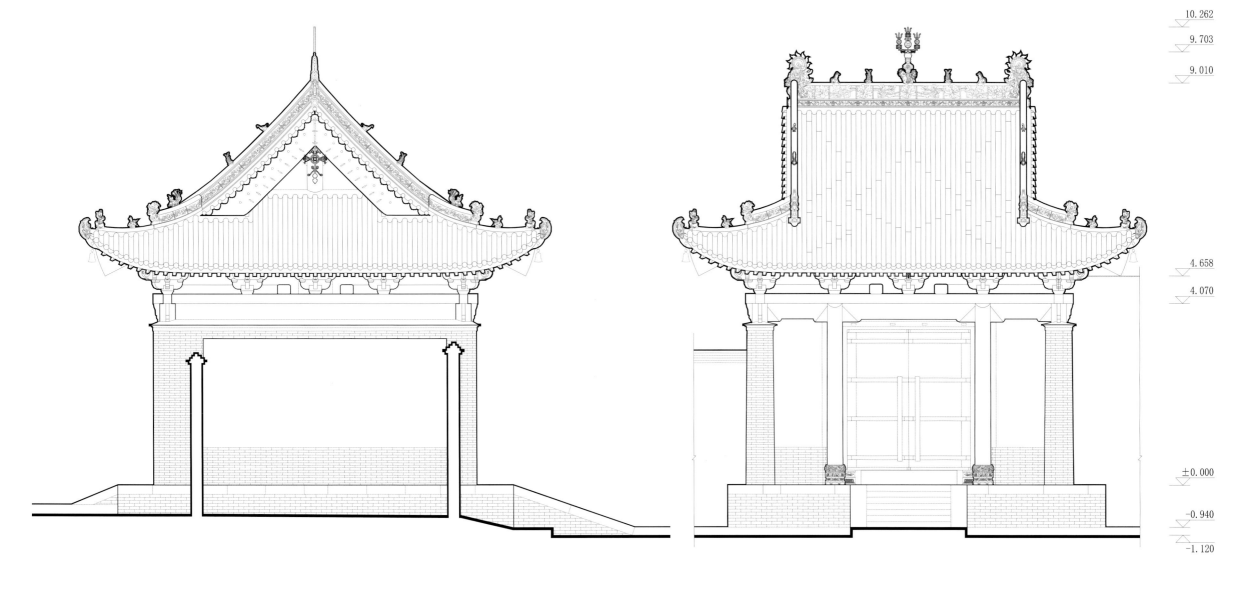

10.262

9.703

9.010

4.658

4.070

±0.000

-0.940

-1.120

1075 1160 3420 3420 1140 2805

13020

文经门东立面图
East elevation of the Wenjing Gate

750 1732 3425 1732 1060

8699

0 0.5 1m

文经门北立面图
North elevation of the Wenjing Gate

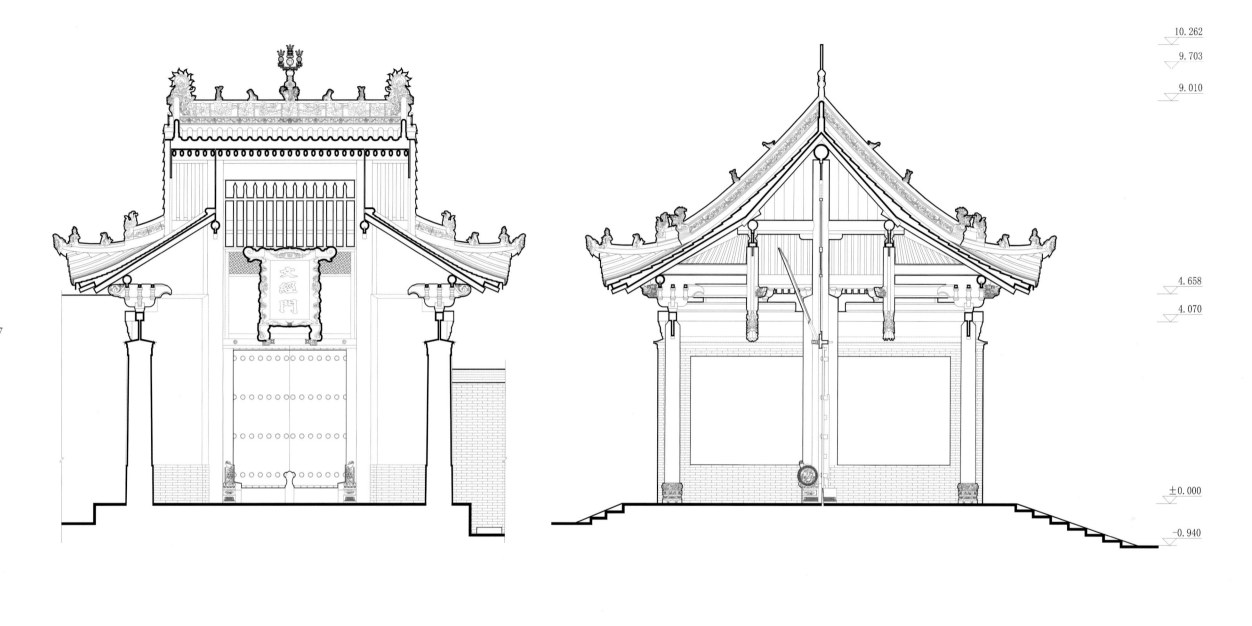

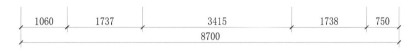

文经门纵剖面图
Longitudinal section of the Wenjing Gate

文经门横剖面图
Cross-section of the Wenjing Gate

中国古建筑测绘大系·祠庙建筑——解州关帝庙

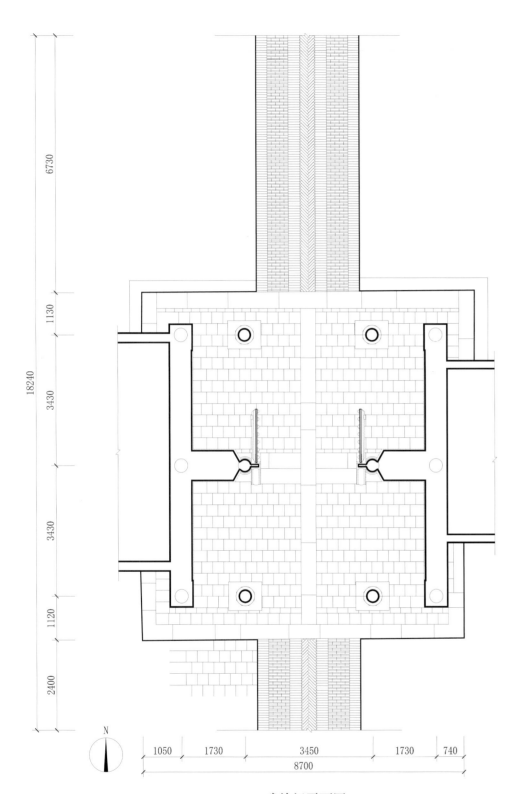

武纬门平面图
Plan of the Wuwei Gate

南立面图区
10.262
9.702
9.010

4.676
4.070

±0.000
-0.400

1050　1730　3450　1730　740
8700

武纬门南立面图
South elevation of the Wuwei Gate

0　0.5　1m

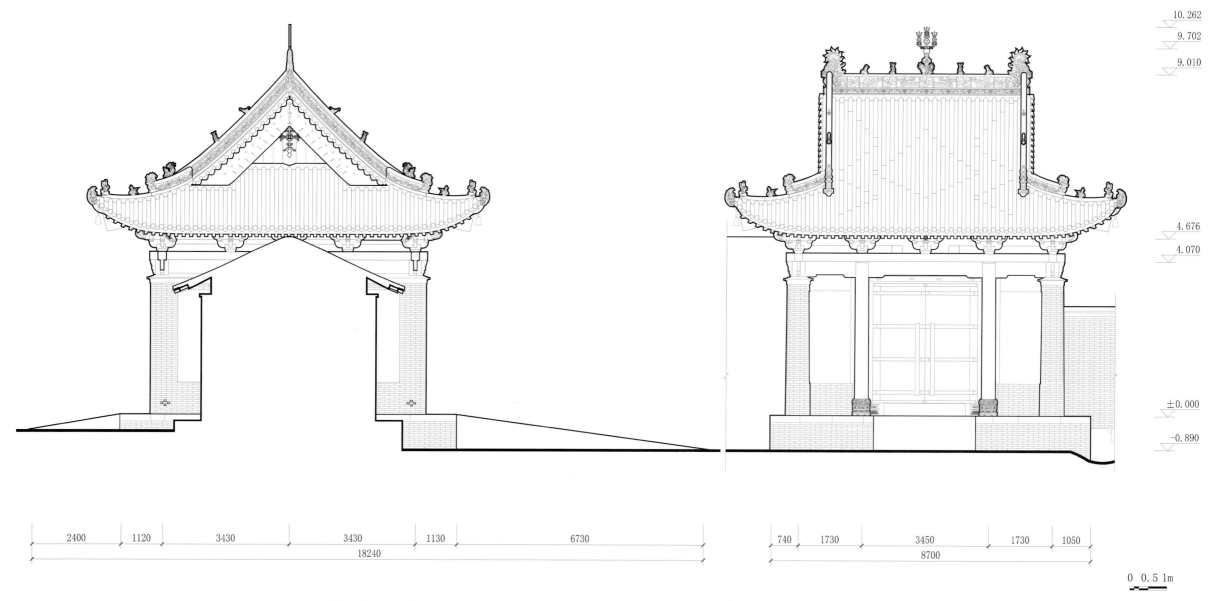

10.262
9.702
9.010
4.676
4.070
±0.000
-0.890

2400　1120　　3430　　　3430　1130　　　6730
18240

740　1730　　3450　　1730　1050
8700

0　0.5　1m

武纬门东立面图
East elevation of the Wuwei Gate

武纬门北立面图
North elevation of the Wuwei Gate

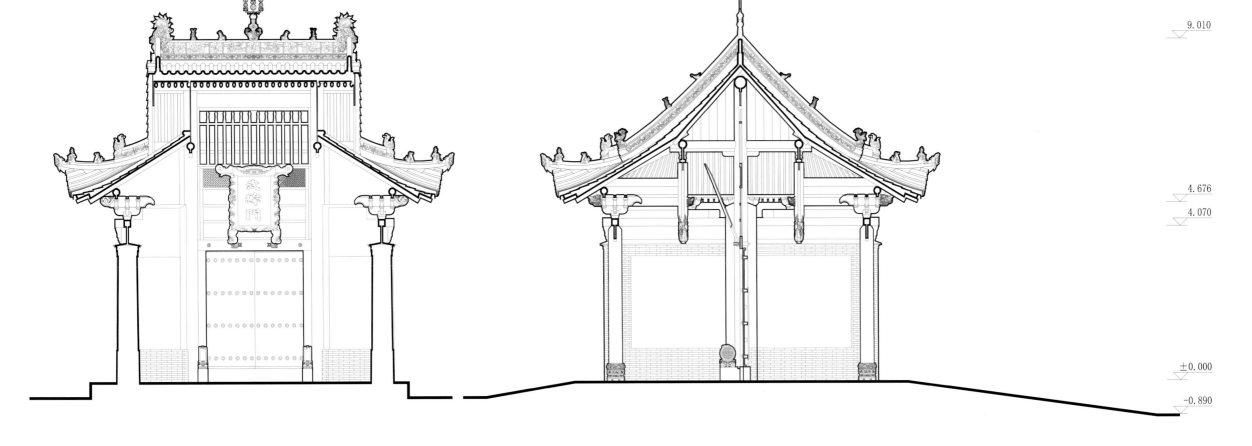

10.262

9.010

060

4.676

4.070

±0.000

-0.890

| 1050 | 1730 | 3450 | 1730 | 740 |
8700

| 2400 | 1120 | 3430 | 3430 | 1130 | 6730 |
18240

0　0.5 1m

武纬门纵剖面图
Longitudinal section of the Wuwei Gate

武纬门横剖面图
Cross-section of the Wuwei Gate

追风伯祠
The Zhuifengbo Temple

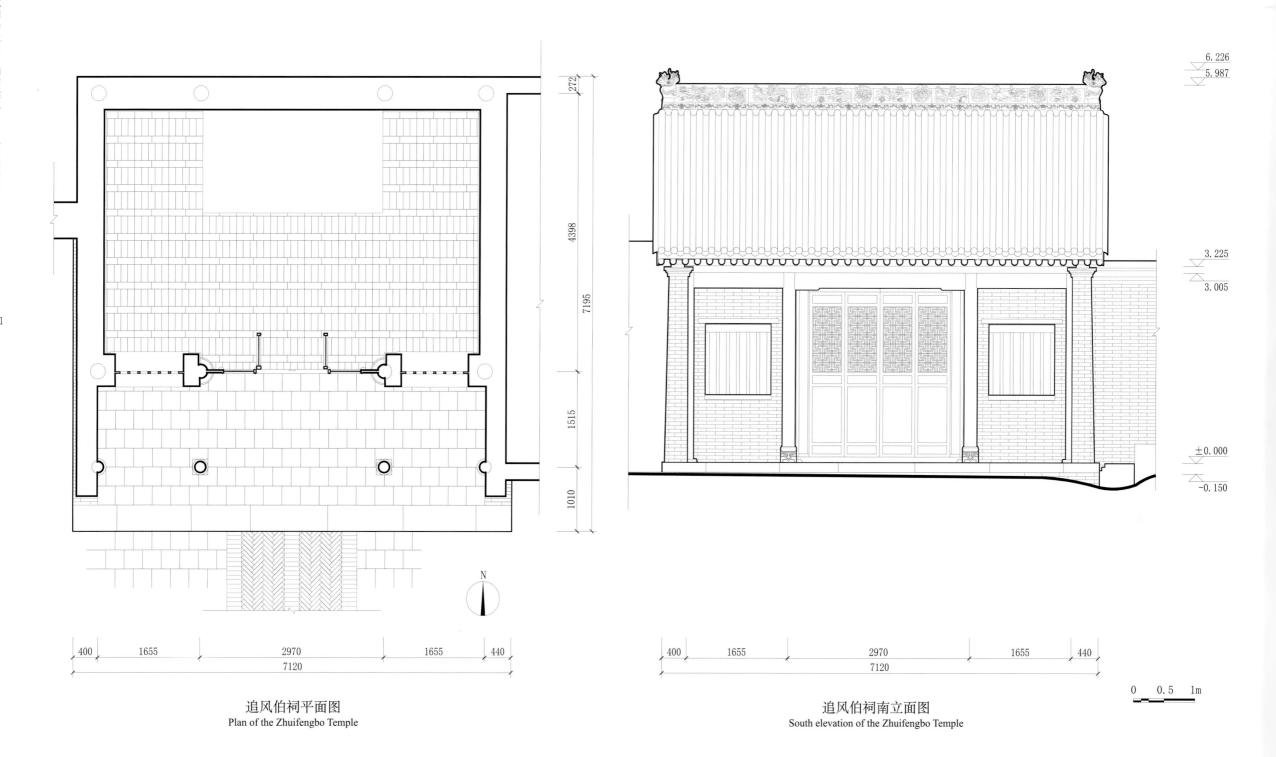

追风伯祠平面图
Plan of the Zhuifengbo Temple

追风伯祠南立面图
South elevation of the Zhuifengbo Temple

中国古建筑测绘大系·祠庙建筑——解州关帝庙

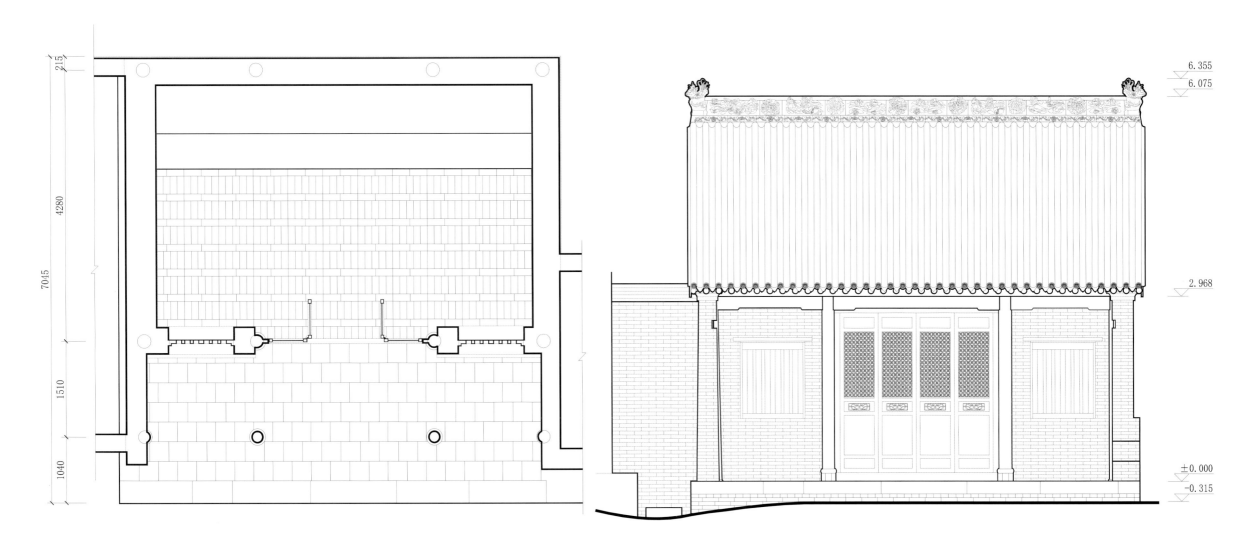

部将祠平面图
Plan of the Bujiang Temple

部将祠南立面图
South elevation of the Bujiang Temple

中国古建筑测绘大系·祠庙建筑——解州关帝庙

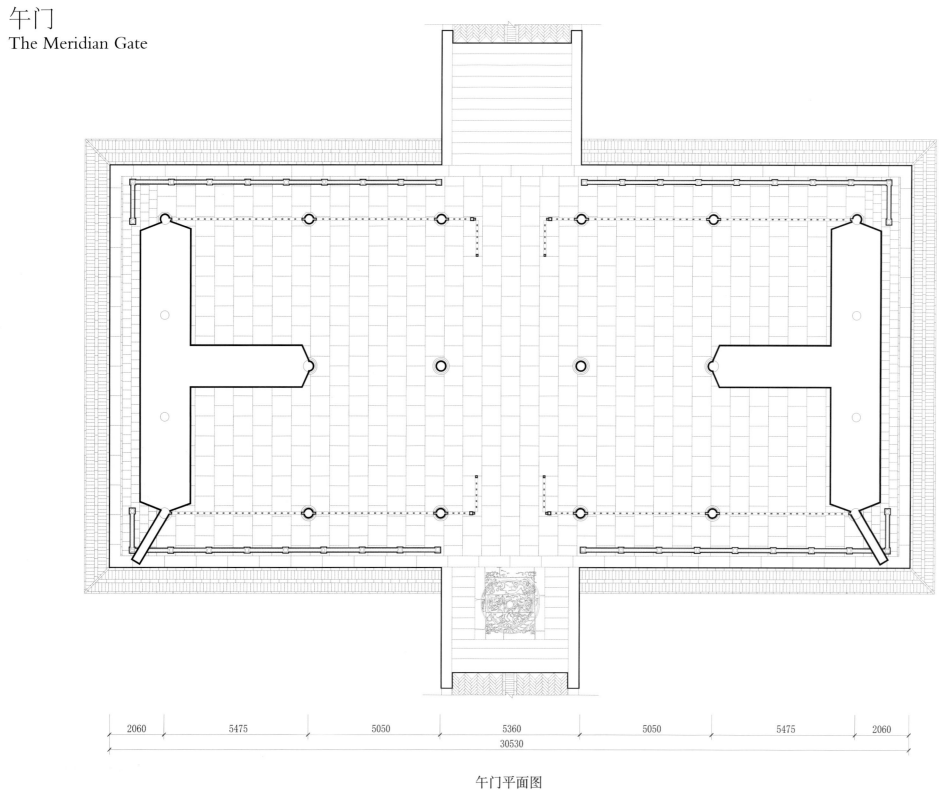

4973

1995

3560

1885

24323

1885

3560

1995

4470

2060　5475　5050　5360　5050　5475　2060

30530

N

0　1　2m

午门平面图
Plan of the Meridian Gate

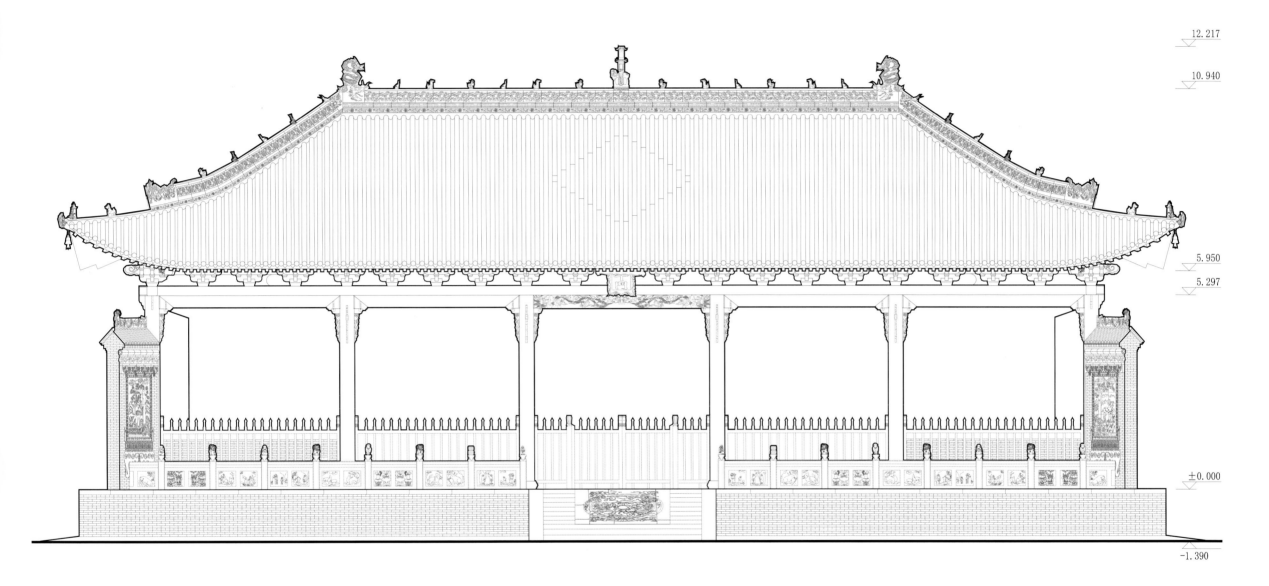

12.217

10.940

5.950

5.297

±0.000

−1.390

2060　　5475　　5050　　5360　　5050　　5475　　2060

30530

0　1　2m

午门南立面图

South elevation of the Meridian Gate

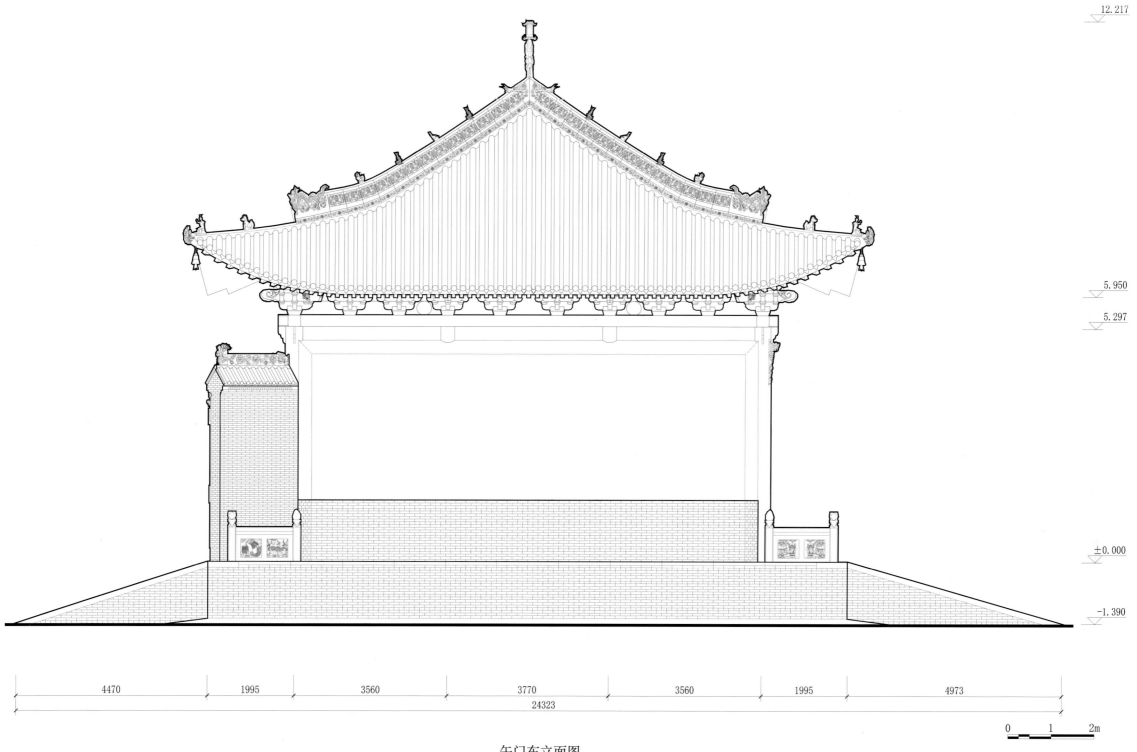

12.217

5.950

5.297

±0.000

-1.390

4470　1995　3560　3770　3560　1995　4973

24323

0　1　2m

午门东立面图
East elevation of the Meridian Gate

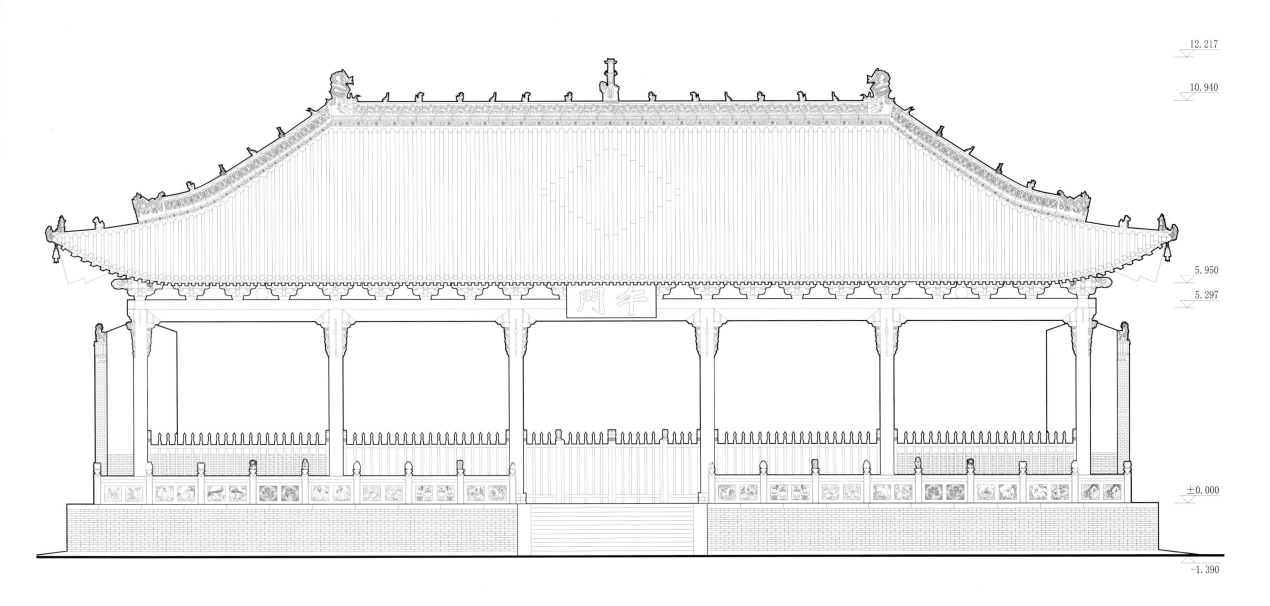

12.217

10.940

5.950

5.297

±0.000

-1.390

| 2060 | 5475 | 5050 | 5360 | 5050 | 5475 | 2060 |

30530

0　1　2m

午门北立面图
North elevation of the Meridian Gate

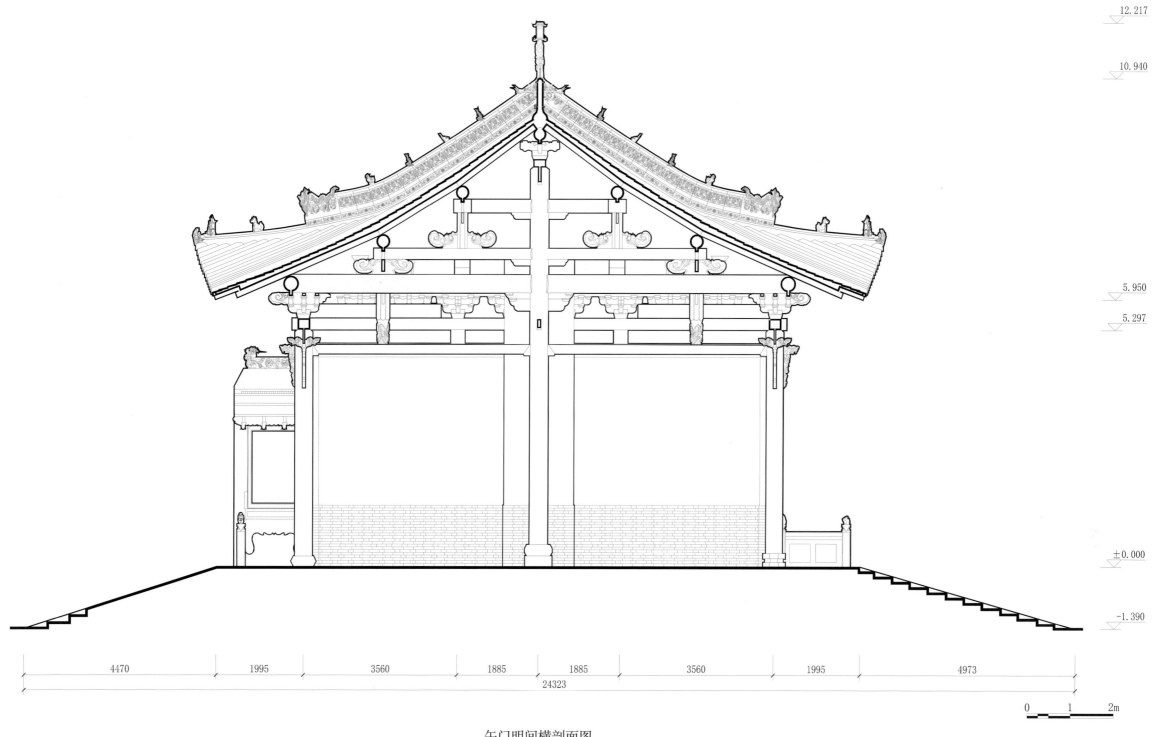

12.217

10.940

5.950

5.297

±0.000

-1.390

4470   1995   3560   1885   1885   3560   1995   4973

24323

0   1   2m

午门明间横剖面图

Central bay section of the Meridian Gate

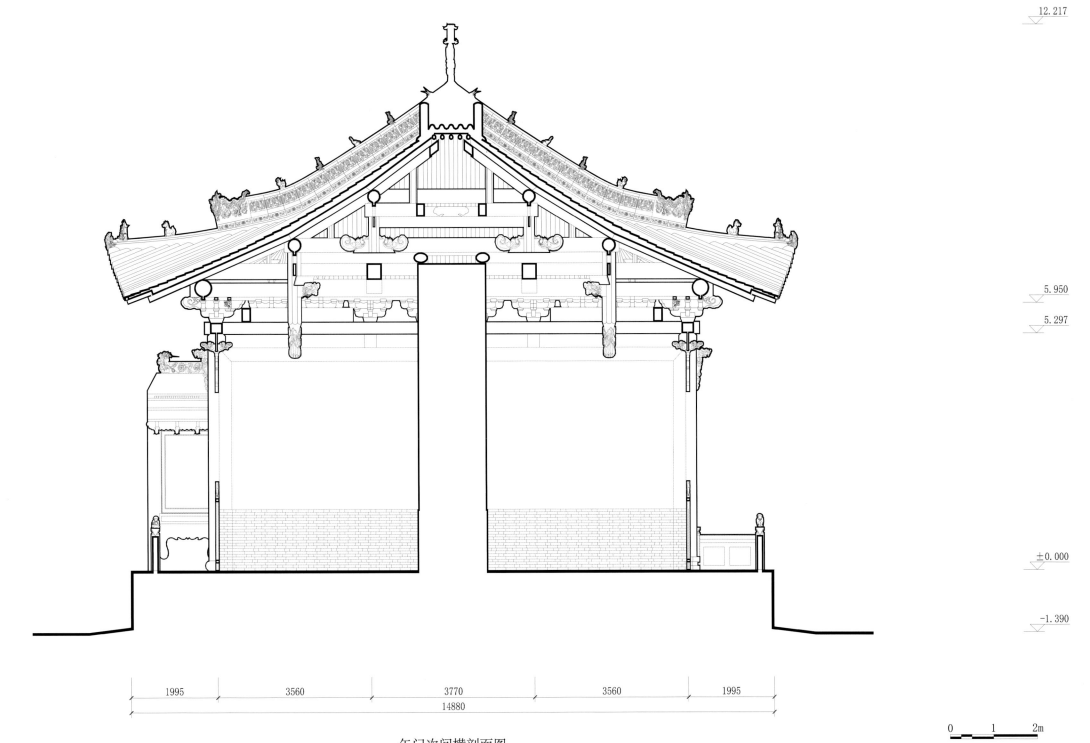

12.217

5.950

5.297

±0.000

−1.390

| 1995 | 3560 | 3770 | 3560 | 1995 |

14880

0 1 2m

午门次间横剖面图
Side bay section of the Meridian Gate

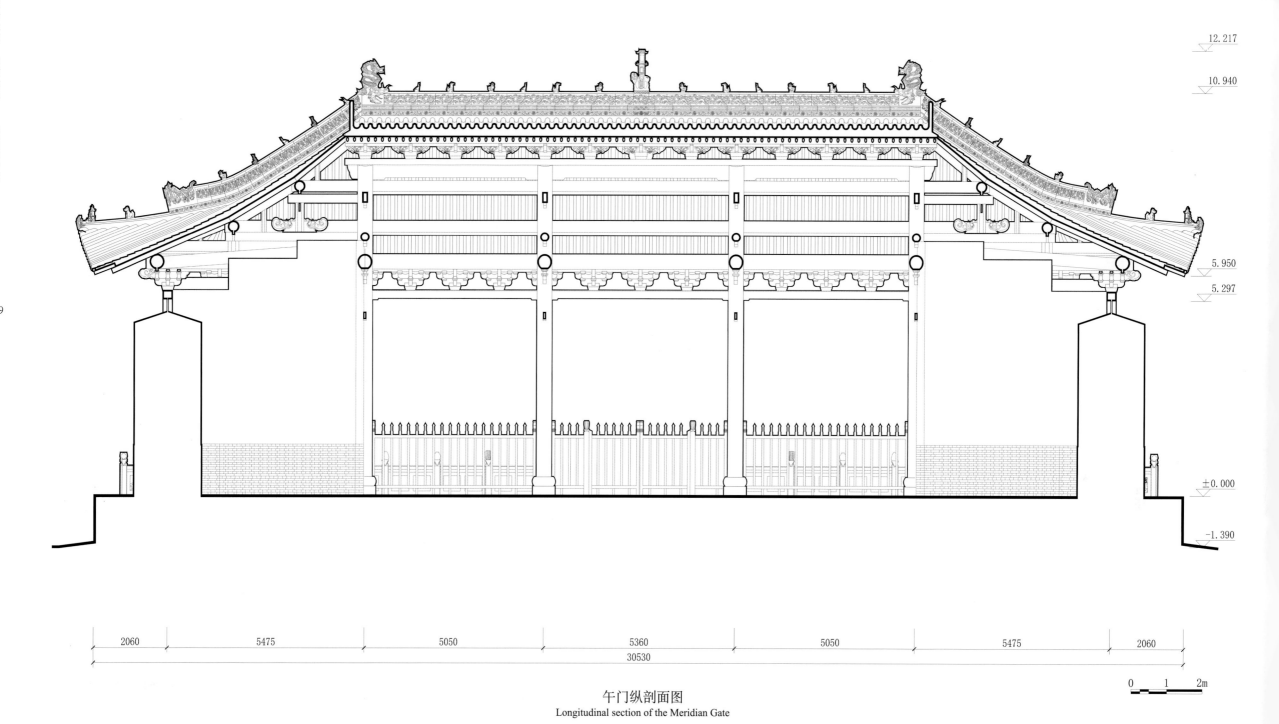

12.217

10.940

5.950

5.297

±0.000

-1.390

2060　5475　5050　5360　5050　5475　2060

30530

0　1　2m

午门纵剖面图
Longitudinal section of the Meridian Gate

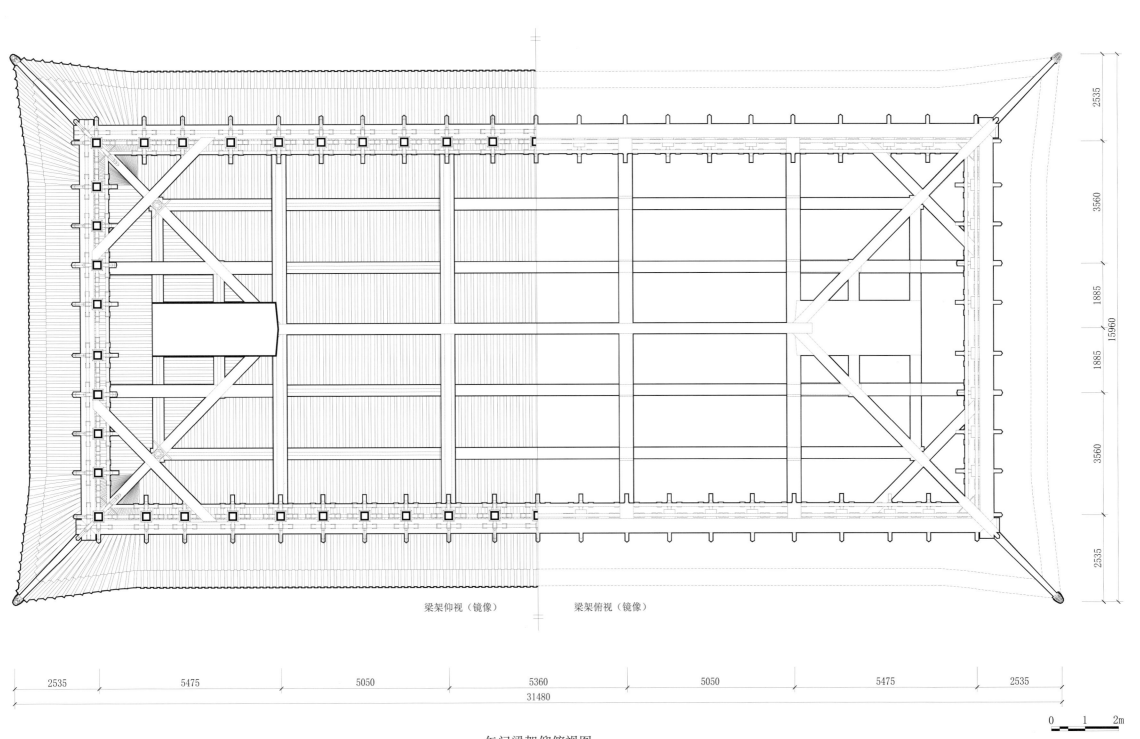

2535
3560
1885
15960
1885
3560
2535

梁架仰视（镜像）　　梁架俯视（镜像）

2535　5475　5050　5360　5050　5475　2535
31480

0　1　2m

午门梁架仰俯视图
Top and bottom views of the beam frame of the Meridian Gate

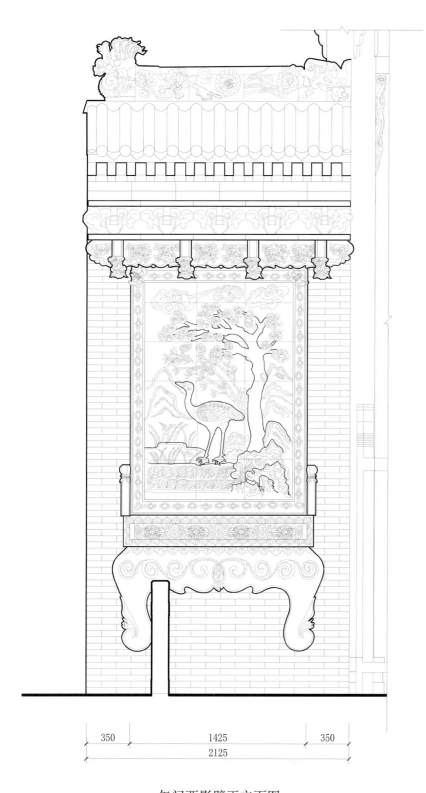

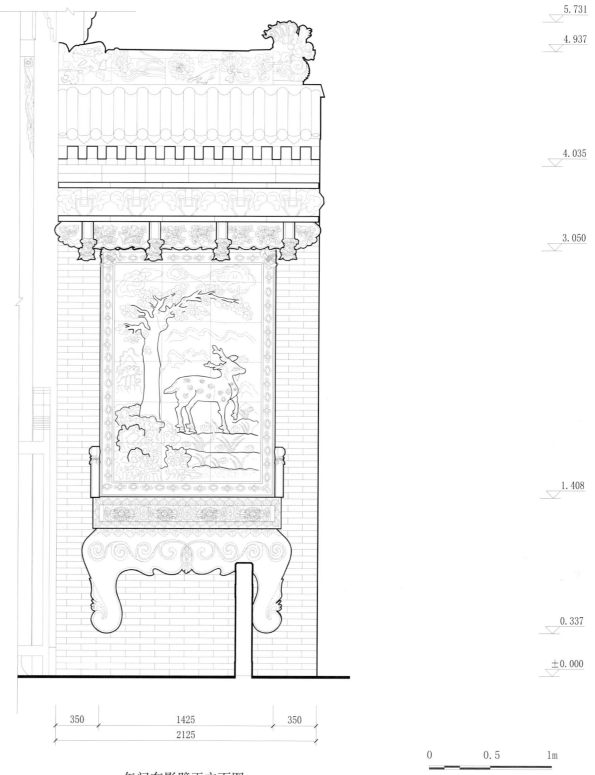

5.731

4.937

4.035

3.050

1.408

0.337

±0.000

| 350 | 1425 | 350 |
2125

| 350 | 1425 | 350 |
2125

0   0.5   1m

午门西影壁正立面图
Front elevation of west screen wall of the Meridian Gate

午门东影壁正立面图
Front elevation of east screen wall of the Meridian Gate

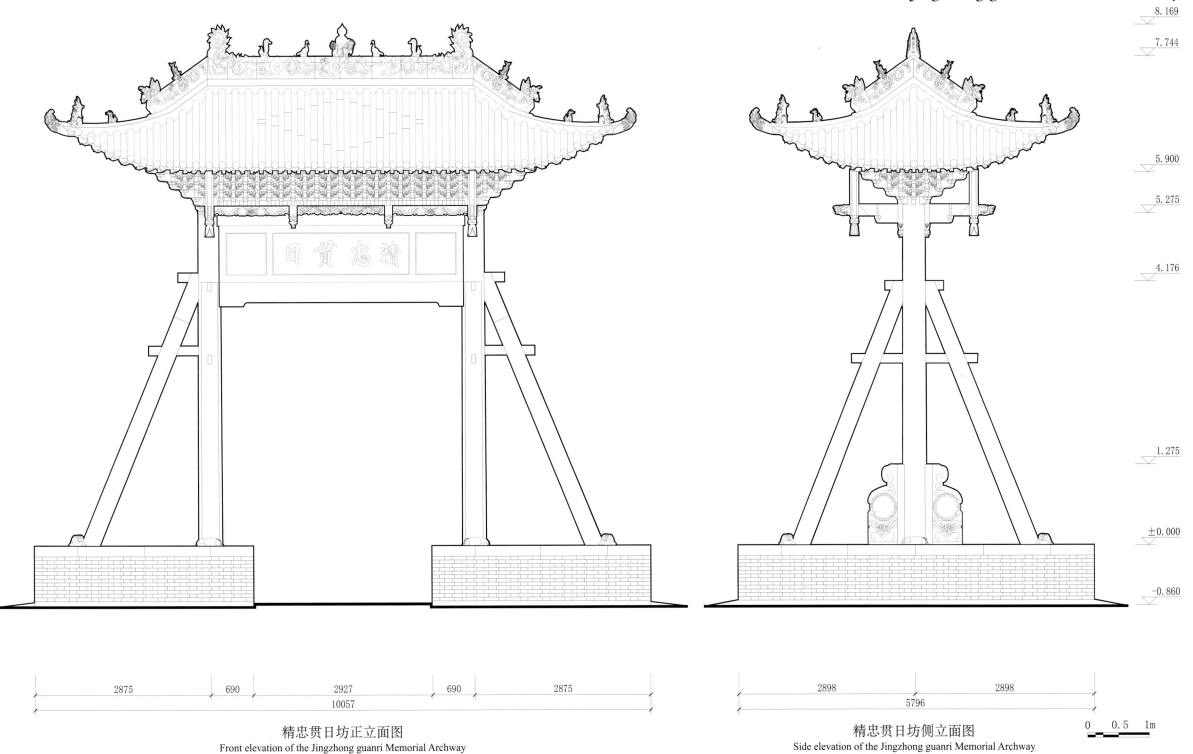

精忠贯日坊

The Jingzhong guanri Memorial Archway

8.169

7.744

5.900

5.275

4.176

1.275

±0.000

-0.860

2875　690　2927　690　2875

10057

2898　2898

5796

精忠贯日坊正立面图

Front elevation of the Jingzhong guanri Memorial Archway

精忠贯日坊侧立面图

Side elevation of the Jingzhong guanri Memorial Archway

0　0.5　1m

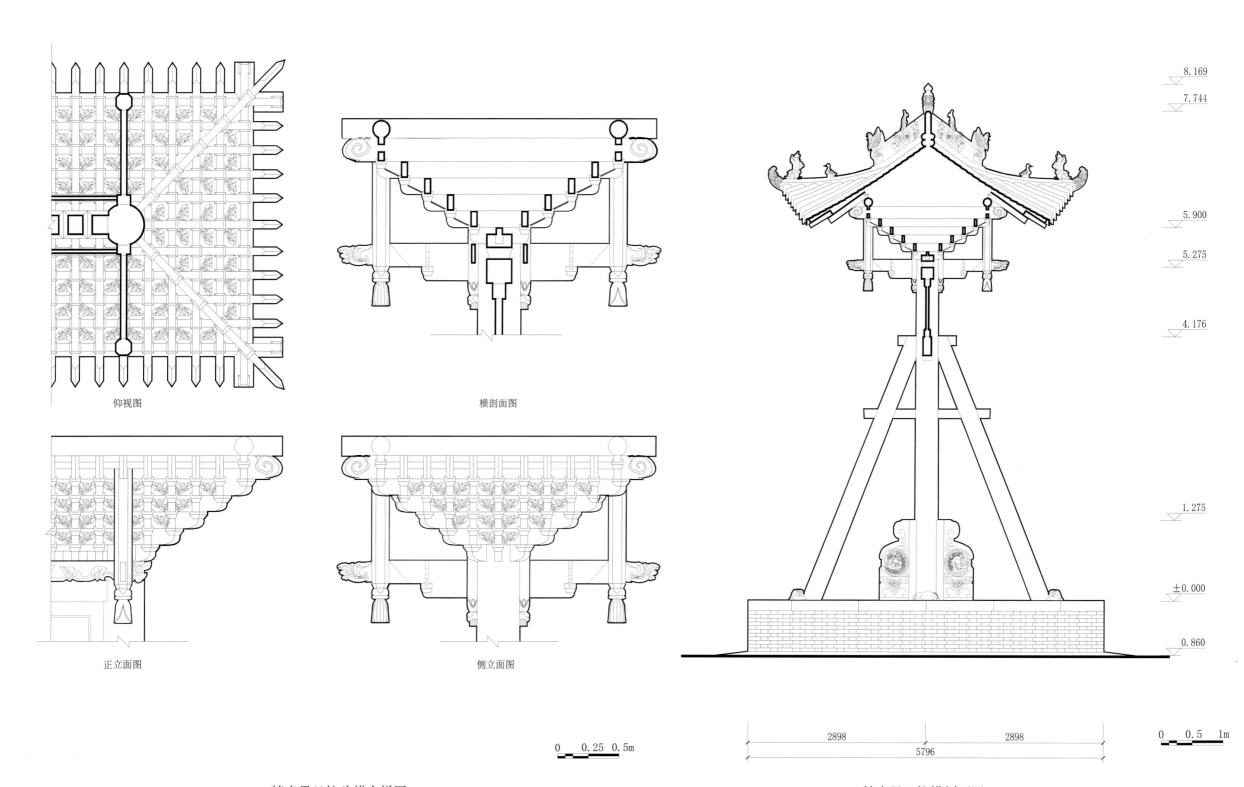

仰视图

横剖面图

正立面图

侧立面图

8.169

7.744

5.900

5.275

4.176

1.275

±0.000

0.860

0 0.25 0.5m

2898 2898

5796

0 0.5 1m

精忠贯日坊斗栱大样图
Expanded view of the bracket set of the Jingzhong guanri Memorial Archway

精忠贯日坊横剖面图
Cross-section of the Jingzhong guanri Memorial Archway

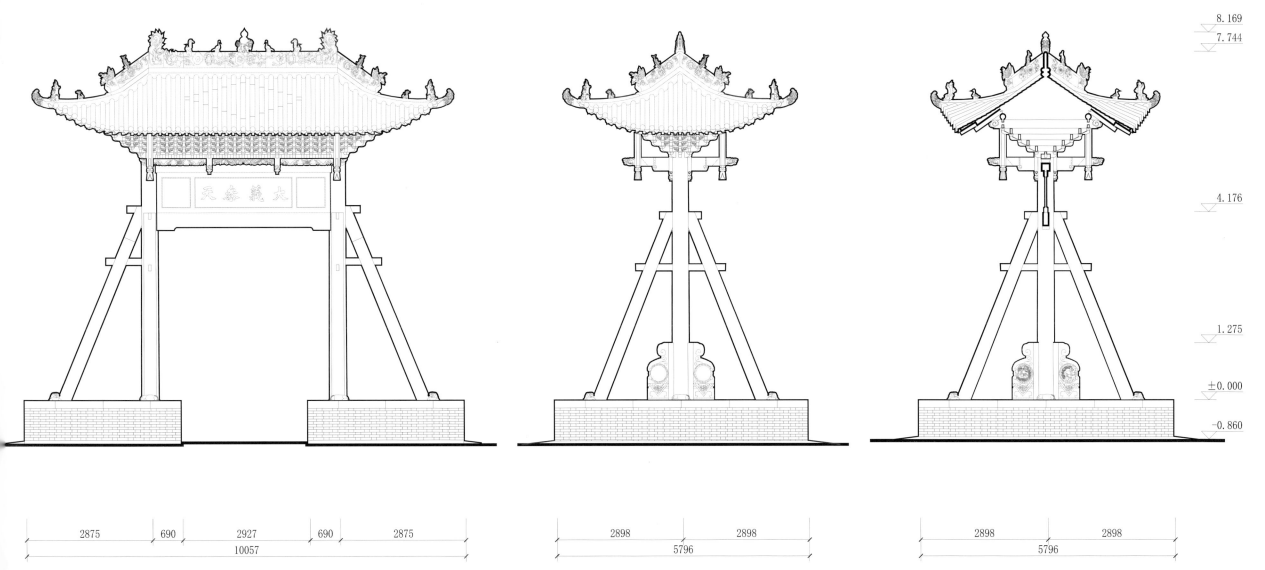

大义参天坊
The Dayi cantian Memorial Archway

中国古建筑测绘大系·祠庙建筑——解州关帝庙

074

8.169
7.744

4.176

1.275

±0.000

-0.860

| 2875 | 690 | 2927 | 690 | 2875 |
10057

| 2898 | 2898 |
5796

| 2898 | 2898 |
5796

大义参天坊正立面图
Front elevation of the Dayi cantian Memorial Archway

大义参天坊侧立面图
Side elevation of the Dayi cantian Memorial Archway

大义参天坊横剖面图
Cross-section of the Dayi cantian Memorial Archway

0 0.5 1m

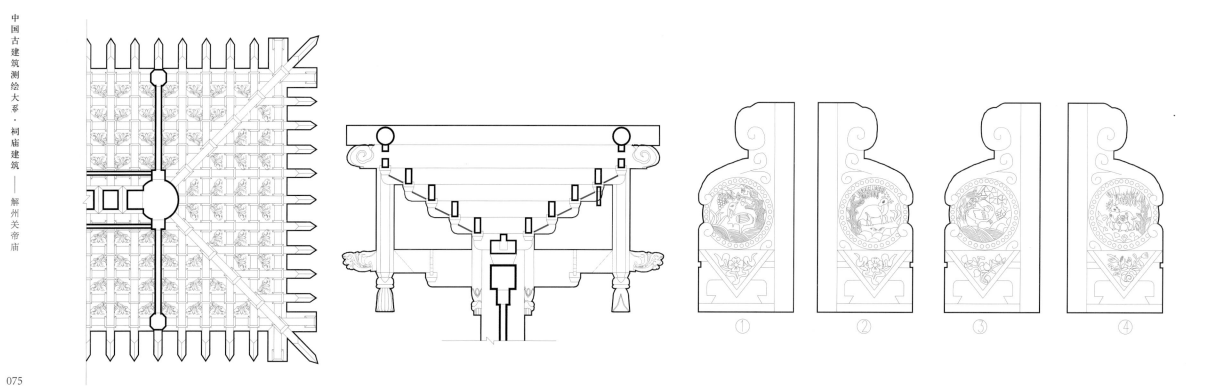

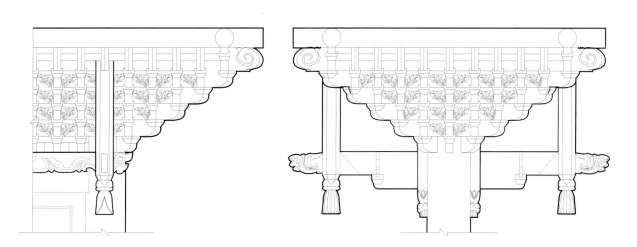

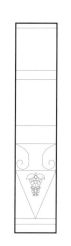

大义参天坊斗栱大样图
Expanded view of the bracket set of the Dayi cantian Memorial Archway

大义参天坊抱鼓石大样图
Expanded view of the drum-shaped bearing stone of the Dayi cantian Memorial Archway

0   0.25   0.5m

中国古建筑测绘大系·祠庙建筑——解州关帝庙

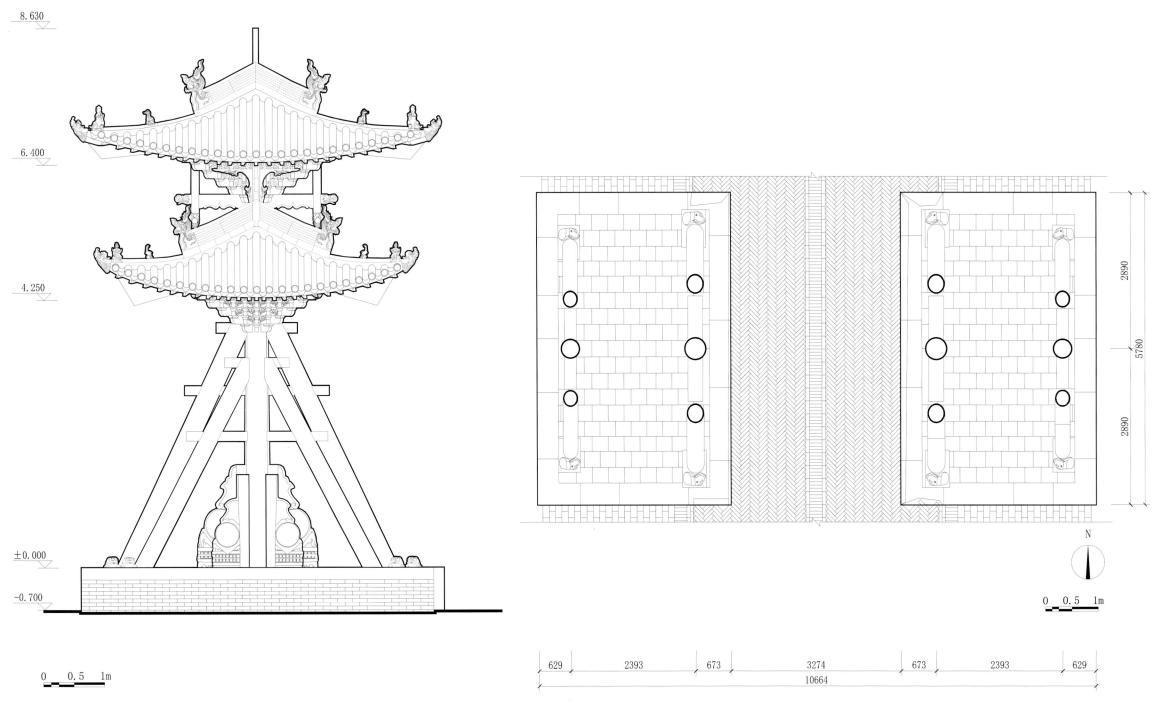

8.630

6.400

4.250

±0.000

-0.700

0 0.5 1m

N

0 0.5 1m

2890

5780

2890

629 2393 673 3274 673 2393 629
10664

山海钟灵坊侧立面图
Side elevation of the Shanhai zhongling Memorial Archway

山海钟灵坊平面图
Plan of the Shanhai zhongling Memorial Archway

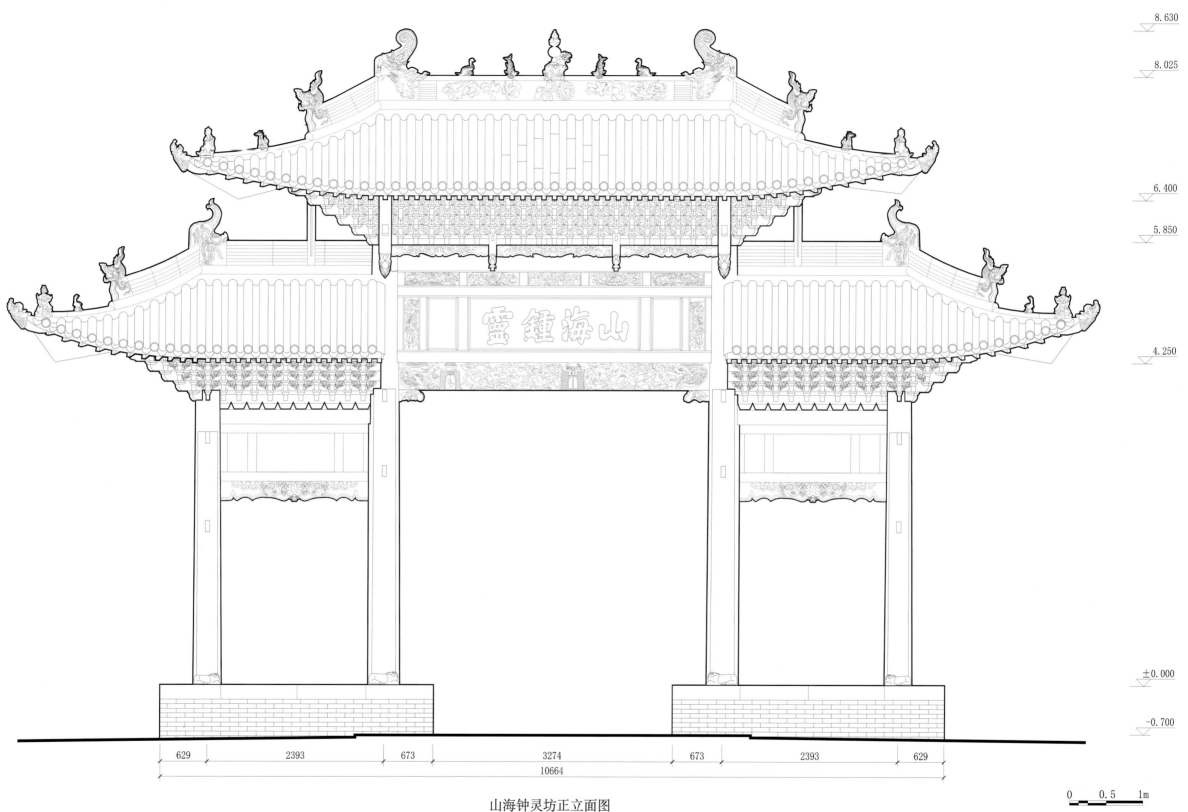

8.630

8.025

6.400

5.850

4.250

±0.000

-0.700

| 629 | 2393 | 673 | 3274 | 673 | 2393 | 629 |

10664

山海钟灵坊正立面图

Front elevation of the Shanhai zhongling Memorial Archway

0   0.5   1m

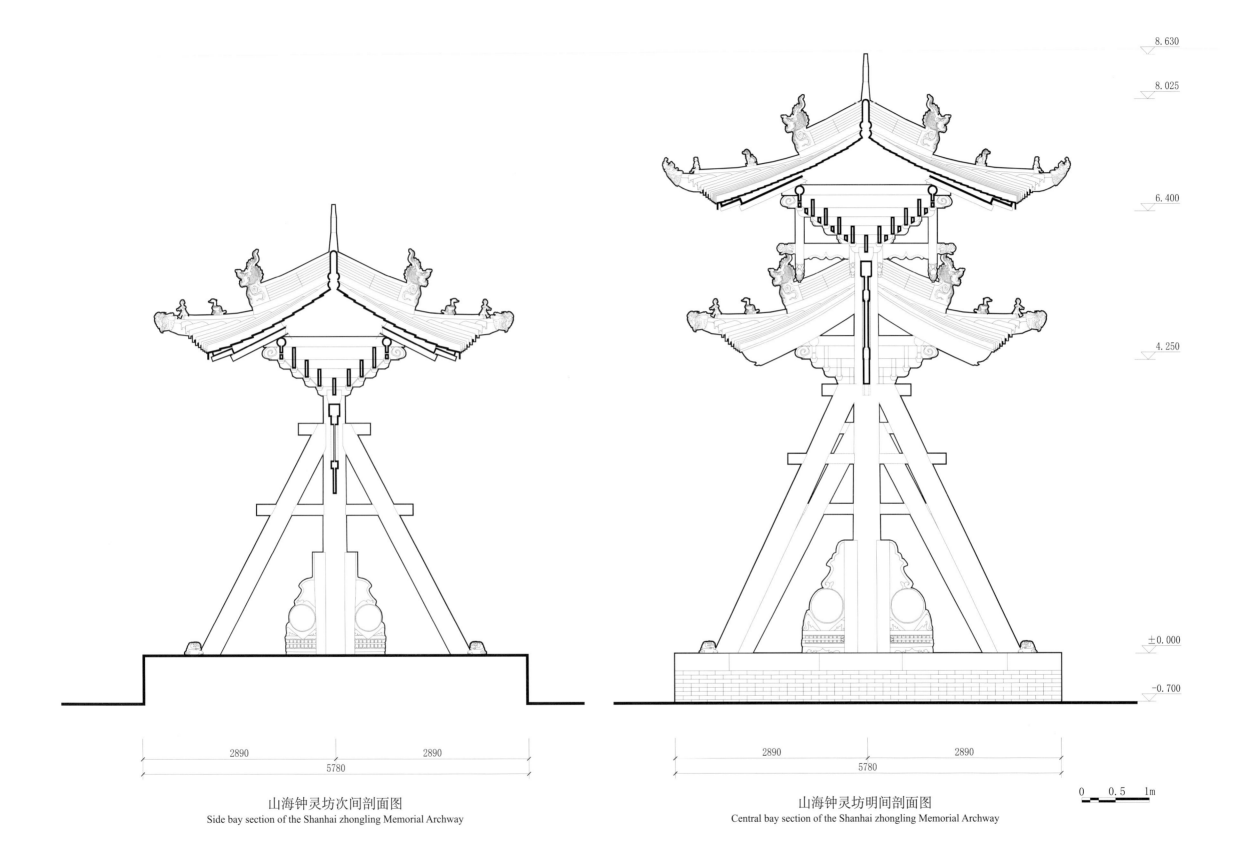

8.630

8.025

6.400

4.250

±0.000

-0.700

2890　　　2890

5780

山海钟灵坊次间剖面图

Side bay section of the Shanhai zhongling Memorial Archway

2890　　　2890

5780

山海钟灵坊明间剖面图

Central bay section of the Shanhai zhongling Memorial Archway

0　　0.5　　1m

御书楼
The Yushu Storied Building

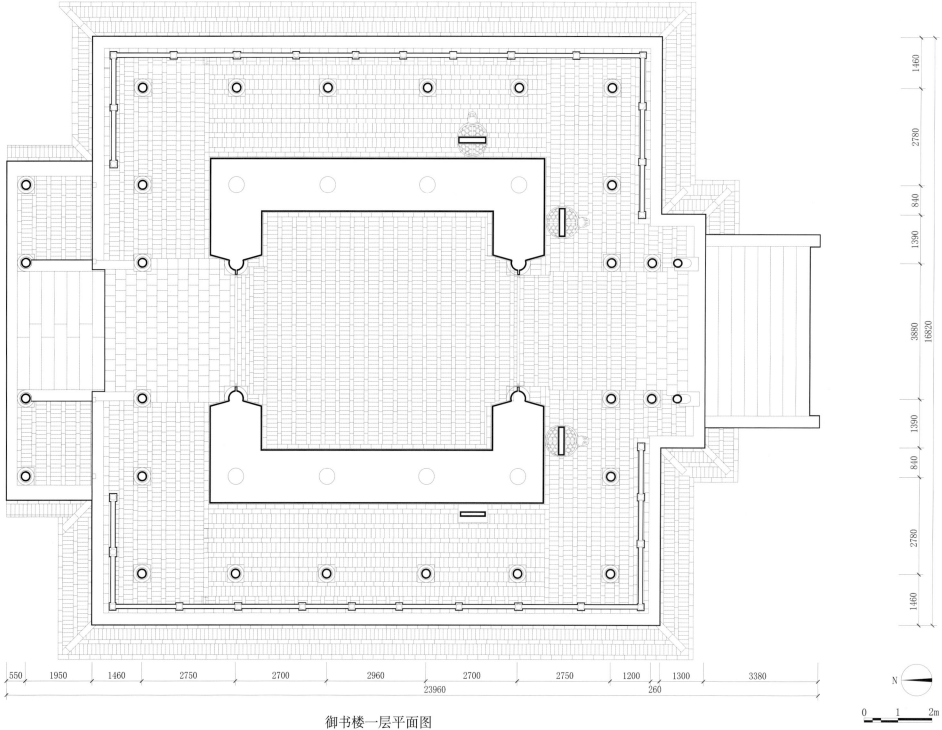

御书楼一层平面图
First Floor Plan of the Yushu Storied Building

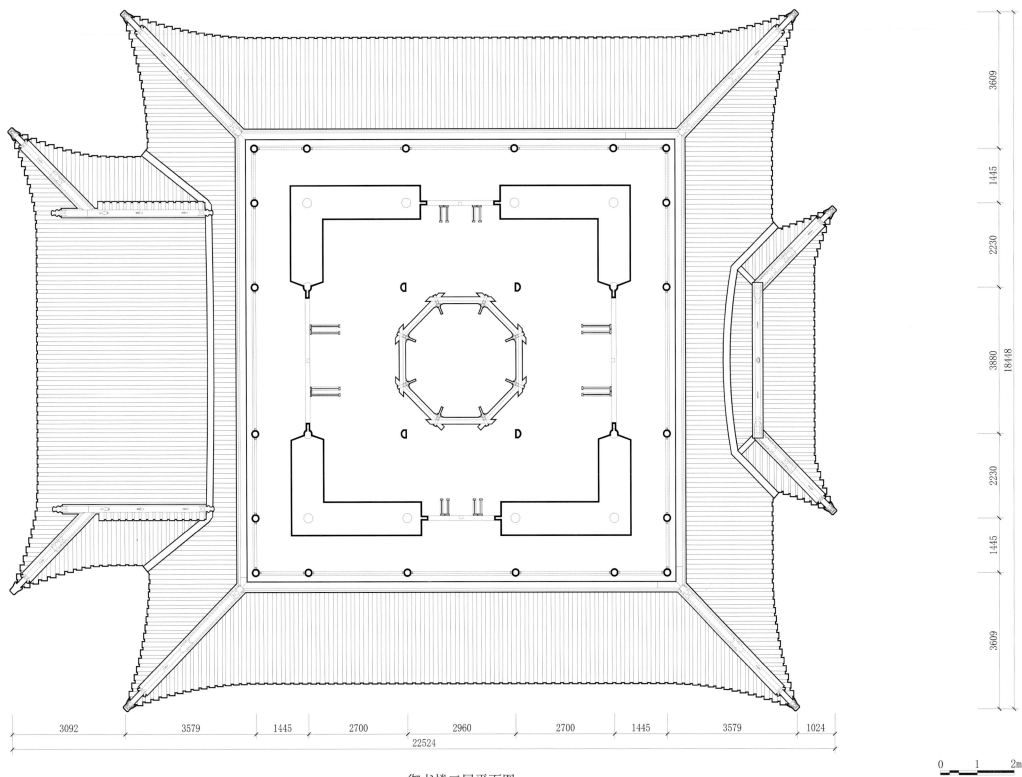

3609

1445

2230

3880 | 18448

2230

1445

3609

3092 | 3579 | 1445 | 2700 | 2960 | 2700 | 1445 | 3579 | 1024

22524

御书楼二层平面图
Second Floor Plan of the Yushu Storied Building

0 1 2m

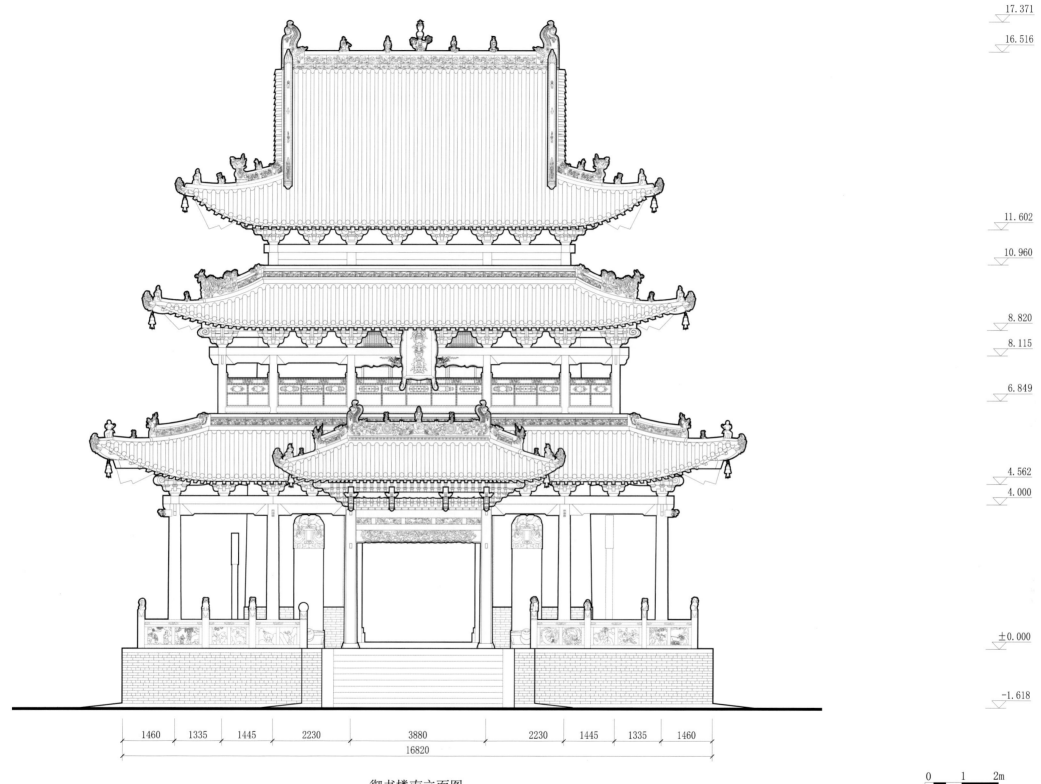

17.371

16.516

11.602

10.960

8.820

8.115

6.849

4.562

4.000

±0.000

-1.618

| 1460 | 1335 | 1445 | 2230 | 3880 | 2230 | 1445 | 1335 | 1460 |

16820

御书楼南立面图
South elevation of the Yushu Storied Building

0 1 2m

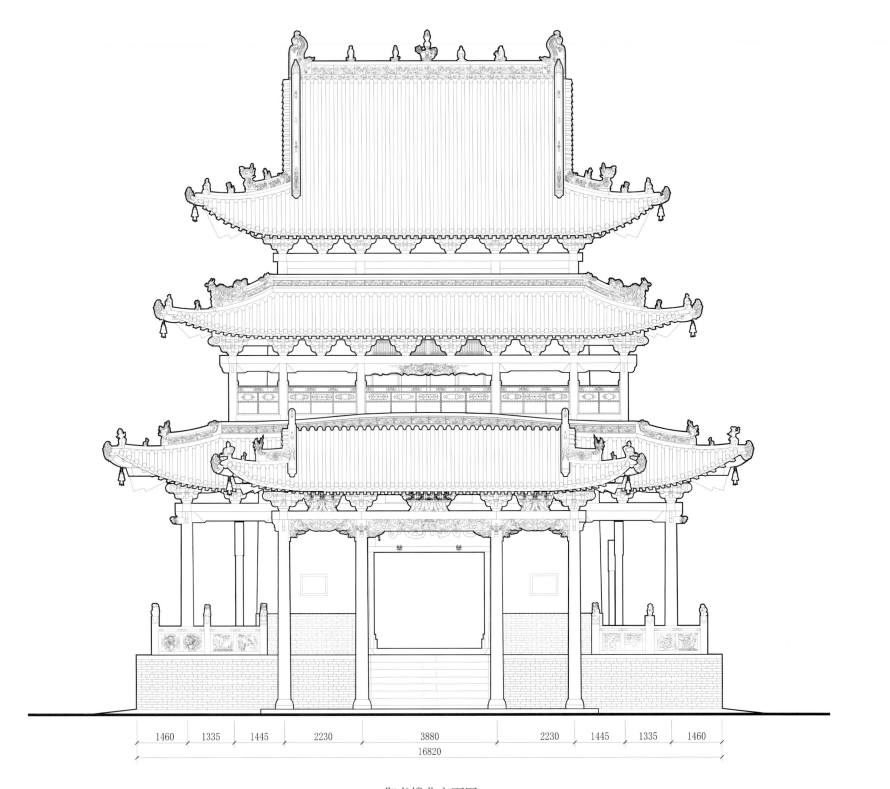

17.371

16.516

11.602

10.960

8.820

8.115

6.849

4.562

4.000

±0.000

-1.618

1460　1335　1445　2230　3880　2230　1445　1335　1460

16820

御书楼北立面图

North elevation of the Yushu Storied Building

0　1　2m

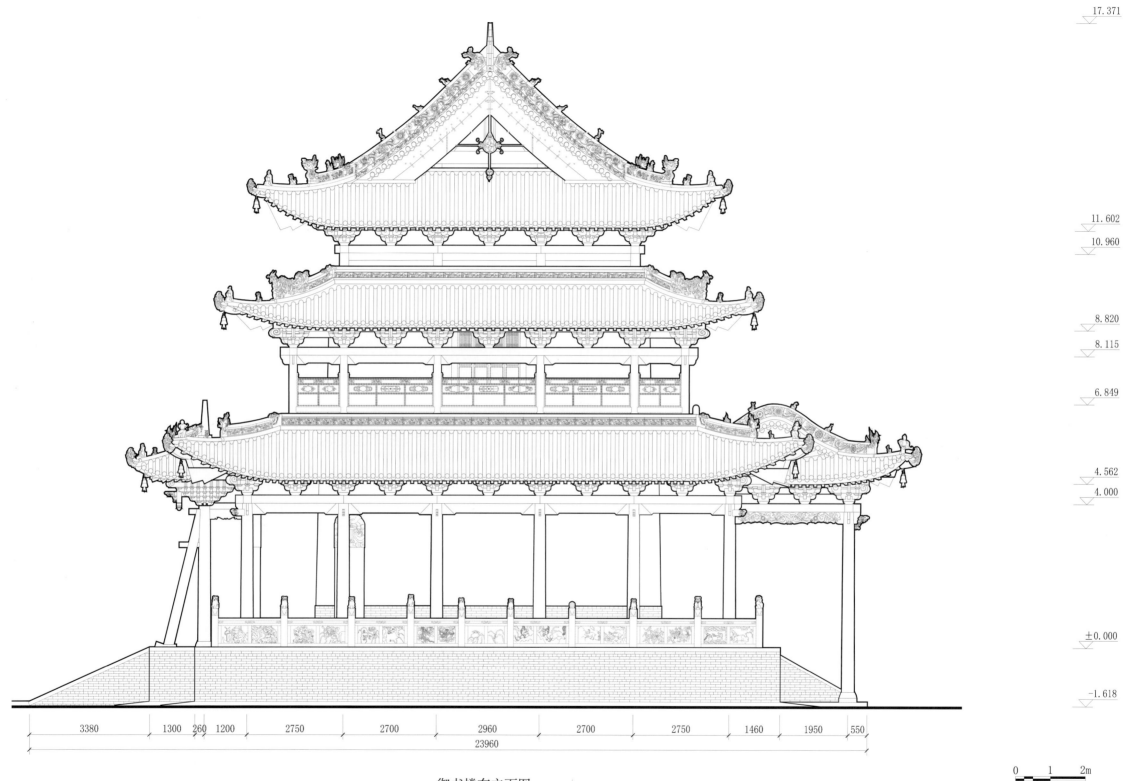

17.371

11.602

10.960

8.820

8.115

6.849

4.562

4.000

±0.000

-1.618

| 3380 | 1300 | 260 | 1200 | 2750 | 2700 | 2960 | 2700 | 2750 | 1460 | 1950 | 550 |

23960

御书楼东立面图
East elevation of the Yushu Storied Building

0 1 2m

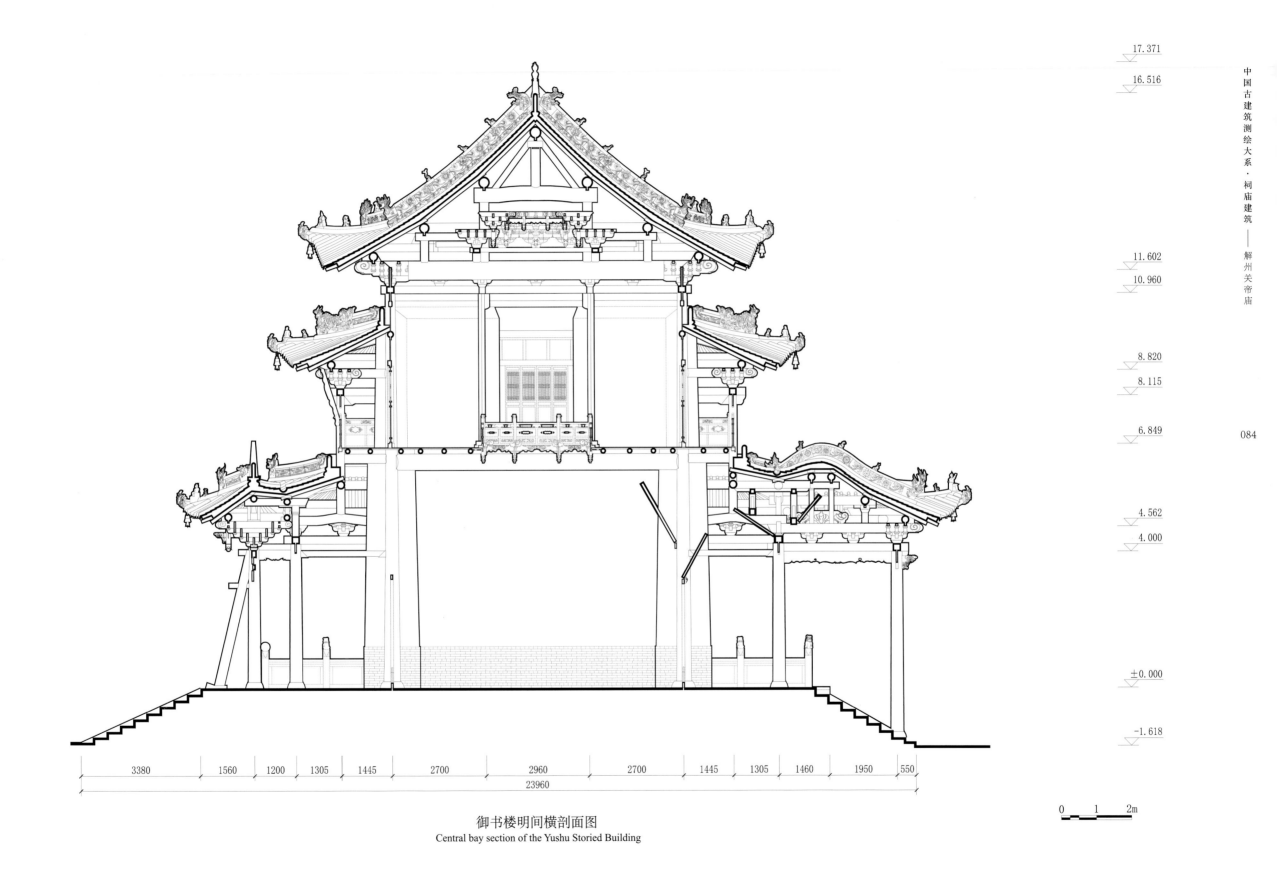

17.371

16.516

11.602

10.960

8.820

8.115

6.849

4.562

4.000

±0.000

-1.618

| 3380 | 1560 | 1200 | 1305 | 1445 | 2700 | 2960 | 2700 | 1445 | 1305 | 1460 | 1950 | 550 |

23960

0 1 2m

御书楼明间横剖面图
Central bay section of the Yushu Storied Building

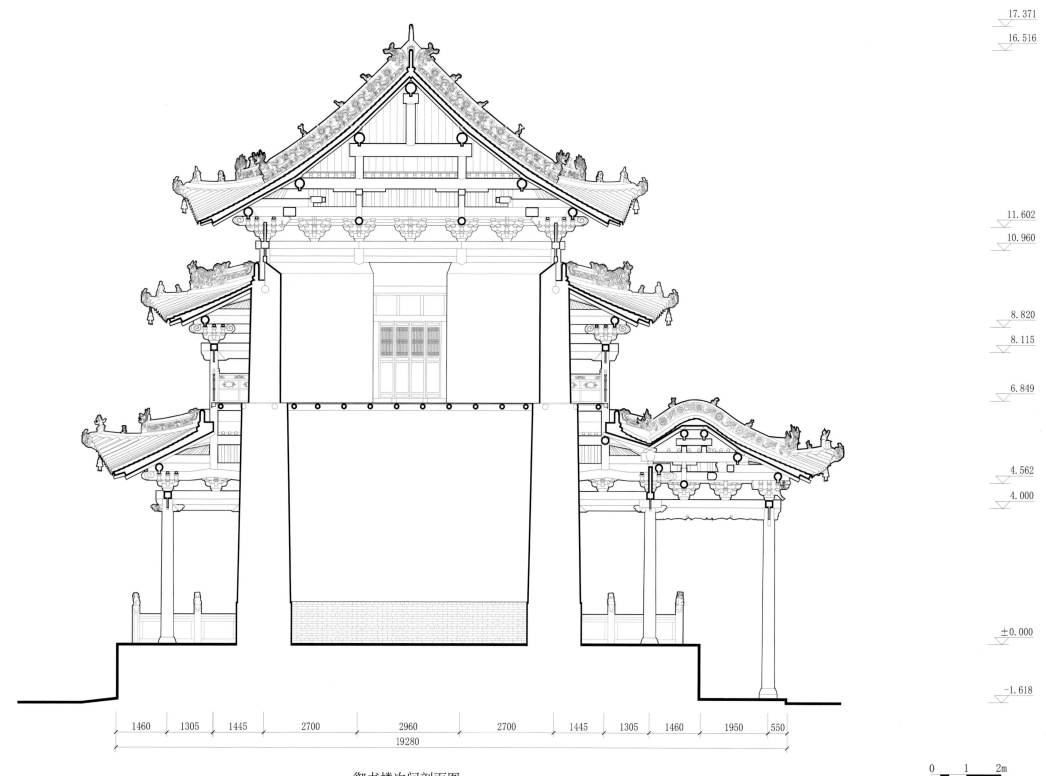

17.371

16.516

11.602

10.960

8.820

8.115

6.849

4.562

4.000

±0.000

-1.618

| 1460 | 1305 | 1445 | 2700 | 2960 | 2700 | 1445 | 1305 | 1460 | 1950 | 550 |

19280

御书楼次间剖面图
Side bay section of the Yushu Storied Building

0 1 2m

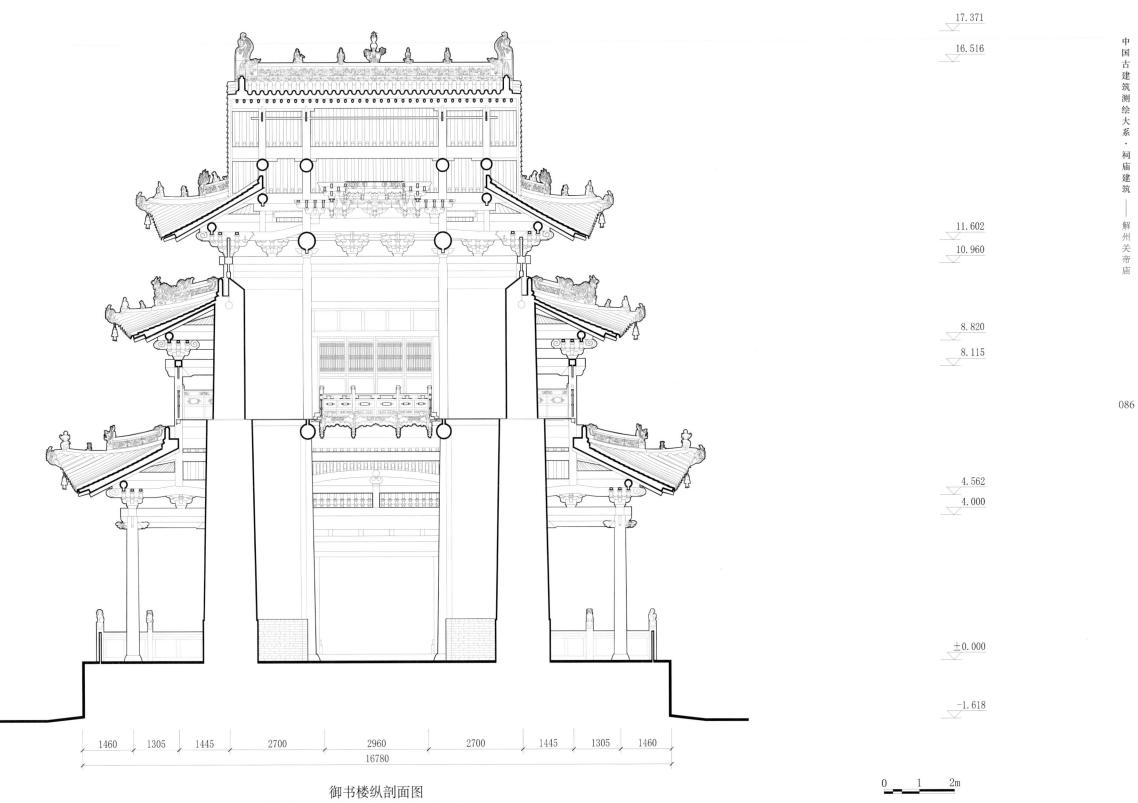

17. 371

16. 516

11. 602

10. 960

8. 820

8. 115

4. 562

4. 000

±0. 000

-1. 618

| 1460 | 1305 | 1445 | 2700 | 2960 | 2700 | 1445 | 1305 | 1460 |

16780

御书楼纵剖面图

Longitudinal section of the Yushu Storied Building

0   1   2m

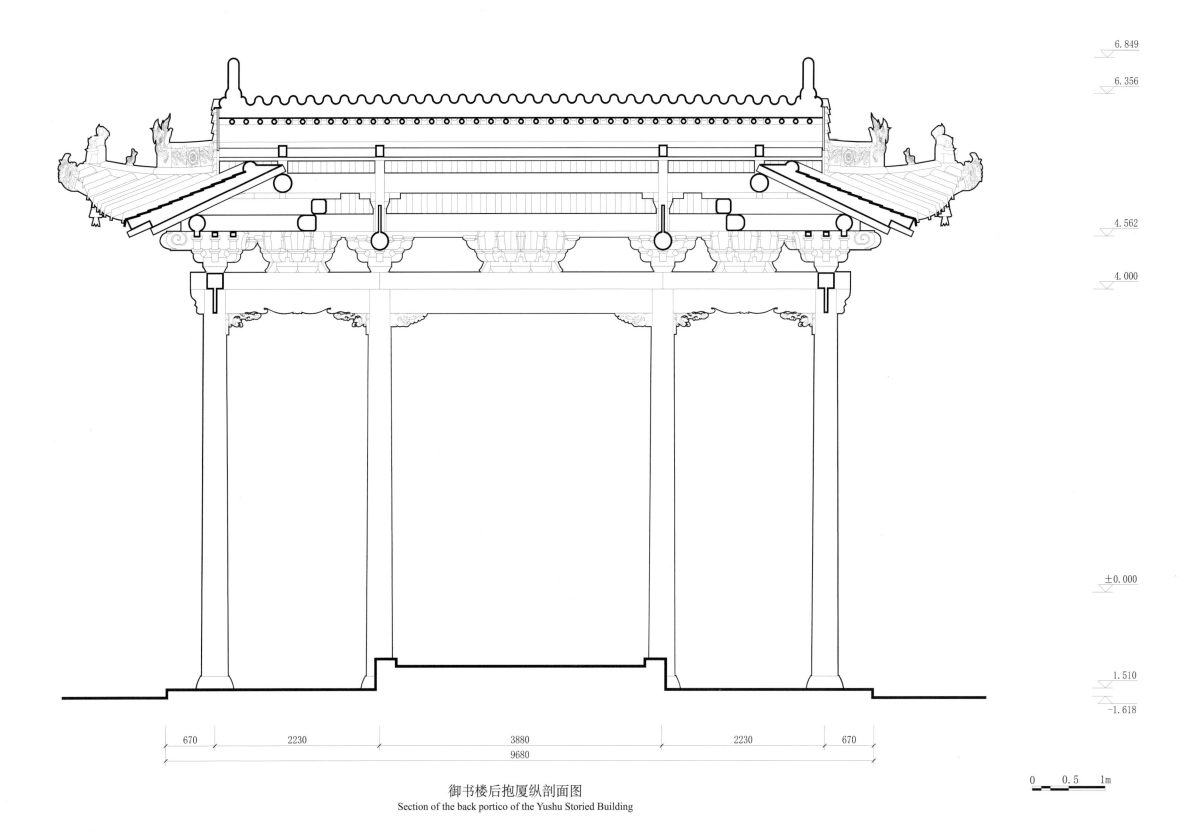

6.849

6.356

4.562

4.000

±0.000

1.510

-1.618

670 2230 3880 2230 670

9680

0 0.5 1m

御书楼后抱厦纵剖面图
Section of the back portico of the Yushu Storied Building

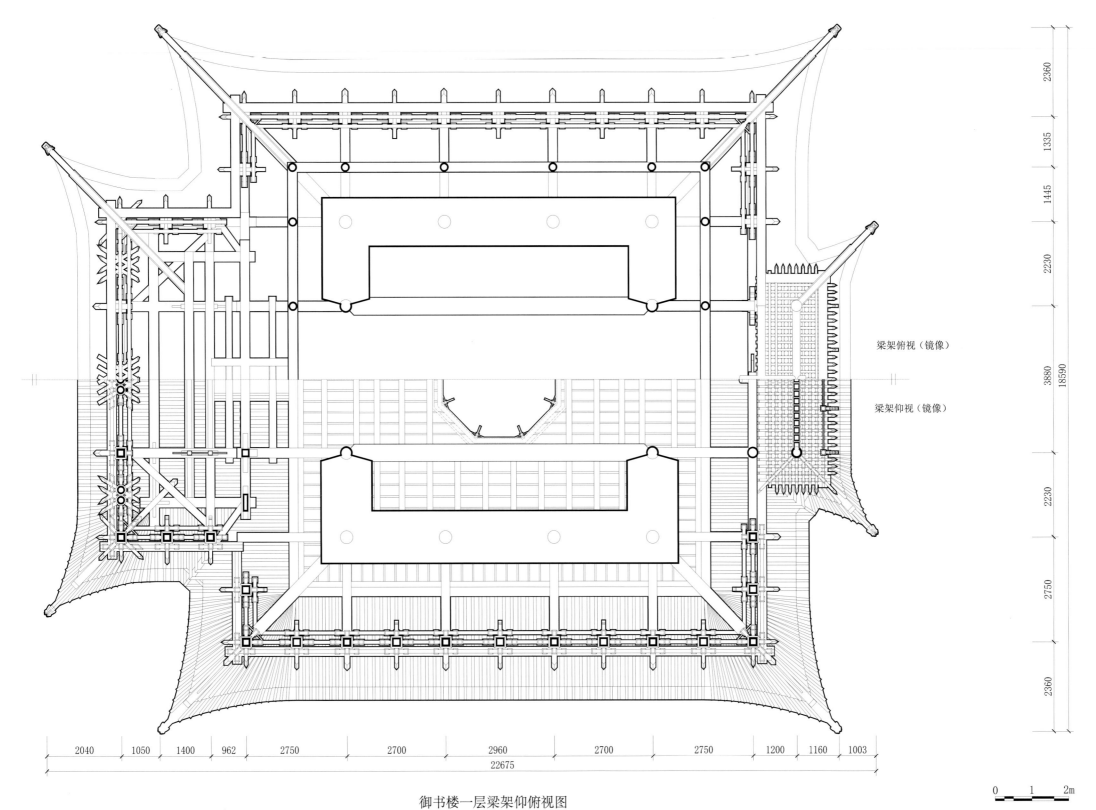

梁架俯视（镜像）

梁架仰视（镜像）

御书楼一层梁架仰俯视图
Top and bottom views of the beam frame and the eaves of the first-floor of the Yushu Storied Building

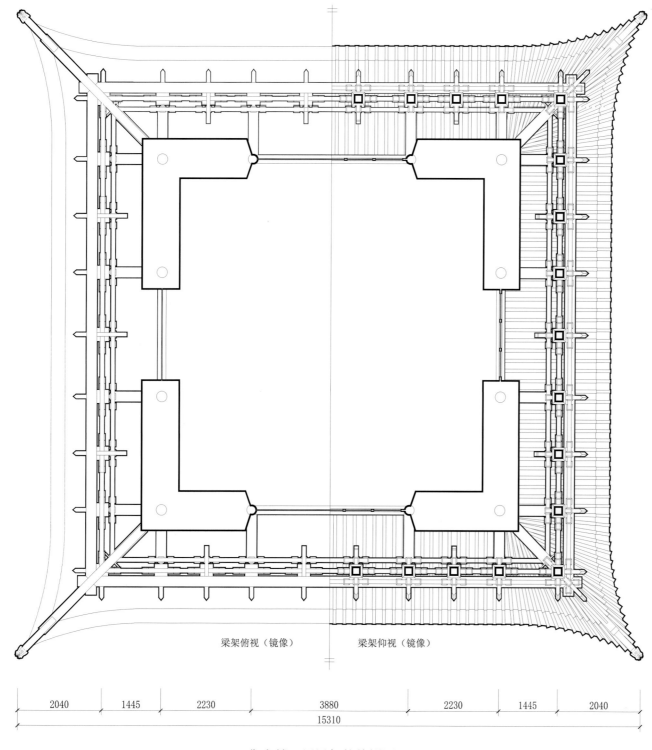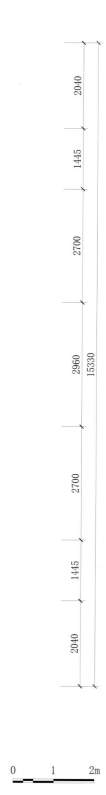

梁架俯视（镜像）   梁架仰视（镜像）

御书楼二层梁架仰俯视图

Top and bottom views of the beam frame and the eaves of the second-floor of the Yushu Storied Building

2040 1445 2230 3880 2230 1445 2040

15310

2040 1445 2700 2960 2700 1445 2040

15330

0   1   2m

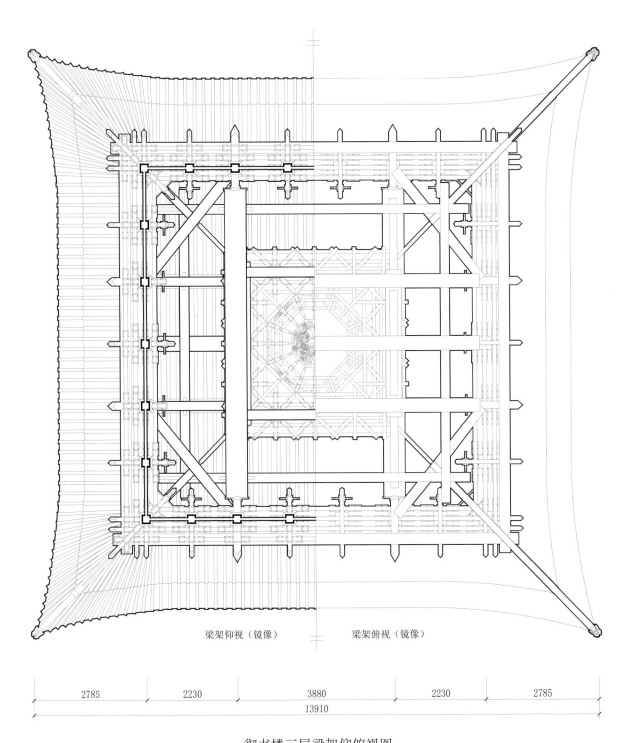

梁架仰视（镜像）　　　梁架俯视（镜像）

御书楼三层梁架仰俯视图

Top and bottom views of the beam frame and the eaves of the third-floor of the Yushu Storied Building

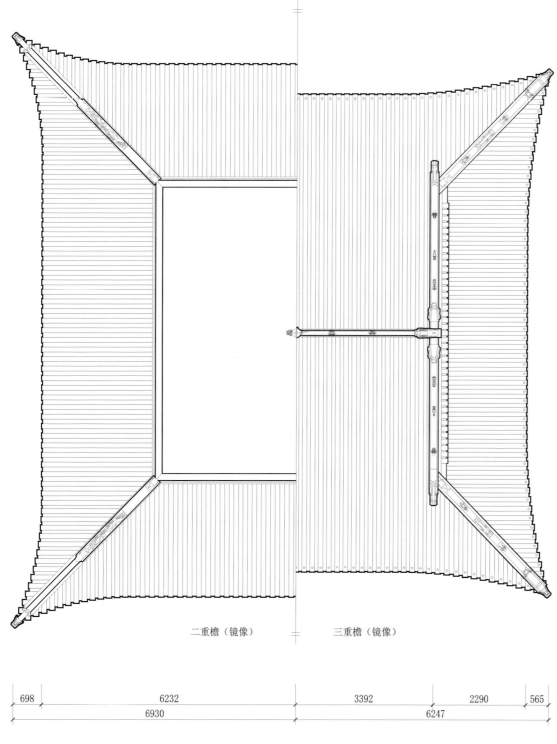
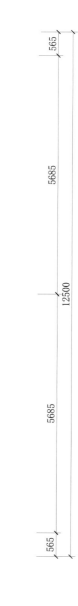
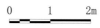

二重檐（镜像）　　　三重檐（镜像）

御书楼屋顶平面图
Roof plan of the Yushu Storied Building

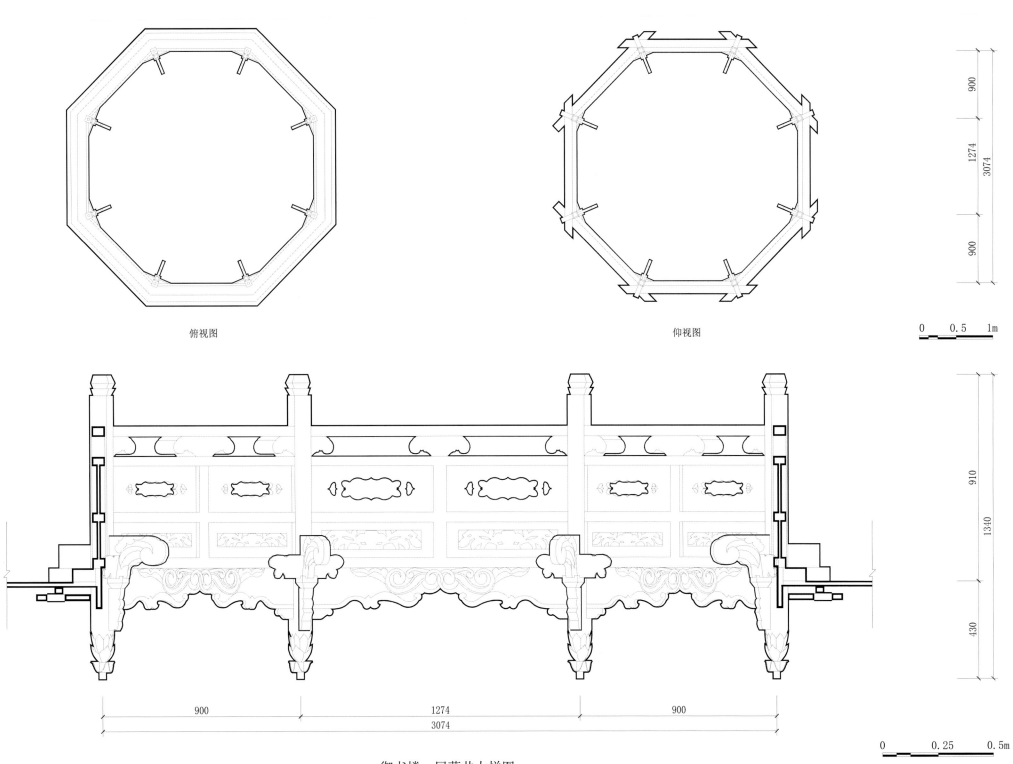

俯视图

仰视图

900

1274

3074

900

910

1340

430

900

1274

900

3074

0　0.5　1m

0　0.25　0.5m

御书楼一层藻井大样图
Details of caisson ceiling of the first-floor of the Yushu Storied Building

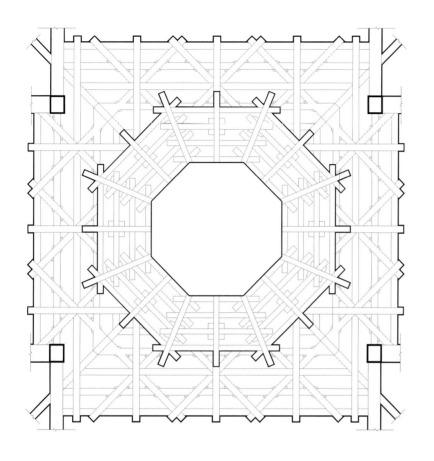

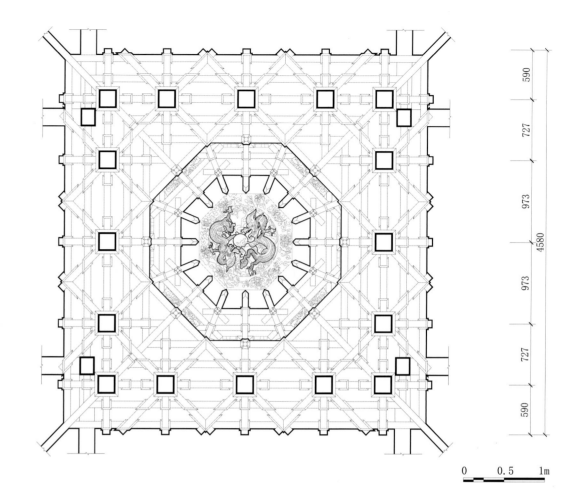

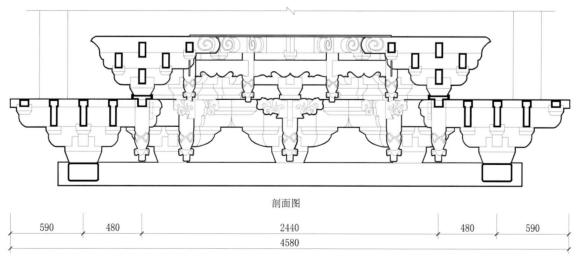

剖面图

御书楼二层藻井大样图
Details of caisson ceiling of the second-floor of the Yushu Storied Building

東华门南廊
The South Corridor of Donghua Gate

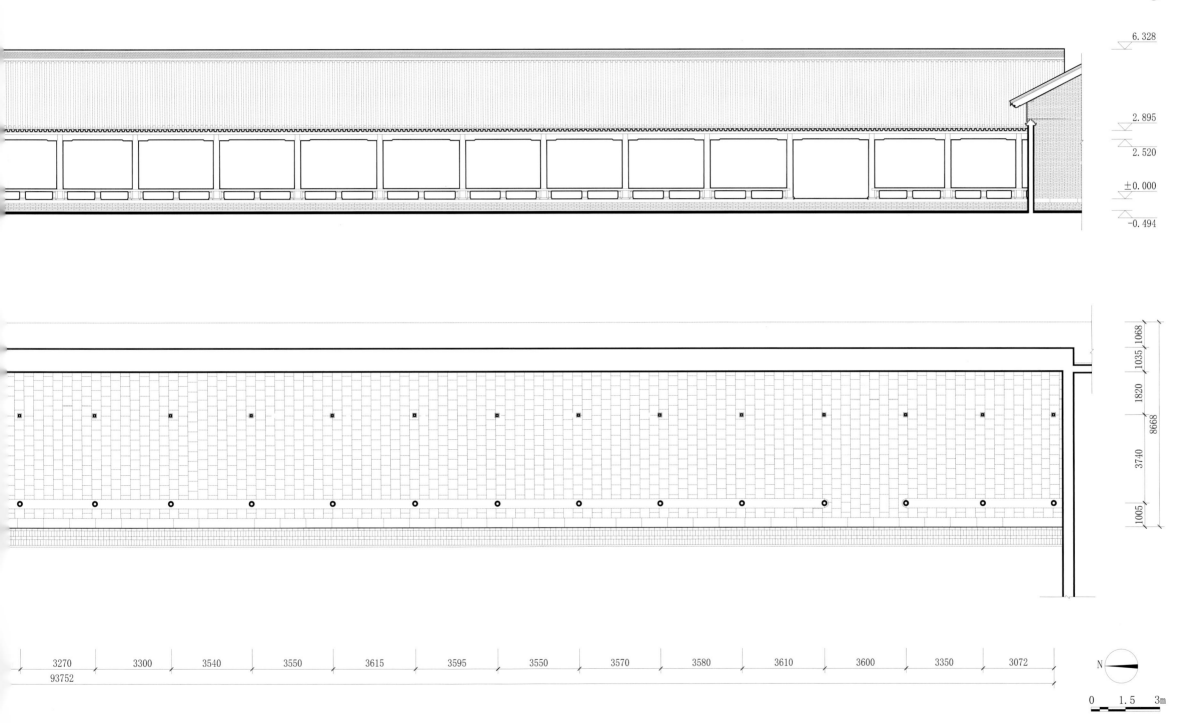

6.328

2.895

2.520

±0.000

−0.494

1068
1035
1820
8668
3740
1005

3270　3300　3540　3550　3615　3595　3550　3570　3580　3610　3600　3350　3072
93752

N

0　1.5　3m

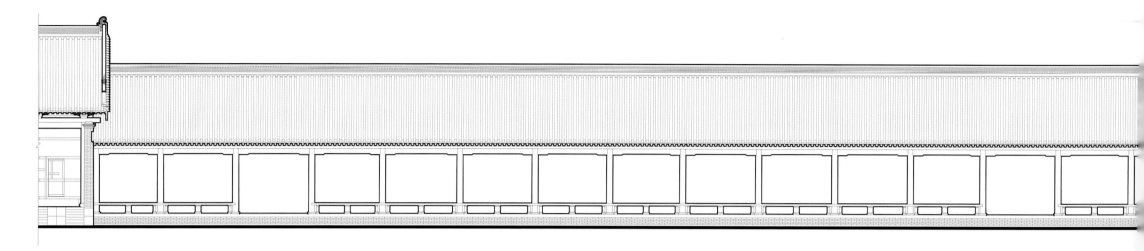

东华门南廊正立面图
Front elevation of the South Corridor of Donghua Gate

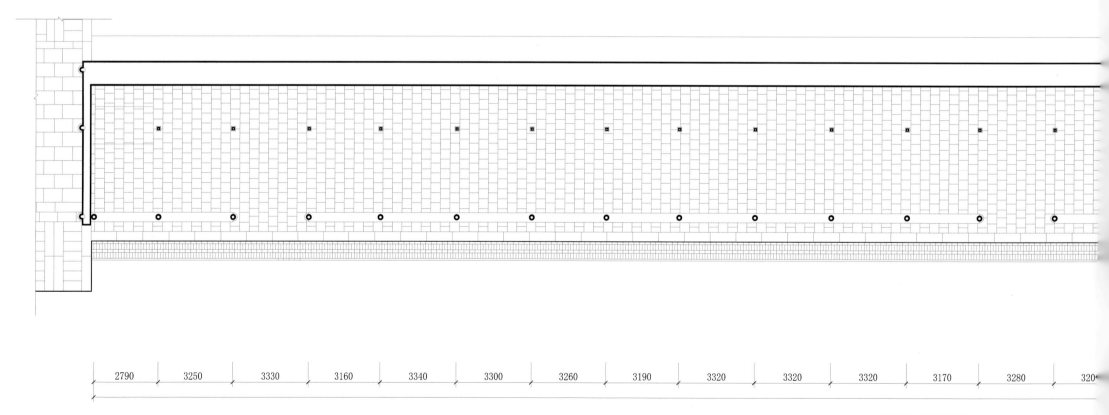

| 2790 | 3250 | 3330 | 3160 | 3340 | 3300 | 3260 | 3190 | 3320 | 3320 | 3320 | 3170 | 3280 | 320 |

东华门南廊平面图
Plan of the South Corridor of Donghua Gate

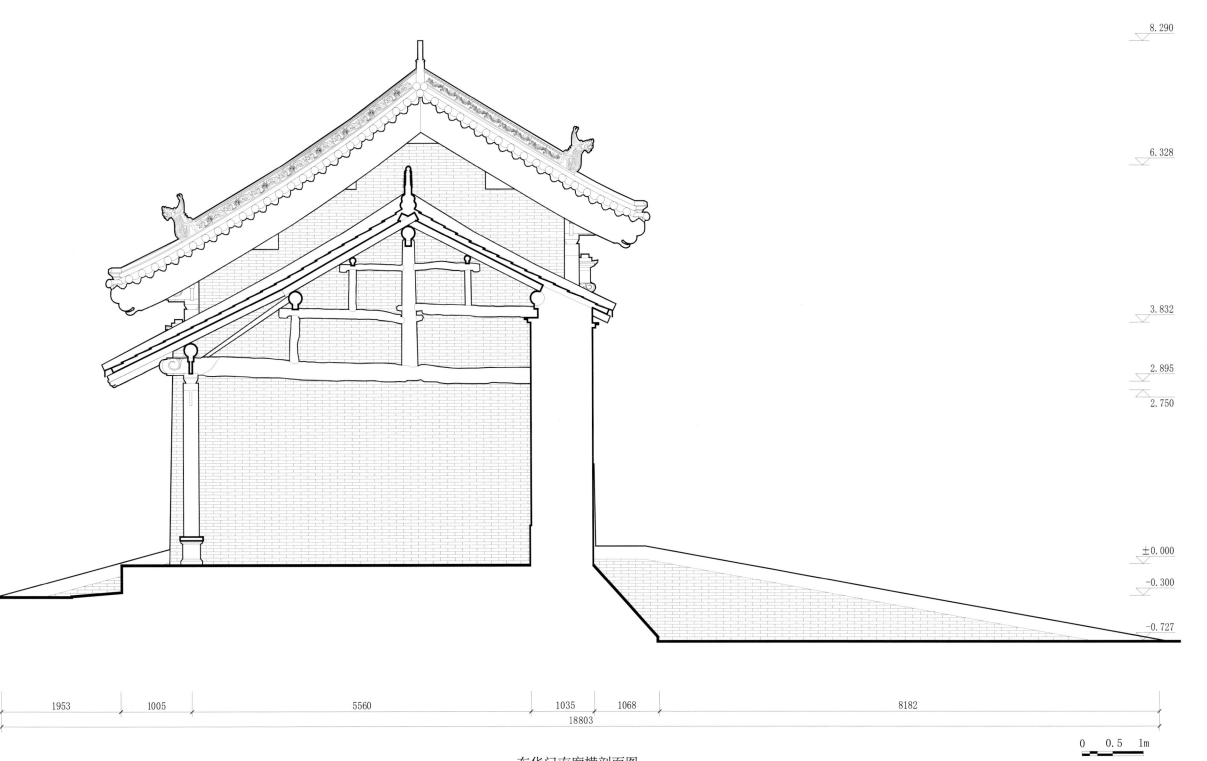

8.290

6.328

3.832

2.895

2.750

±0.000

-0.300

-0.727

1953　1005　5560　1035　1068　8182

18803

0　0.5　1m

东华门南廊横剖面图
Cross-section of the South Corridor of Donghua Gate

东华门
The Donghua Gate

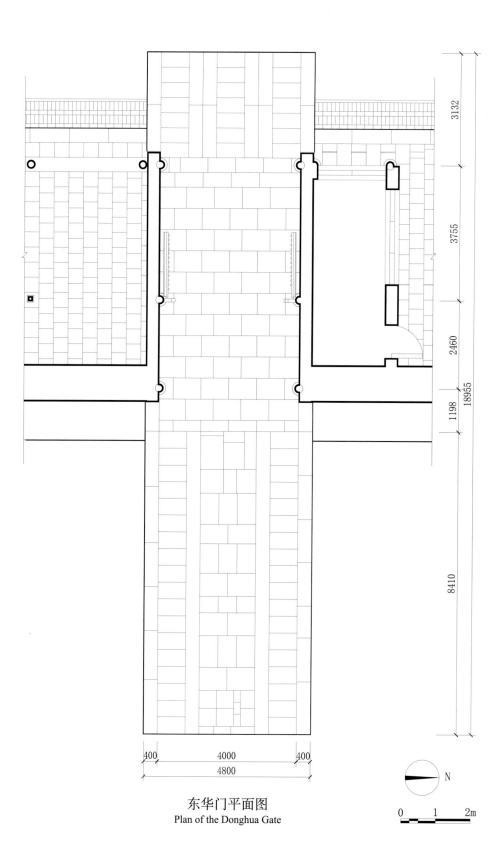

3132

3755

2460

18955

1198

8410

400　4000　400
4800

N

0　1　2m

东华门平面图
Plan of the Donghua Gate

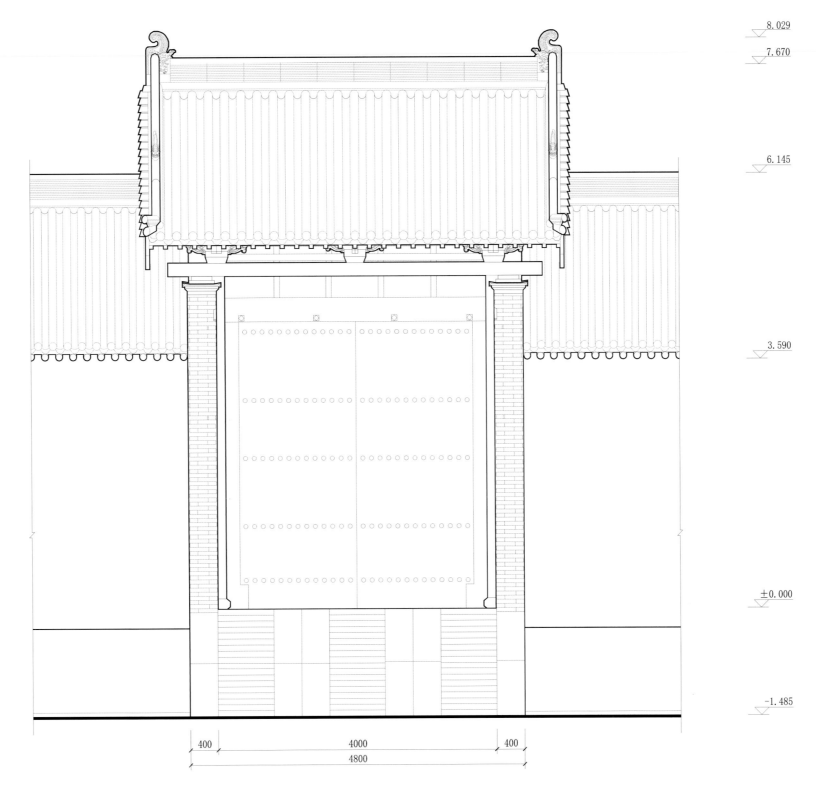

8.029

7.670

6.145

3.590

±0.000

-1.485

400   4000   400

4800

东华门正立面图

Front elevation of the Donghua Gate

0   0.5   1m

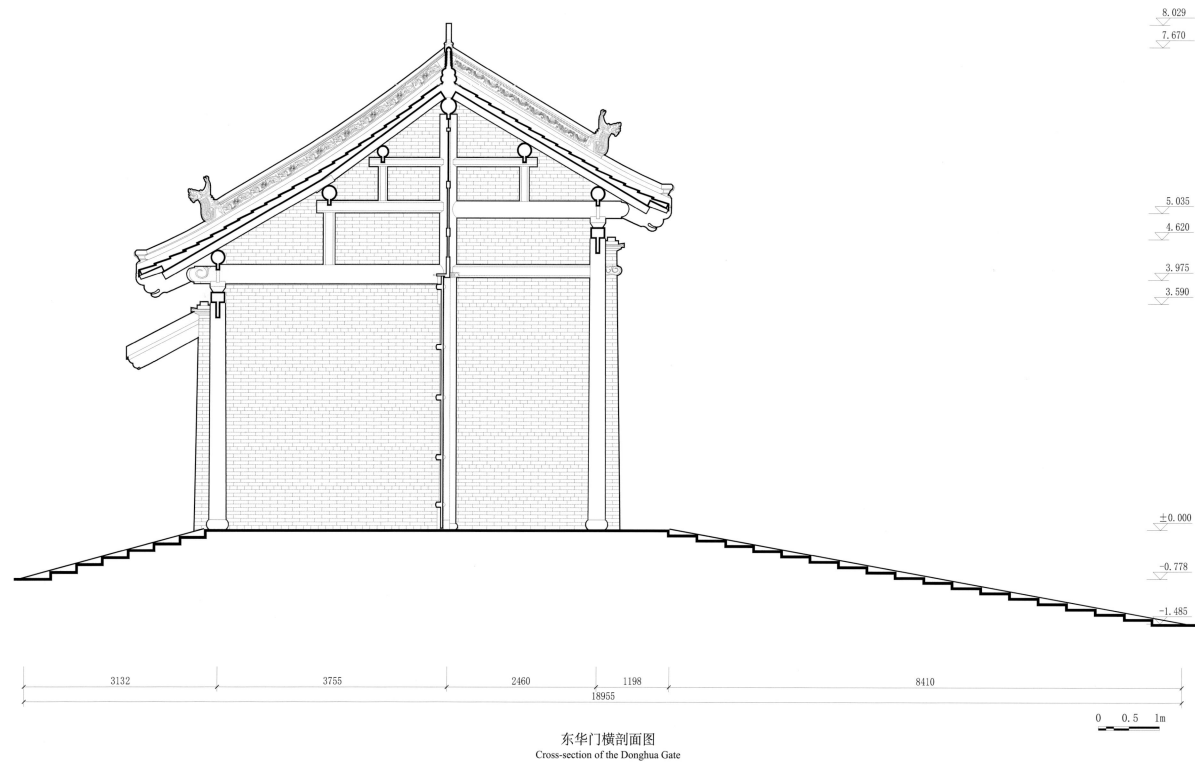

8.029
7.670

5.035
4.620

3.975
3.590

±0.000

-0.778

-1.485

3132   3755   2460   1198   8410
18955

0   0.5   1m

东华门横剖面图
Cross-section of the Donghua Gate

东华门北廊
The North Corridor of Donghua Gate

中国古建筑测绘大系·祠庙建筑——解州关帝庙

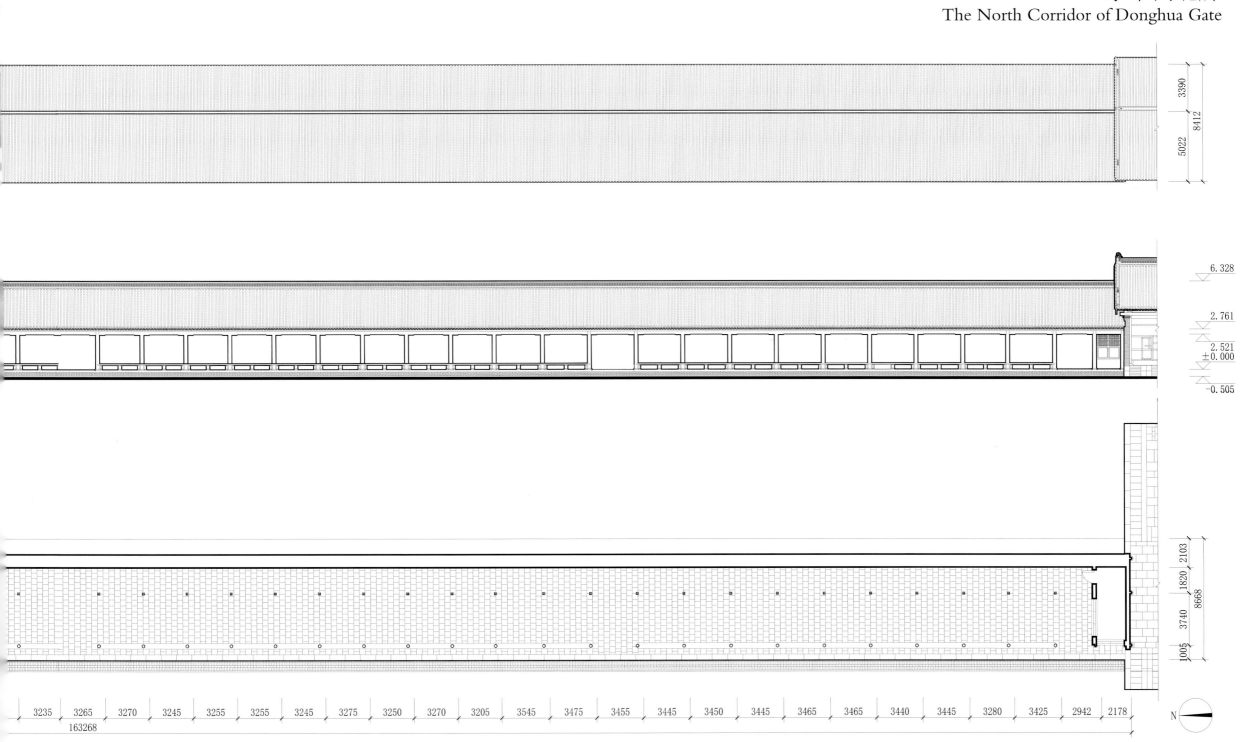

3390
5022 8412

6.328
2.761
2.521
±0.000
−0.505

100

2103
1820
3740 8668
1005

| 3235 | 3265 | 3270 | 3245 | 3255 | 3255 | 3245 | 3275 | 3250 | 3270 | 3205 | 3545 | 3475 | 3455 | 3445 | 3450 | 3445 | 3465 | 3465 | 3440 | 3445 | 3280 | 3425 | 2942 | 2178 |

163268

N

0　　2.5　　5m

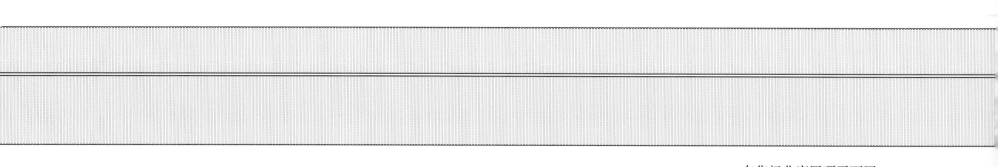

东华门北廊屋顶平面图
Roof plan of the North Corridor of Donghua Gate

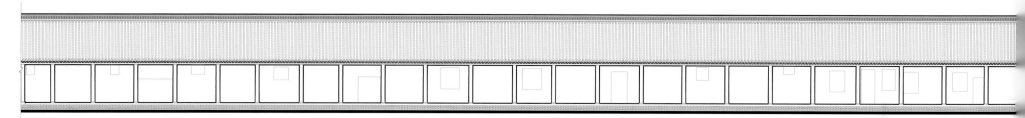

东华门北廊立面图
Front elevation of the North Corridor of Donghua Gate

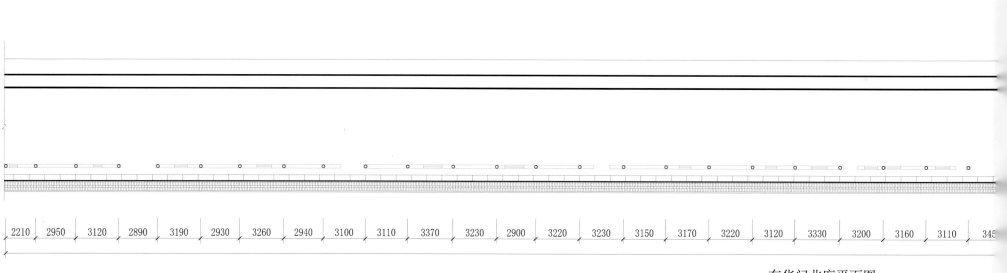

2210	2950	3120	2890	3190	2930	3260	2940	3100	3110	3370	3230	2900	3220	3230	3150	3170	3220	3120	3330	3200	3160	3110	345

东华门北廊平面图
Plan of the North Corridor of Donghua Gate

西华门南廊
The South Corridor of Xihua Gate

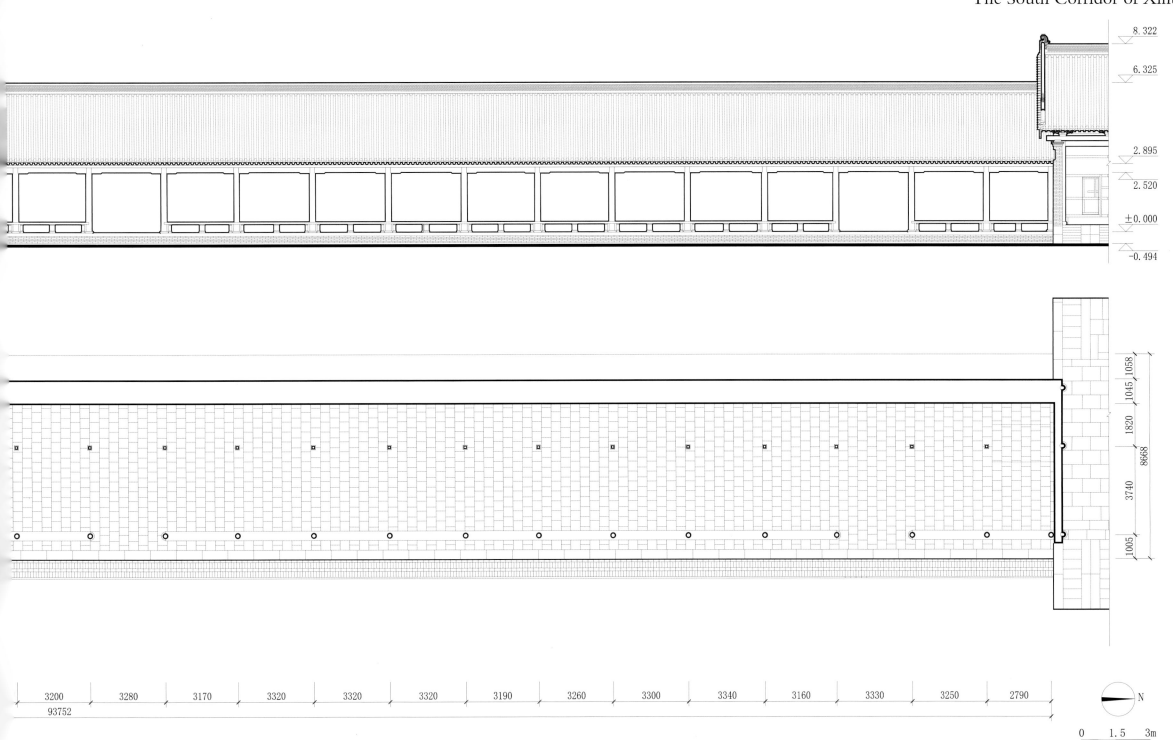

中国古建筑测绘大系·祠庙建筑——解州关帝庙

8.322

6.325

2.895

2.520

±0.000

-0.494

1058
1045
1820
8668
3740
1005

3200 3280 3170 3320 3320 3320 3190 3260 3300 3340 3160 3330 3250 2790

93752

N

0   1.5   3m

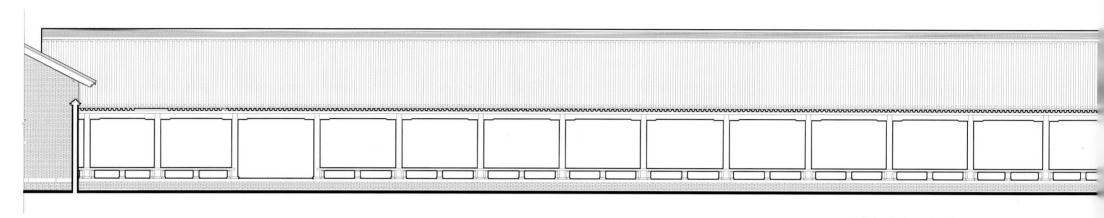

西华门南廊正立面图
Front elevation of the South Corridor of Xihua Gate

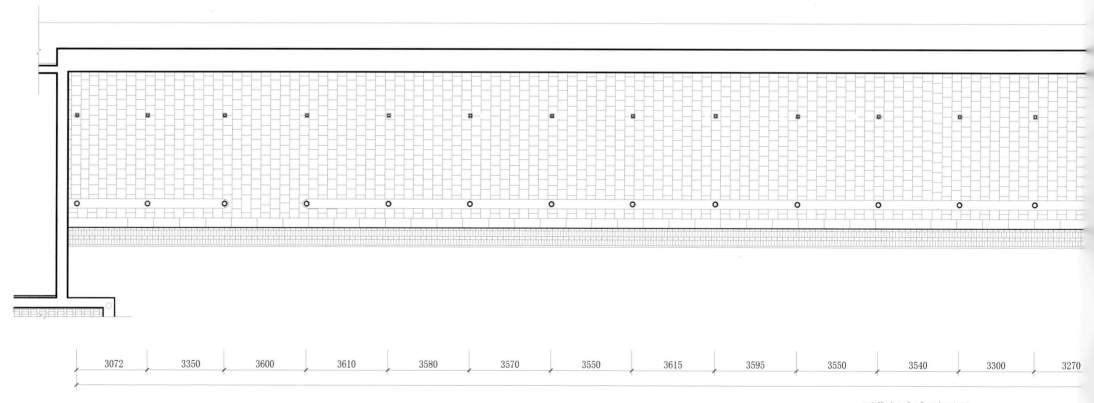

| 3072 | 3350 | 3600 | 3610 | 3580 | 3570 | 3550 | 3615 | 3595 | 3550 | 3540 | 3300 | 3270 |

西华门南廊平面图
Plan of the South Corridor of Xihua Gate

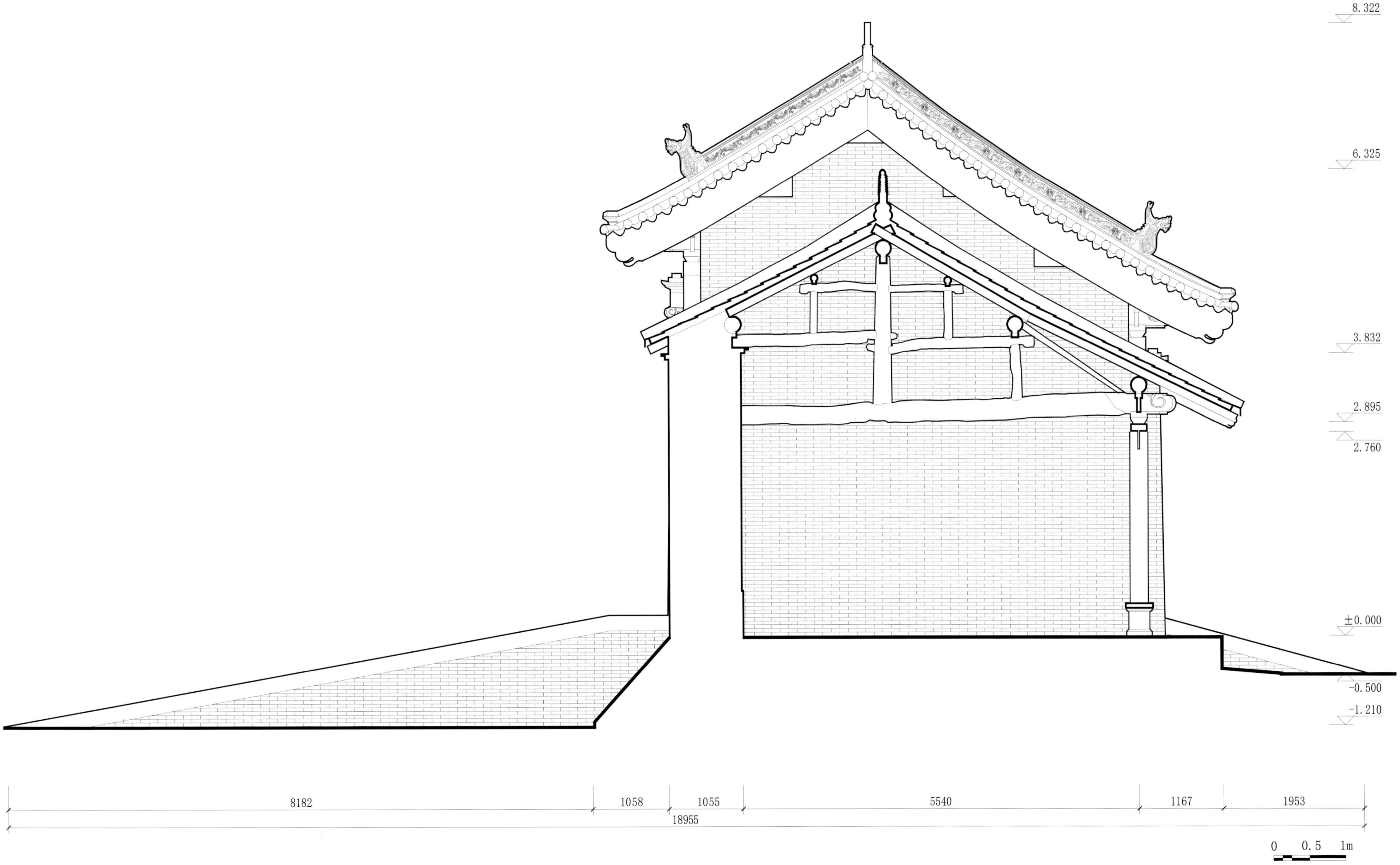

8.322

6.325

3.832

2.895

2.760

±0.000

−0.500

−1.210

| 8182 | 1058 | 1055 | 5540 | 1167 | 1953 |

18955

0   0.5   1m

西华门南廊横剖面图

Cross-section of the South Corridor of Xihua Gate

西华门
The Xihua Gate

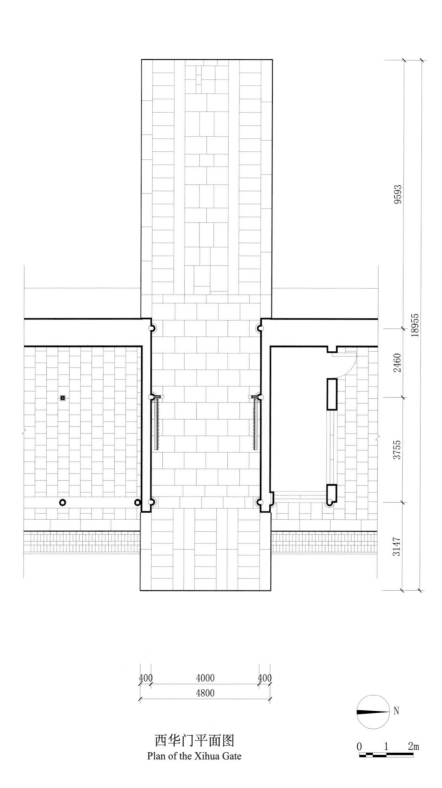

9593

18955

2460

3755

3147

400 4000 400
4800

N

0 1 2m

西华门平面图
Plan of the Xihua Gate

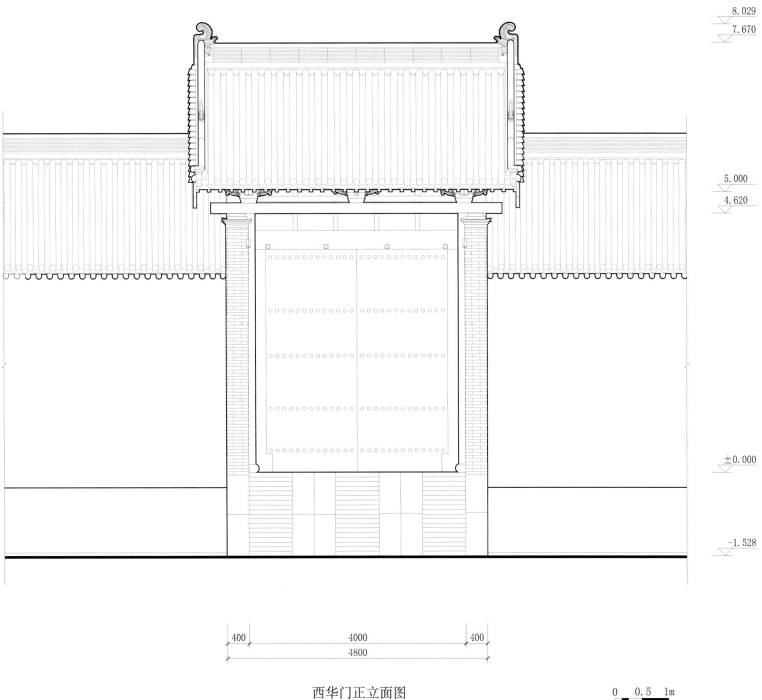

8.029

7.670

5.000

4.620

±0.000

-1.528

400   4000   400
4800

西华门正立面图
Front elevation of the Xihua Gate

0   0.5   1m

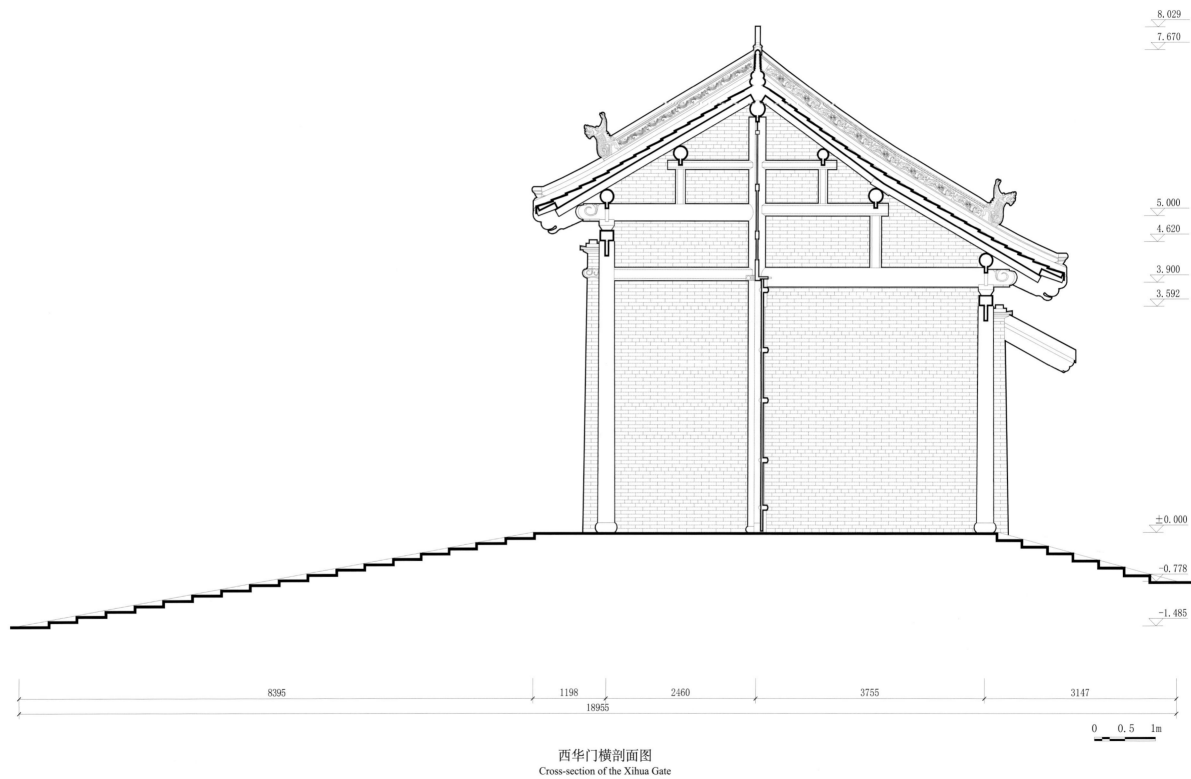

8.029

7.670

5.000

4.620

3.900

3.592

±0.000

−0.778

−1.485

8395　1198　2460　3755　3147

18955

0　0.5　1m

西华门横剖面图

Cross-section of the Xihua Gate

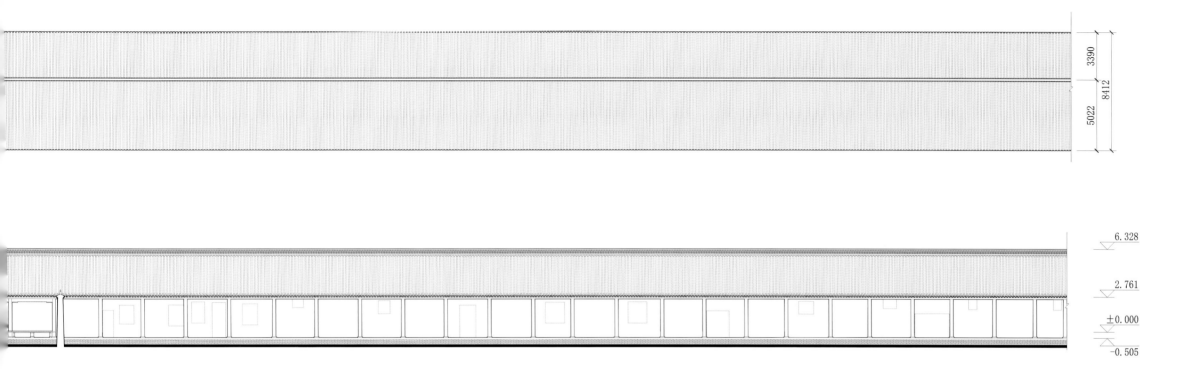

3390
8412
5022

6.328

2.761

±0.000

-0.505

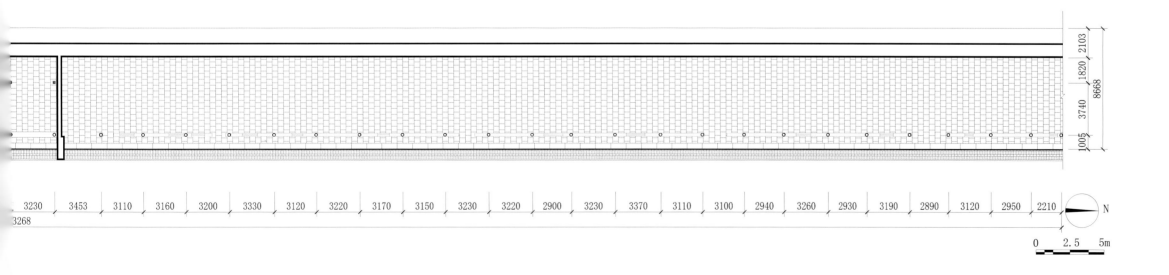

2103
1820
8668
3740
1005

3230 3453 3110 3160 3200 3330 3120 3220 3170 3150 3230 3220 2900 3230 3370 3110 3100 2940 3260 2930 3190 2890 3120 2950 2210
3268

N

0   2.5   5m

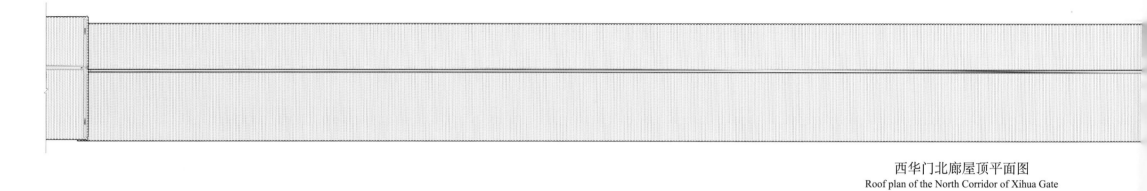

西华门北廊屋顶平面图
Roof plan of the North Corridor of Xihua Gate

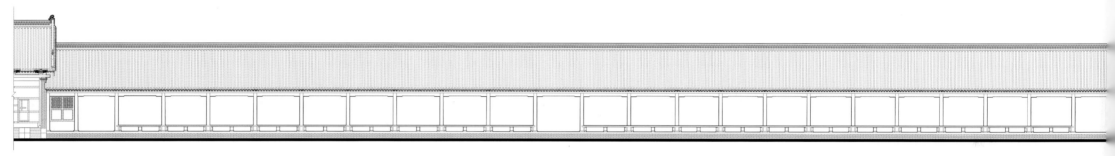

西华门北廊立面图
Front elevation of the North Corridor of Xihua Gate

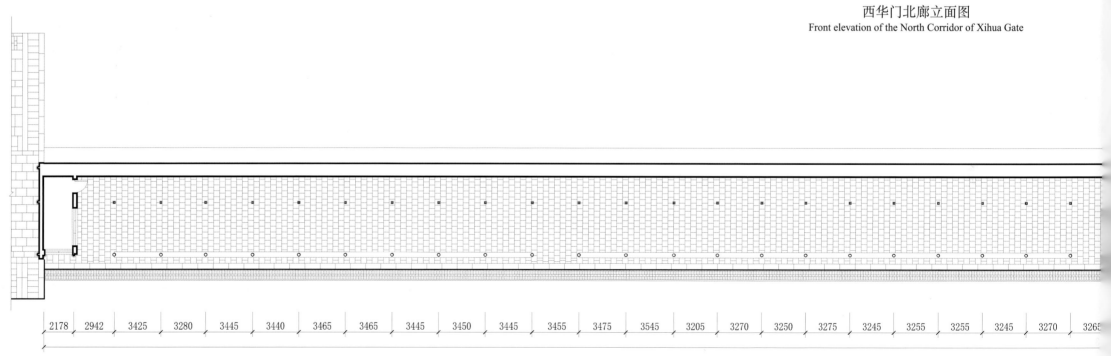

| 2178 | 2942 | 3425 | 3280 | 3445 | 3440 | 3465 | 3465 | 3445 | 3450 | 3445 | 3455 | 3475 | 3545 | 3205 | 3270 | 3250 | 3275 | 3245 | 3255 | 3255 | 3245 | 3270 | 3265 |

西华门北廊平面图
Plan of the North Corridor of Xihua Gate

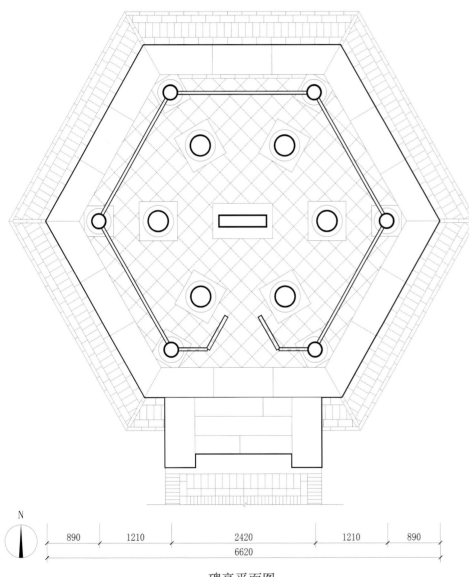

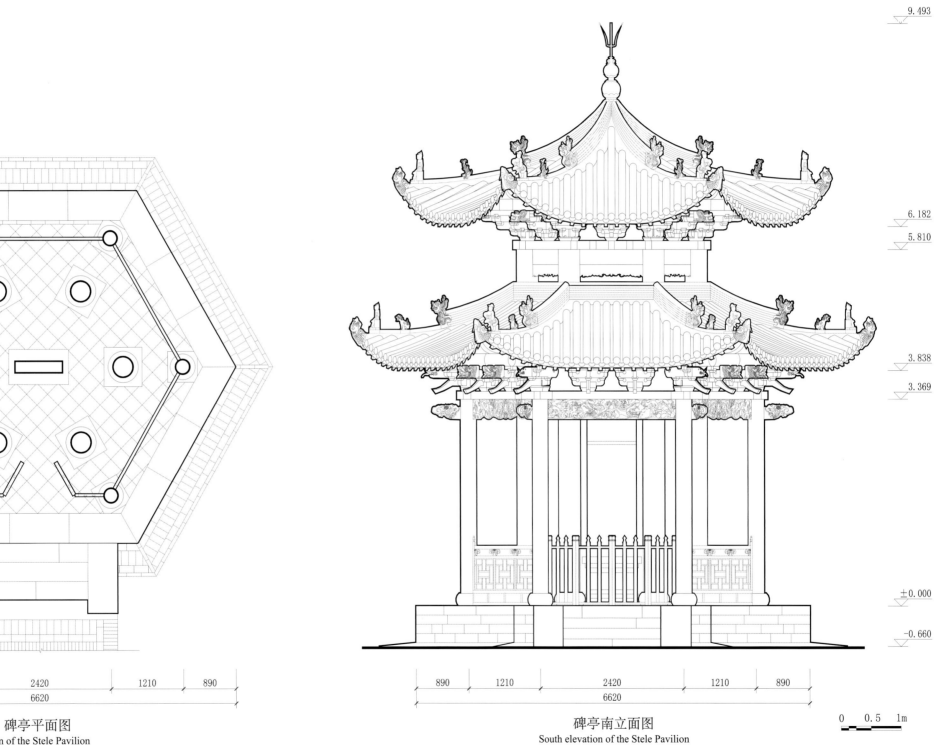

9.493

6.182
5.810

110

3.838
3.369

±0.000

-0.660

碑亭平面图
Plan of the Stele Pavilion

890　1210　2420　1210　890
6620

碑亭南立面图
South elevation of the Stele Pavilion

890　1210　2420　1210　890
6620

0　0.5　1m

9.493

6.182
5.810

3.838
3.369

±0.000

-0.660

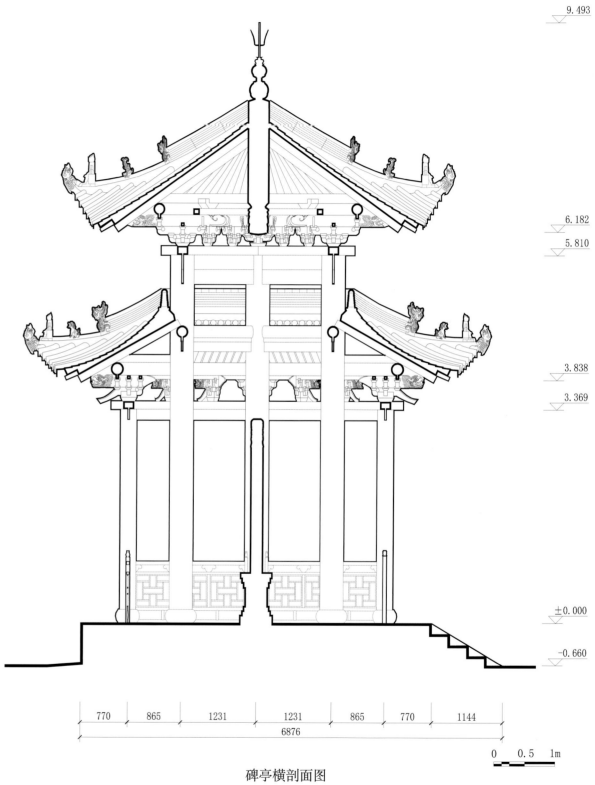

| 770 | 2096 | 2096 | 770 | 1144 |

6876

碑亭西立面图
West elevation of the Stele Pavilion

| 770 | 865 | 1231 | 1231 | 865 | 770 | 1144 |

6876

0   0.5   1m

碑亭横剖面图
Cross-section of the Stele Pavilion

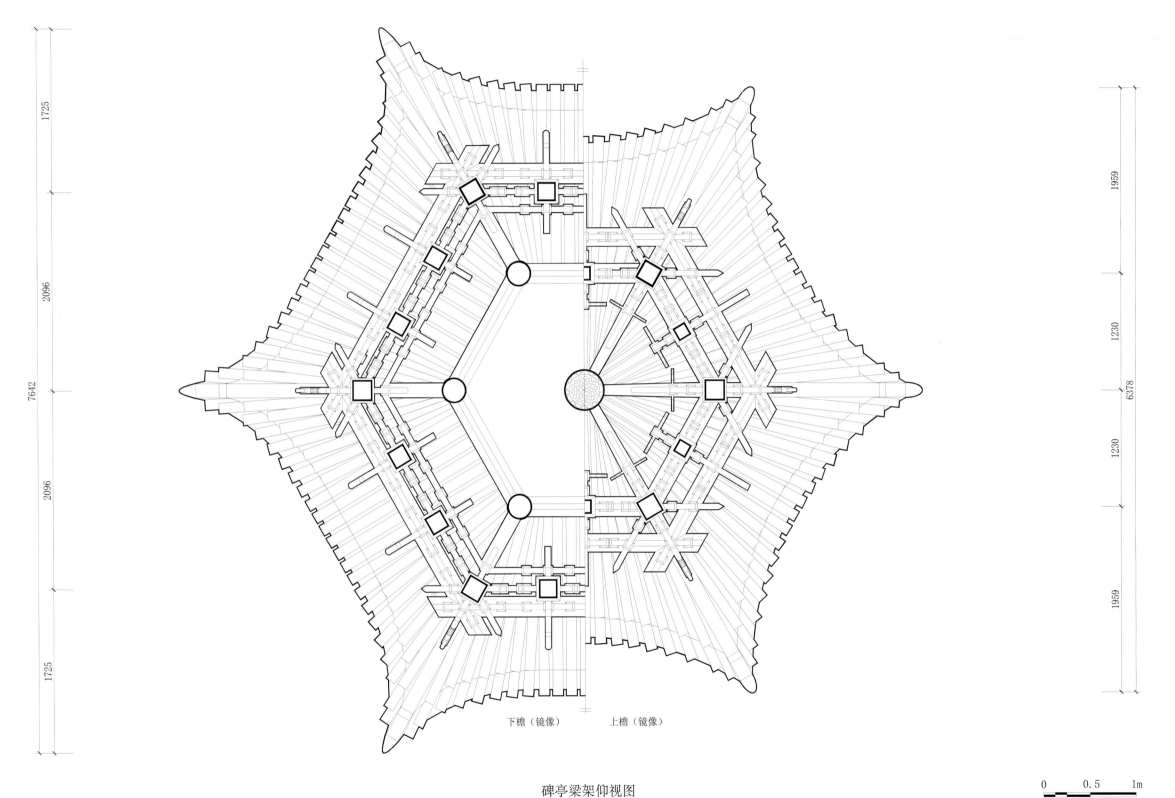

1725　1959

2096　1230

7642　6378

2096　1230

1725　1959

下檐（镜像）　上檐（镜像）

碑亭梁架仰视图
Bottom views of the beam frame of the Stele Pavilion

0　0.5　1m

钟亭
The Bell Pavilion

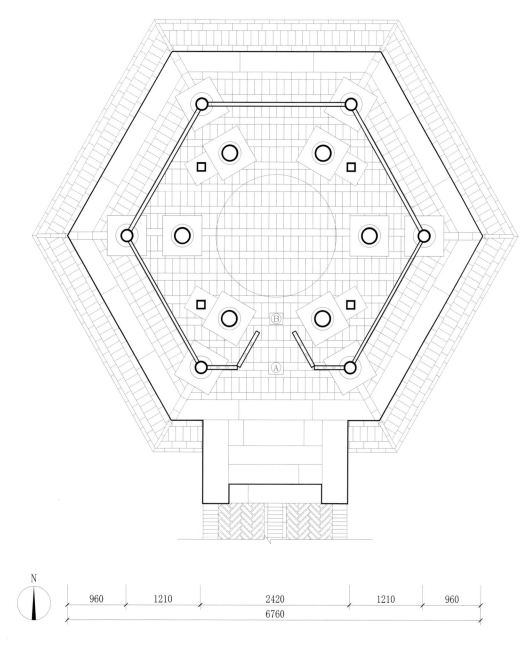

钟亭平面图
Plan of the Bell Pavilion

960　1210　2420　1210　960
6760

N

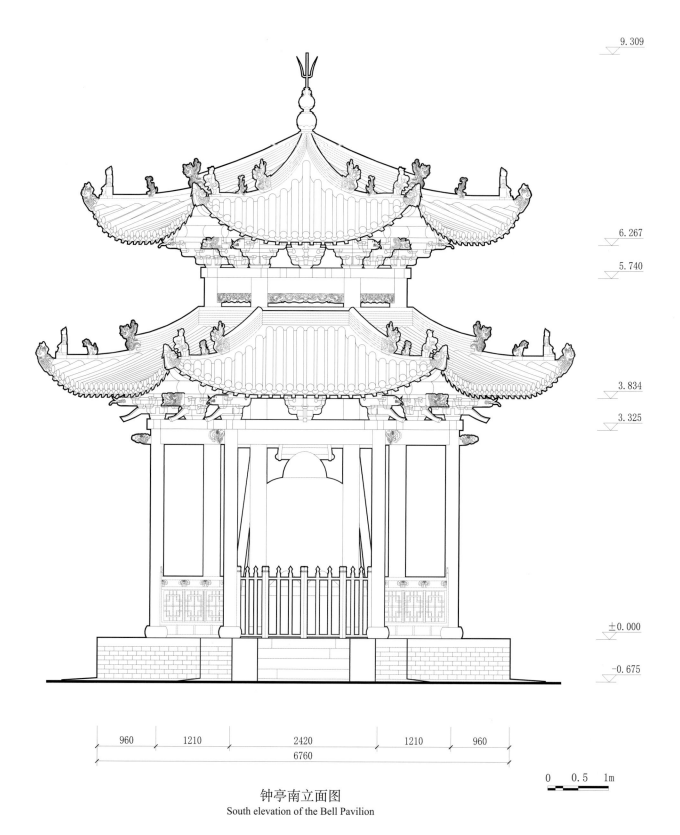

9.309

6.267

5.740

3.834

3.325

±0.000

-0.675

960　1210　2420　1210　960
6760

0　0.5　1m

钟亭南立面图
South elevation of the Bell Pavilion

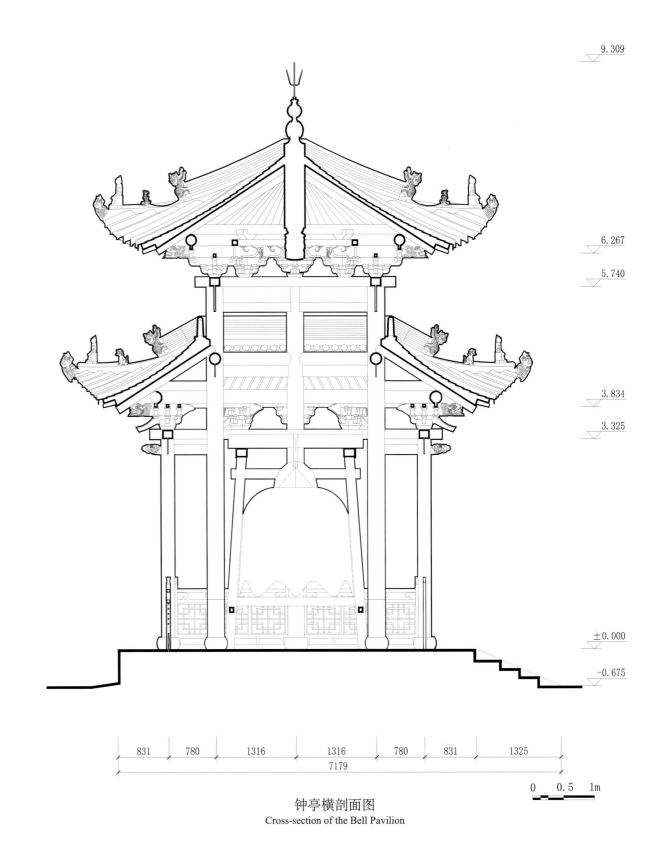

9.309

6.267

5.740

3.834

3.325

±0.000

-0.675

114

| 831 | 2096 | 2096 | 831 | 1325 |

7179

钟亭西立面图
West elevation of the Bell Pavilion

| 831 | 780 | 1316 | 1316 | 780 | 831 | 1325 |

7179

0   0.5   1m

钟亭横剖面图
Cross-section of the Bell Pavilion

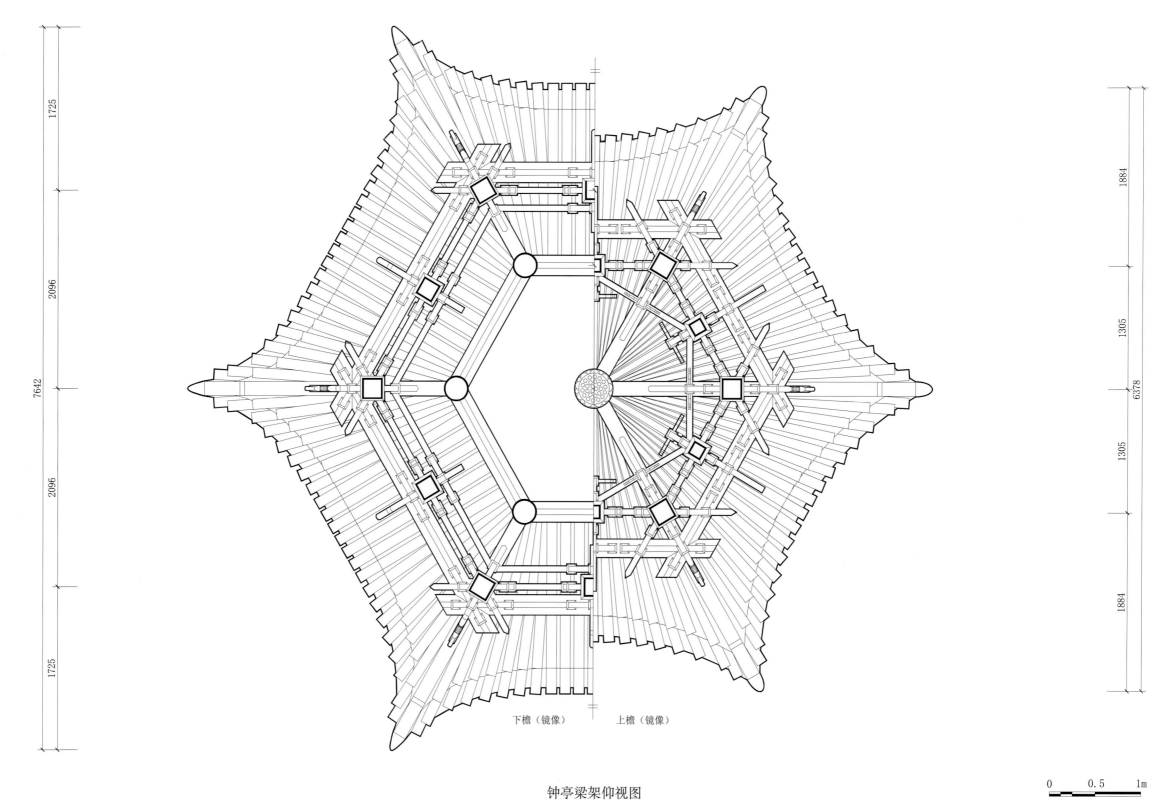

下檐（镜像）　　上檐（镜像）

钟亭梁架仰视图
Top and bottom views of the beam frame of the Bell Pavilion

0　　0.5　　1m

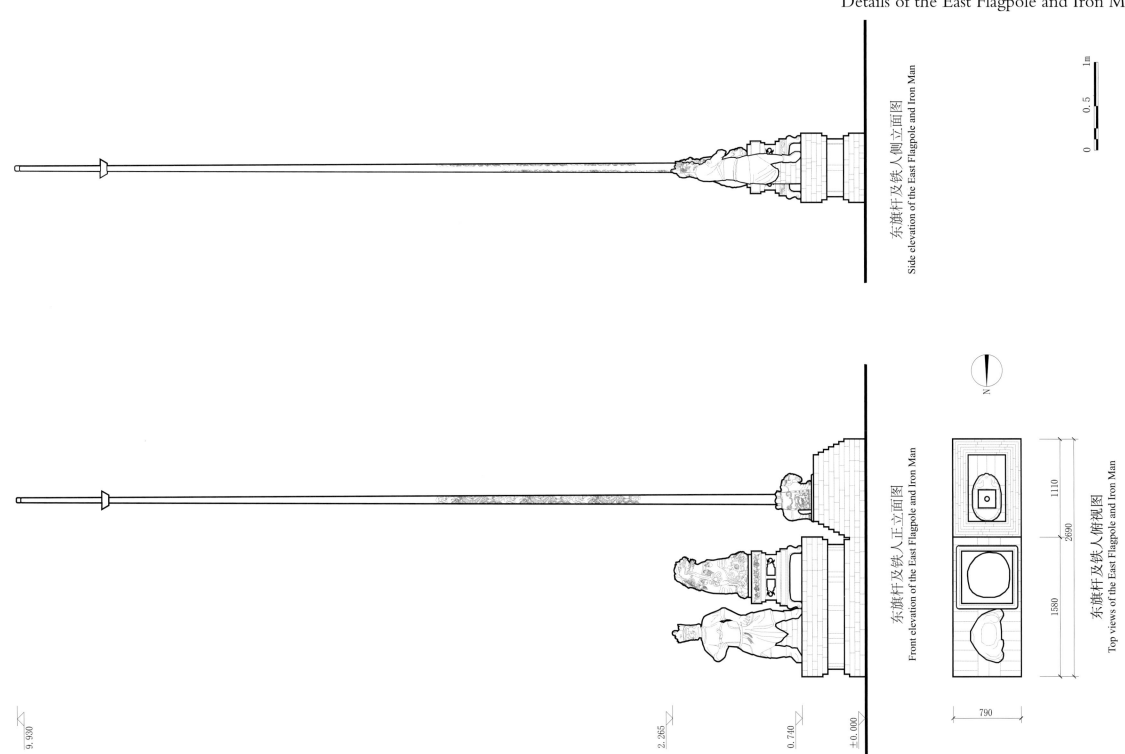

东旗杆及铁人侧立面图
Side elevation of the East Flagpole and Iron Man

0   0.5   1m

东旗杆及铁人正立面图
Front elevation of the East Flagpole and Iron Man

N

9.930

2.265

0.740

±0.000

东旗杆及铁人俯视图
Top views of the East Flagpole and Iron Man

1110

2690

1580

790

东焚帛炉大样图
Details of the East Silk Incinerator

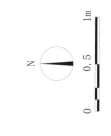

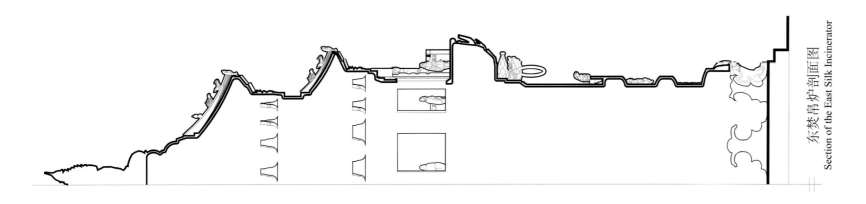

东焚帛炉剖面图
Section of the East Silk Incinerator

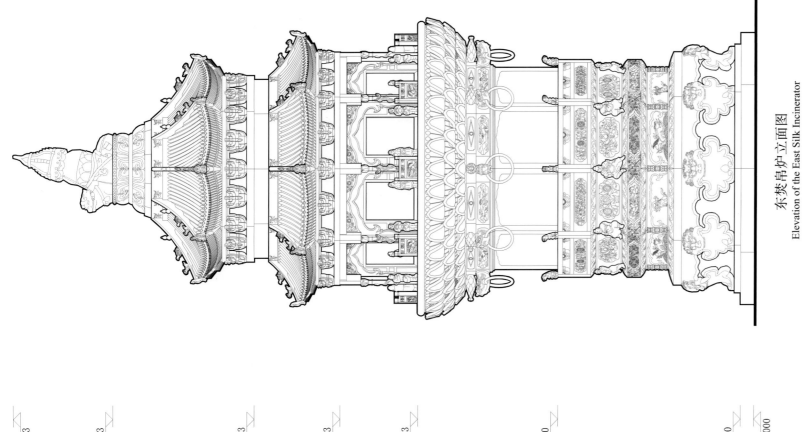

东焚帛炉立面图
Elevation of the East Silk Incinerator

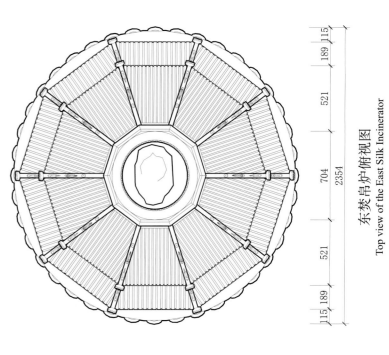

东焚帛炉俯视图
Top view of the East Silk Incinerator

6.063

5.253

4.163

3.453

2.763

1.610

0.160

±0.000

115 189

521

704

521

189 115

2354

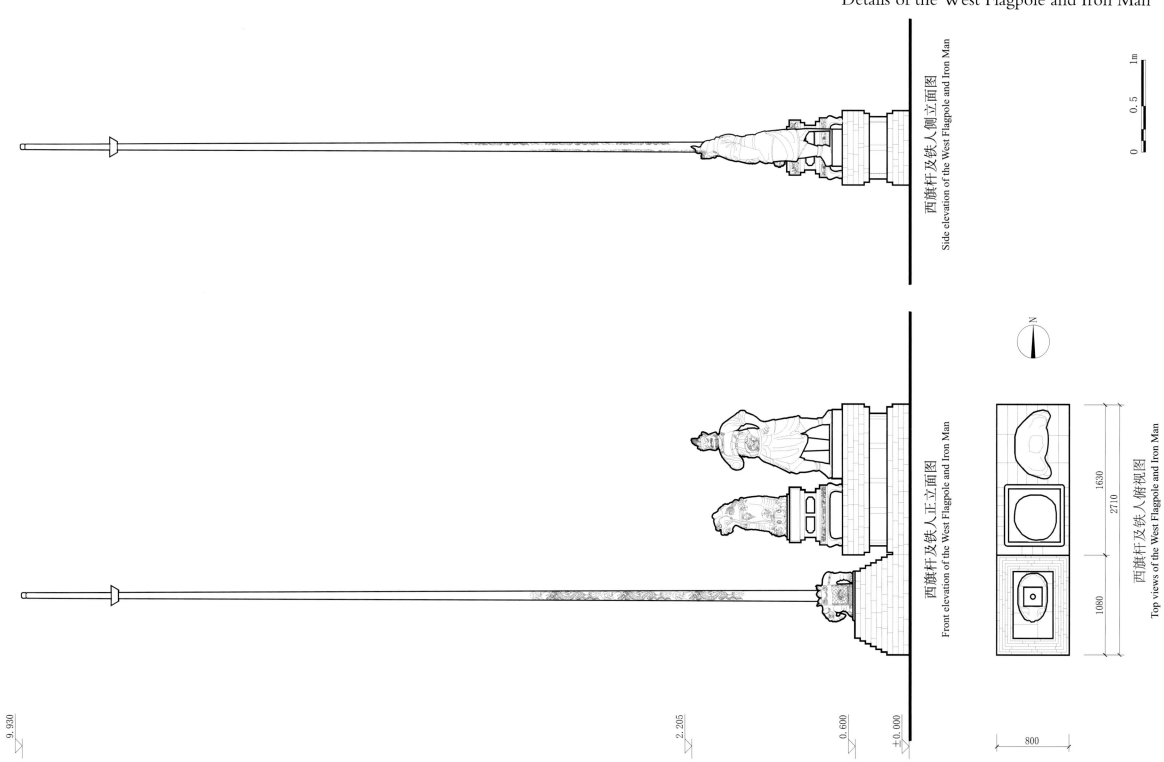

中国古建筑测绘大系·祠庙建筑——解州关帝庙

西旗杆及铁人侧立面图
Side elevation of the West Flagpole and Iron Man

西旗杆及铁人正立面图
Front elevation of the West Flagpole and Iron Man

西旗杆及铁人俯视图
Top views of the West Flagpole and Iron Man

N

118

9.930

2.205

0.600

±0.000

2710
1630
1080

800

0　0.5　1m

西焚帛炉大样图
Details of the West Silk Incinerator

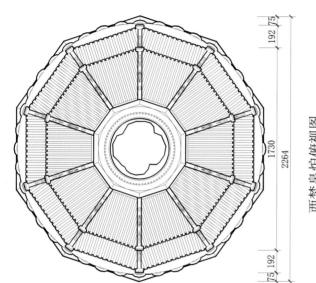

西焚帛炉俯视图
Top views of the West Silk Incinerator

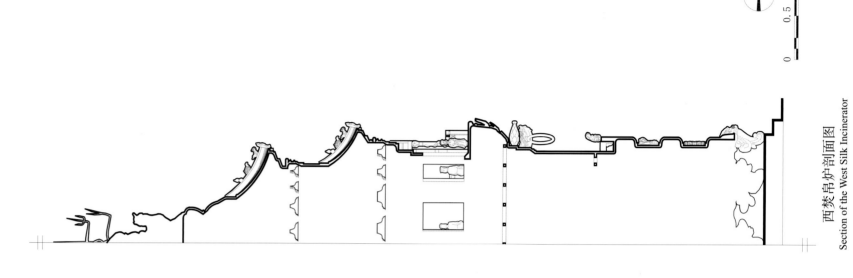

西焚帛炉剖面图
Section of the West Silk Incinerator

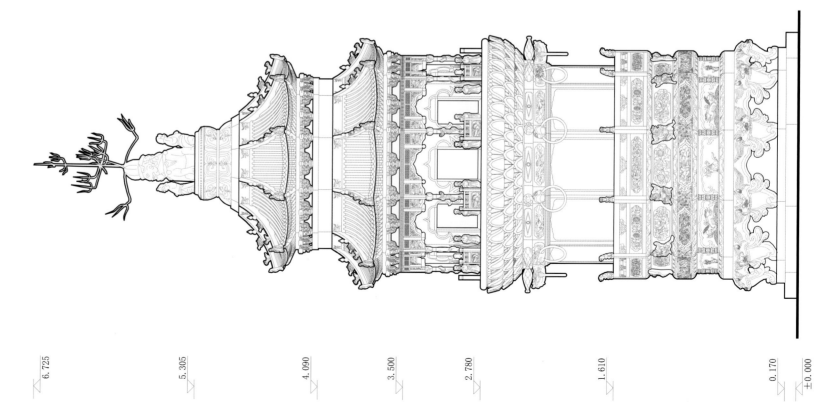

西焚帛炉立面图
Elevation of the West Silk Incinerator

崇宁殿
The Chongning Hall

中国古建筑测绘大系·祠庙建筑——解州关帝庙

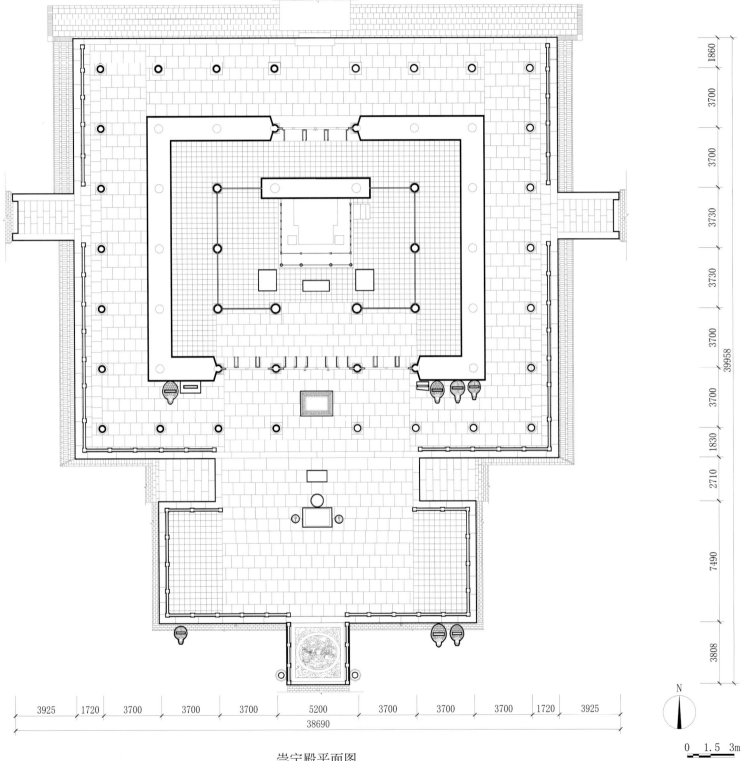

崇宁殿平面图
Plan of the Chongning Hall

0 1.5 3m

N

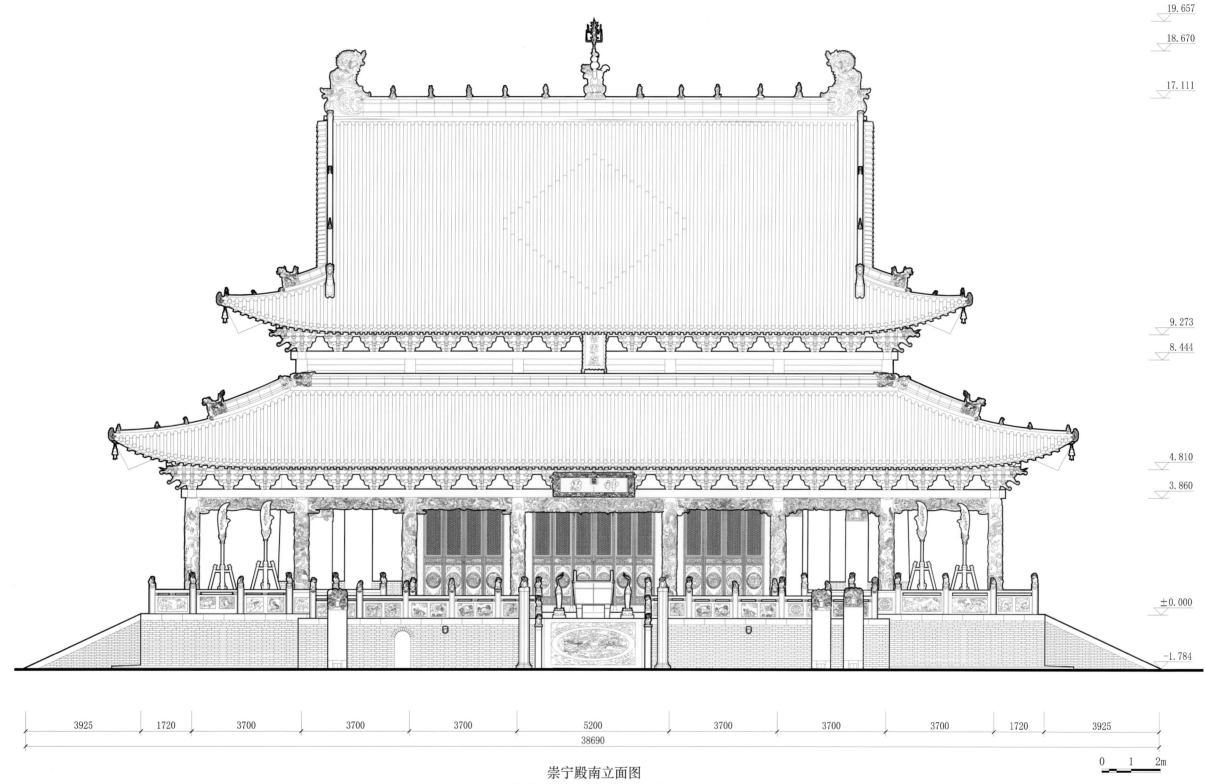

19.657

18.670

17.111

9.273

8.444

4.810

3.860

±0.000

-1.784

3925　1720　3700　3700　3700　5200　3700　3700　3700　1720　3925

38690

崇宁殿南立面图
South elevation of the Chongning Hall

0　1　2m

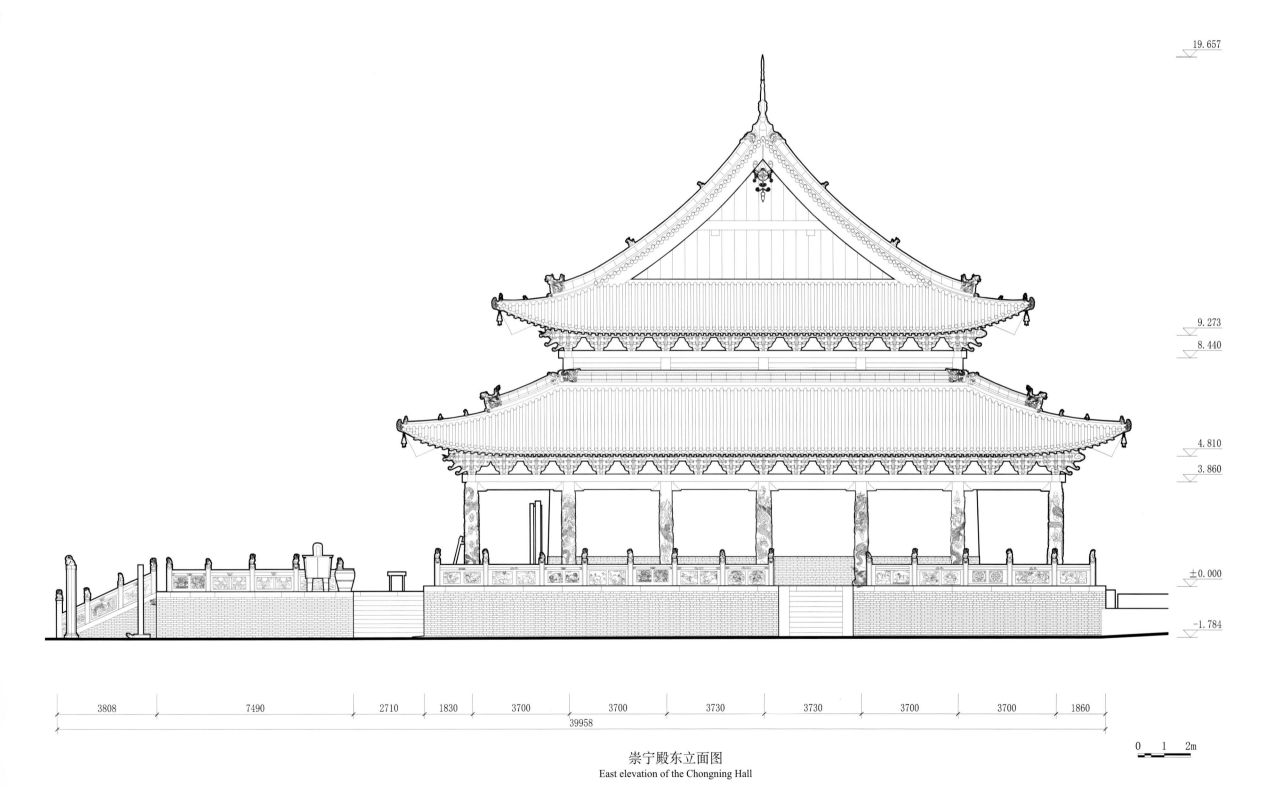

19.657

9.273
8.440

4.810
3.860

±0.000

-1.784

3808　7490　2710　1830　3700　3700　3730　3730　3700　3700　1860

39958

崇宁殿东立面图
East elevation of the Chongning Hall

0　1　2m

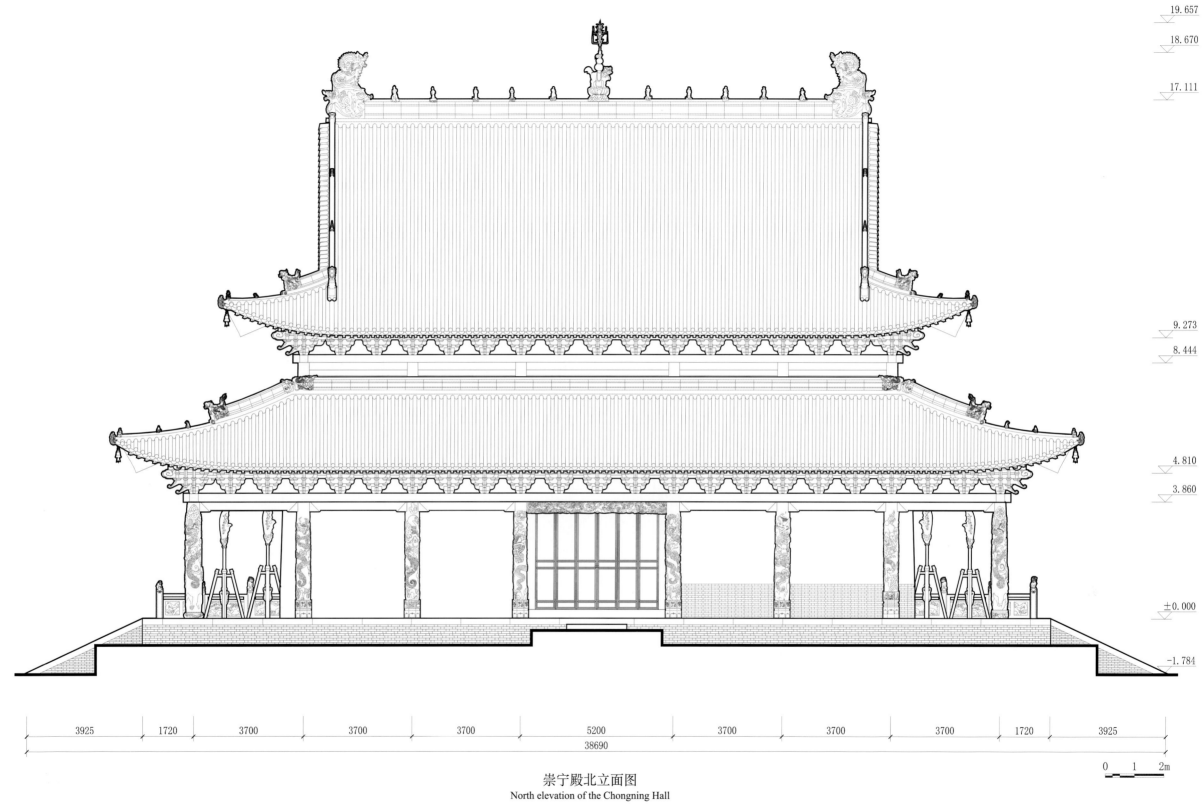

19.657

18.670

17.111

9.273

8.444

4.810

3.860

±0.000

-1.784

3925　1720　3700　3700　3700　5200　3700　3700　3700　1720　3925

38690

0　1　2m

崇宁殿北立面图

North elevation of the Chongning Hall

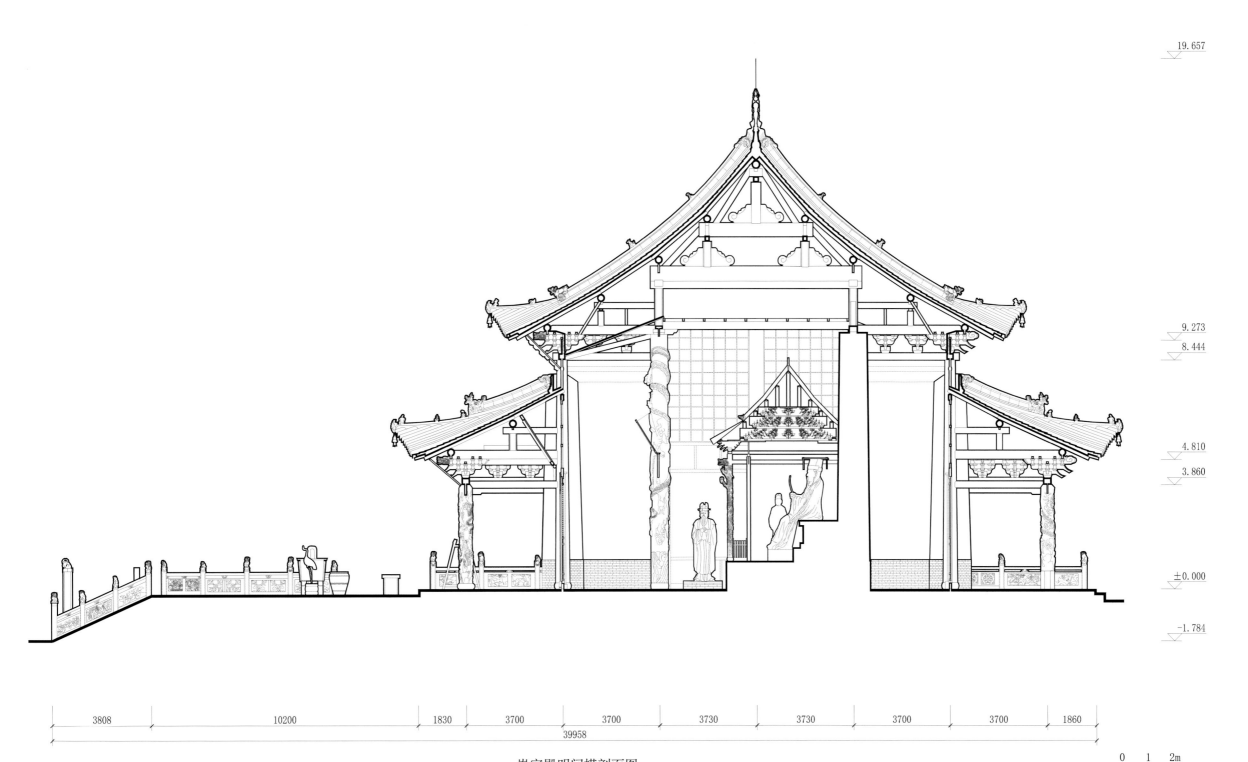

19.657

9.273
8.444

4.810
3.860

±0.000

−1.784

3808　　　　　10200　　　　1830　　3700　　　3700　　　3730　　　3730　　　3700　　　3700　　1860

39958

崇宁殿明间横剖面图
Central bay section of the Chongning Hall

0　　1　　2m

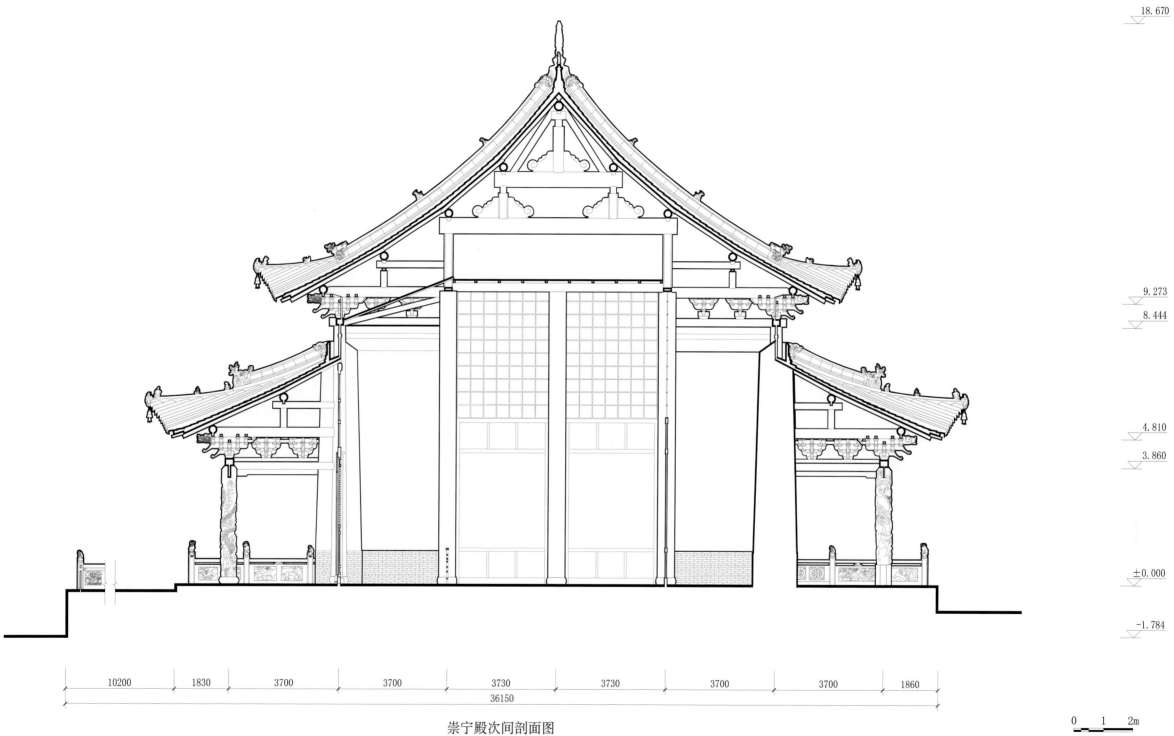

18.670

9.273

8.444

4.810

3.860

±0.000

-1.784

10200 1830 3700 3700 3730 3730 3700 3700 1860

36150

崇宁殿次间剖面图
Second bay section of the Chongning Hall

0 1 2m

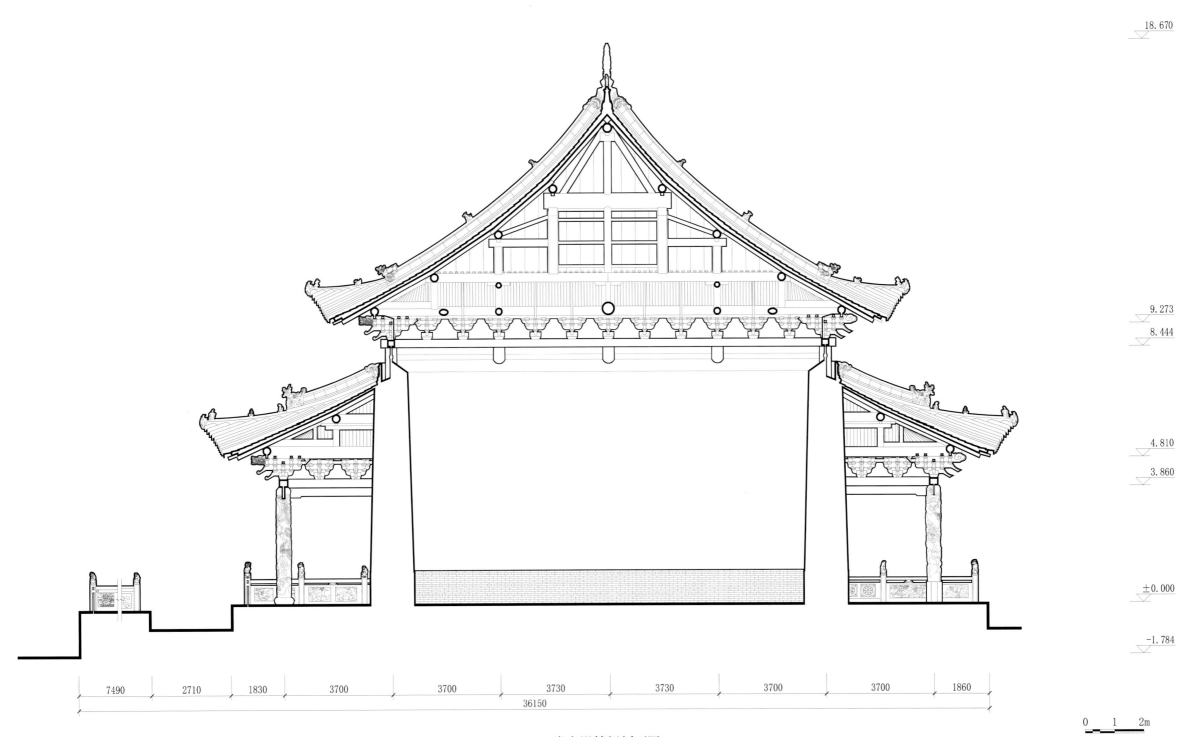

18.670

9.273
8.444

4.810
3.860

±0.000

-1.784

| 7490 | 2710 | 1830 | 3700 | 3700 | 3730 | 3730 | 3700 | 3700 | 1860 |

36150

0 1 2m

崇宁殿梢间剖面图
Side bay section of the Chongning Hall

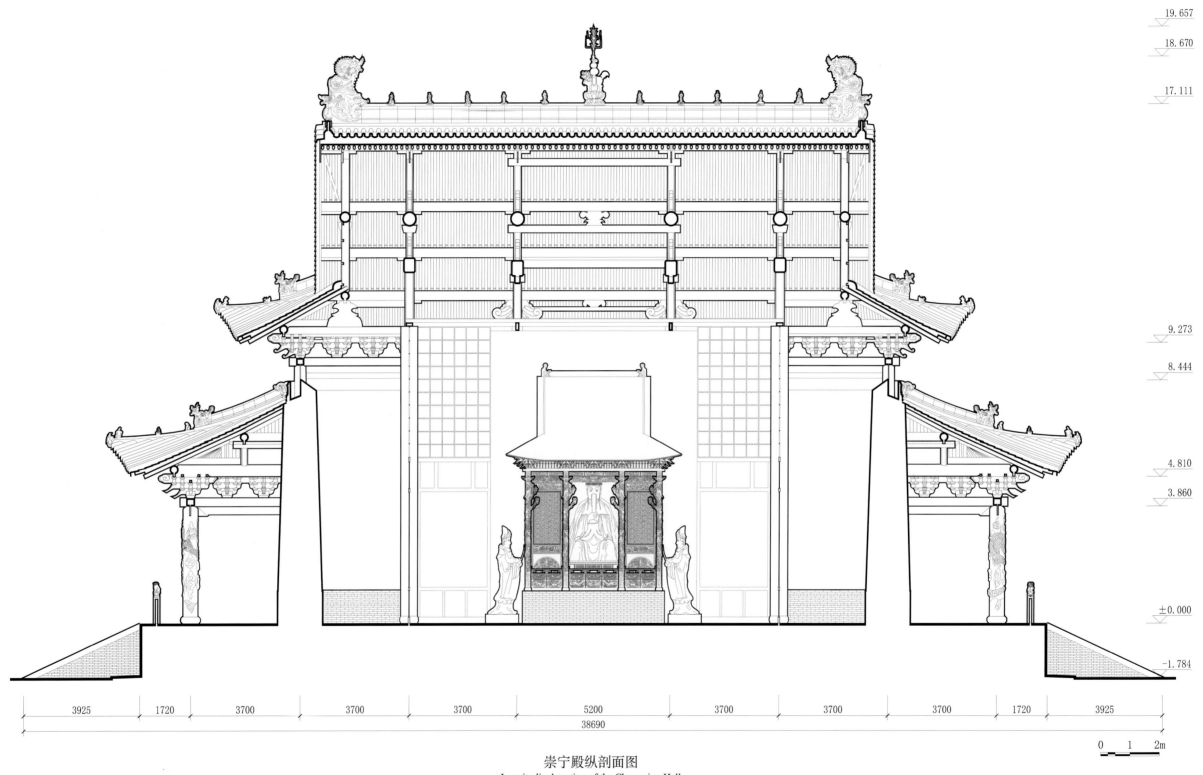

19.657
18.670
17.111
9.273
8.444
4.810
3.860
±0.000
-1.784

3925 1720 3700 3700 3700 5200 3700 3700 3700 1720 3925

38690

0　1　2m

崇宁殿纵剖面图
Longitudinal section of the Chongning Hall

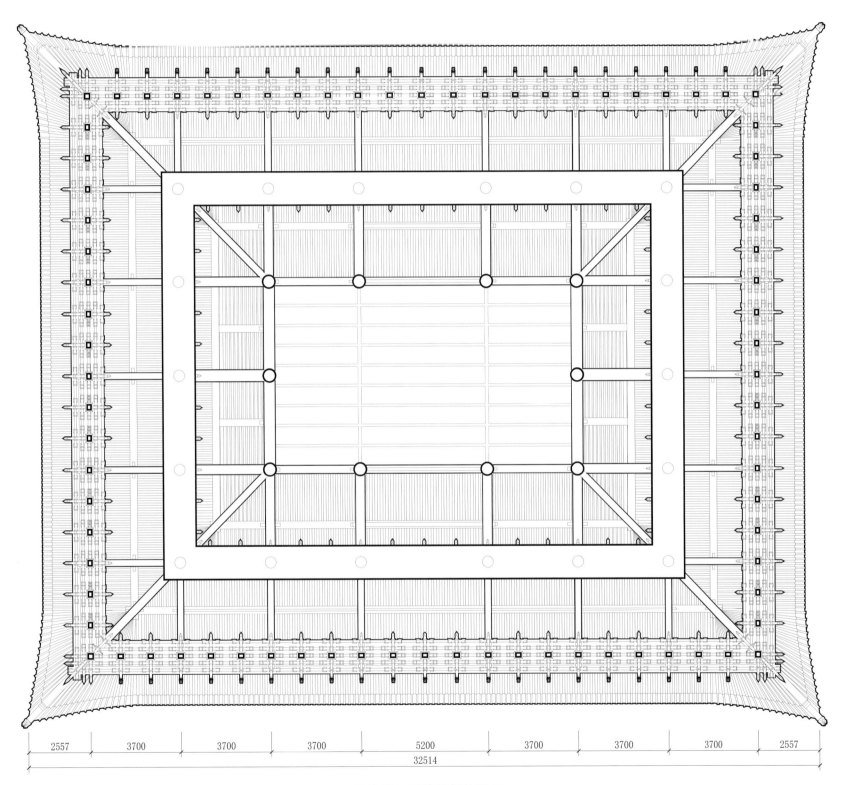

崇宁殿一层檐梁架仰视图
Bottom view of the beam frame and the eaves of the first-floor of the Chongning Hall

0 1 2m

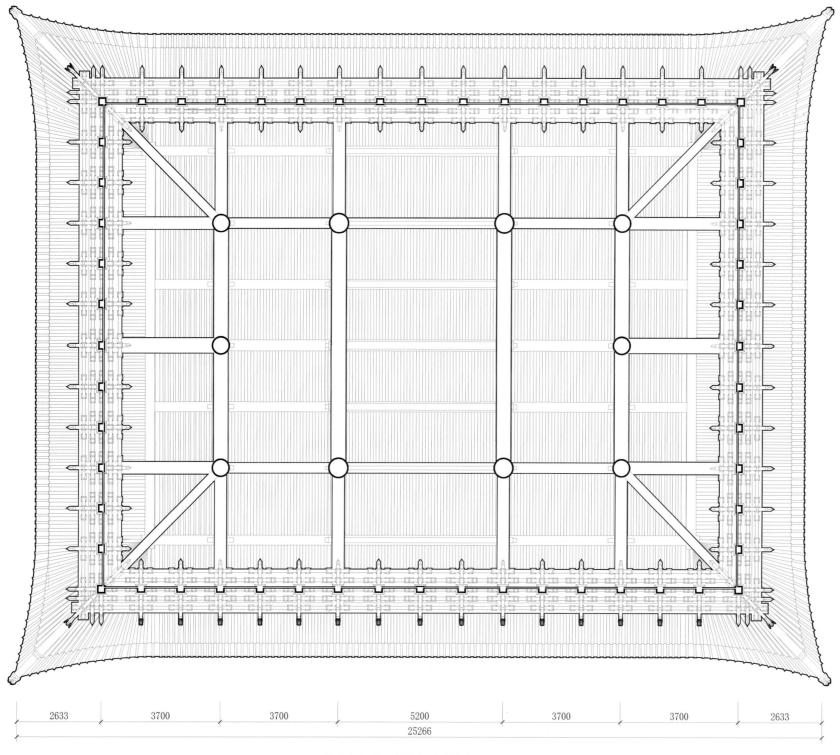

崇宁殿二层檐梁架仰视图
Bottom view of the beam frame and the eaves of the second-floor of the Chongning Hall

0 1 2m

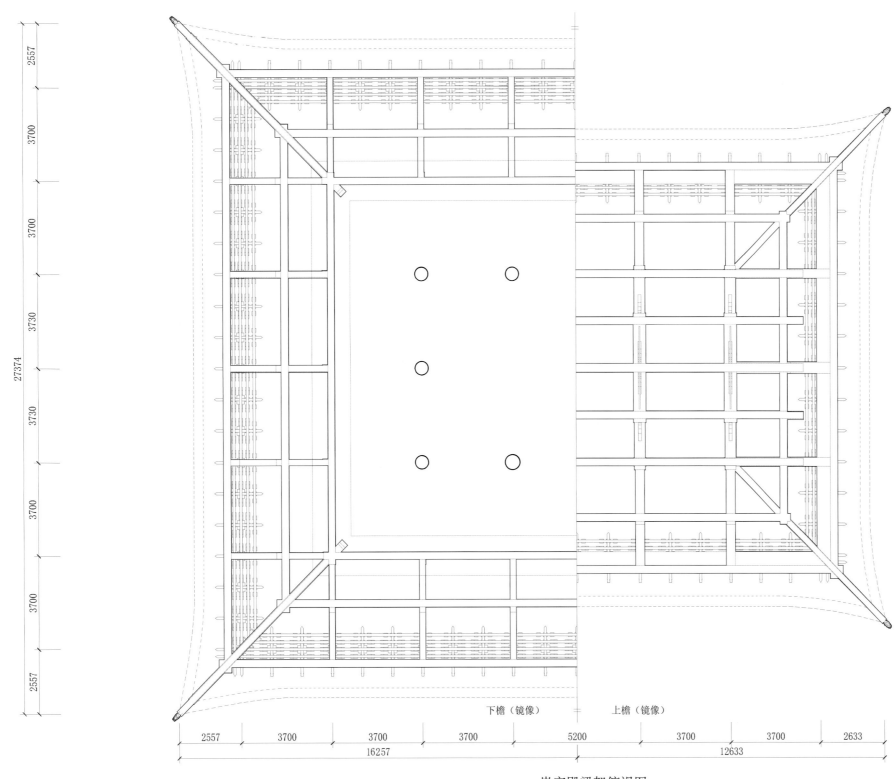

下檐（镜像）　上檐（镜像）

崇宁殿梁架俯视图
Top view of the beam frame of the Chongning Hall

0　1　2m

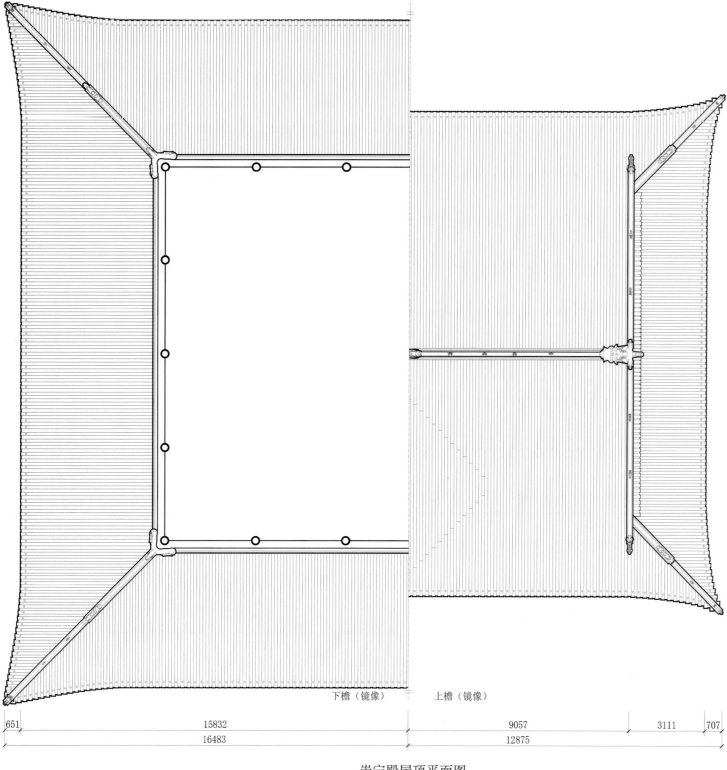

下檐（镜像）　　　上檐（镜像）

崇宁殿屋顶平面图
Roof plan of the Chongning Hall

0　1　2m

N

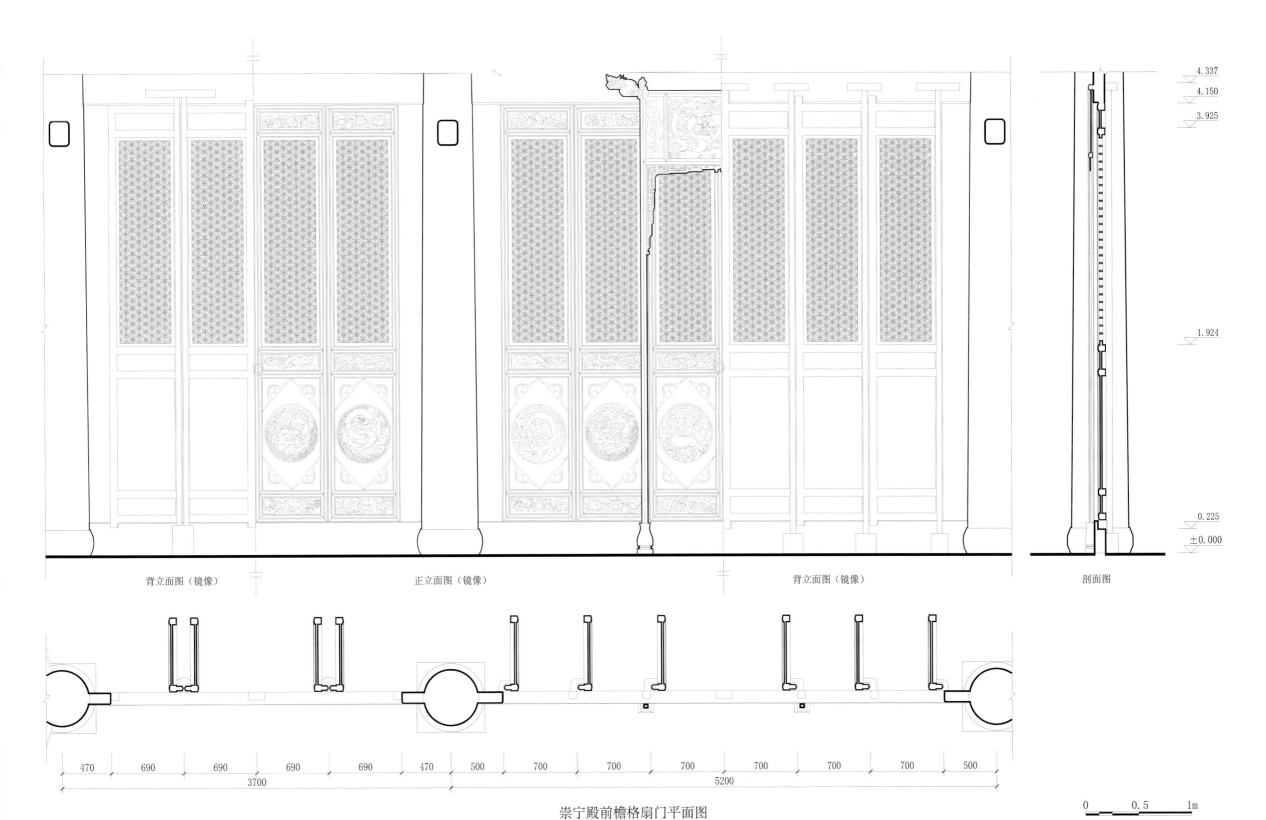

4.337
4.150
3.925

1.924

0.225
±0.000

背立面图（镜像）　　　　　　正立面图（镜像）　　　　　　背立面图（镜像）　　　　　剖面图

470　690　690　690　690　470　500　700　700　700　700　700　700　500

3700　　　　　　　　　5200

崇宁殿前檐格扇门平面图
Plan of the partition door of the front eaves of the Chongning Hall

0　0.5　1m

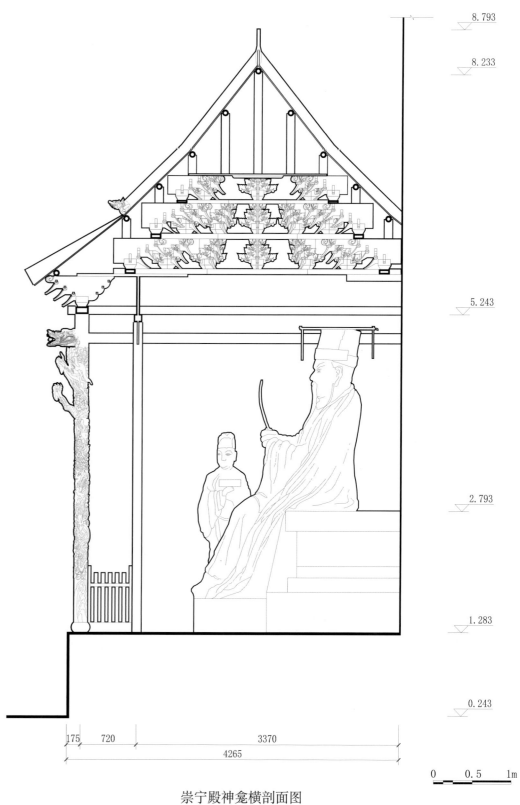

175 1275 1900 1275 175
4800

175 720 3370
4265

8.793
8.233

5.243

2.793

1.283

0.243

0   0.5   1m

崇宁殿神龛正立面图
Front elevation of the niche of the Chongning Hall

崇宁殿神龛横剖面图
Cross-section of the niche of the Chongning Hall

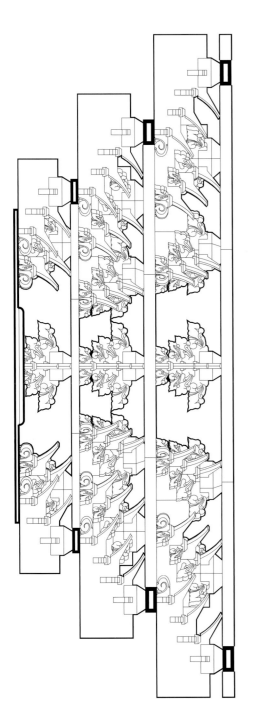

崇宁殿神龛藻井剖面图

Section of the caisson ceiling of the Chongning Hall

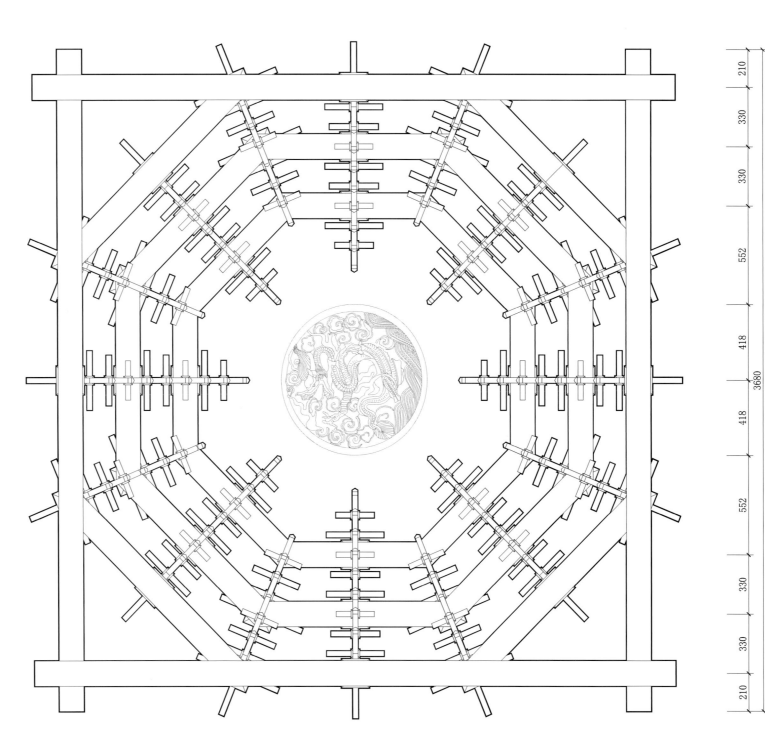

崇宁殿神龛藻井仰视图

Bottom view of the caisson ceiling of the Chongning Hall

东官厅
The Dongguan Hall

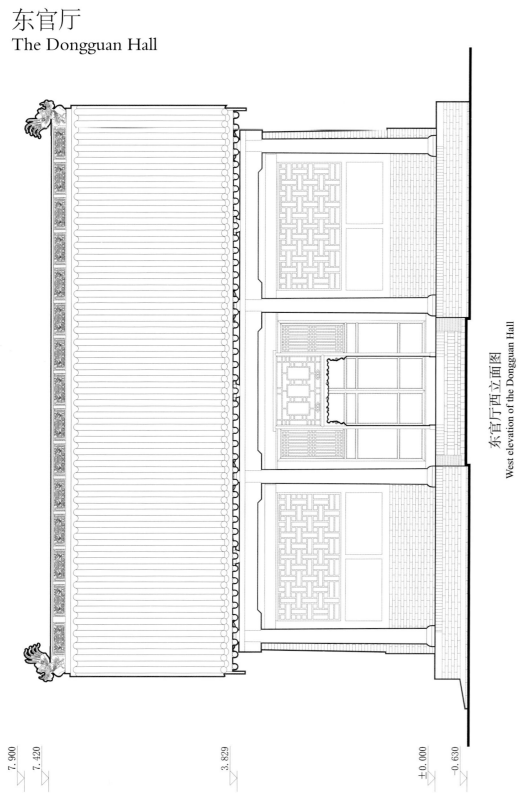

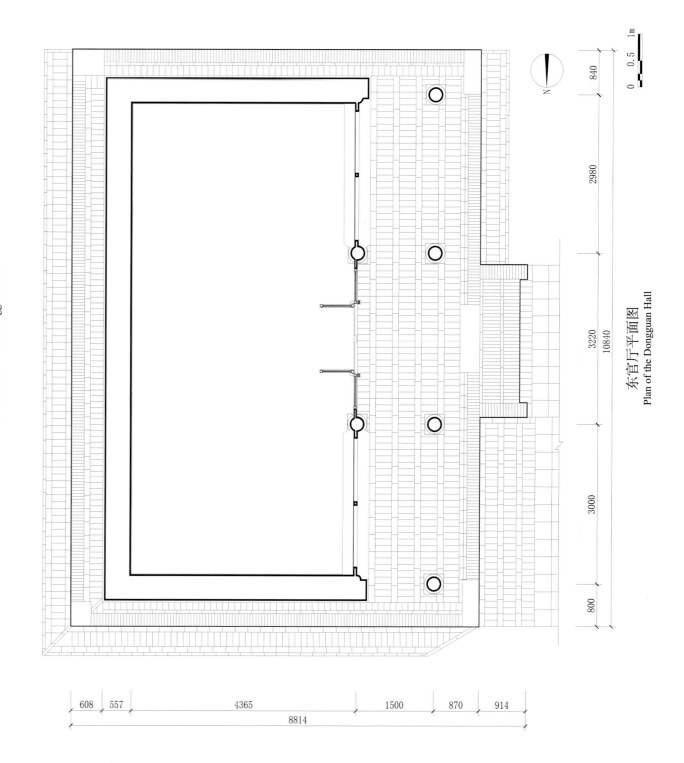

东官厅西立面图
West elevation of the Dongguan Hall

7.900
7.420
3.829
±0.000
-0.630

东官厅平面图
Plan of the Dongguan Hall

840
2980
3220
10840
3000
800

608 557 4365 1500 870 914
8814

0 0.5 1m

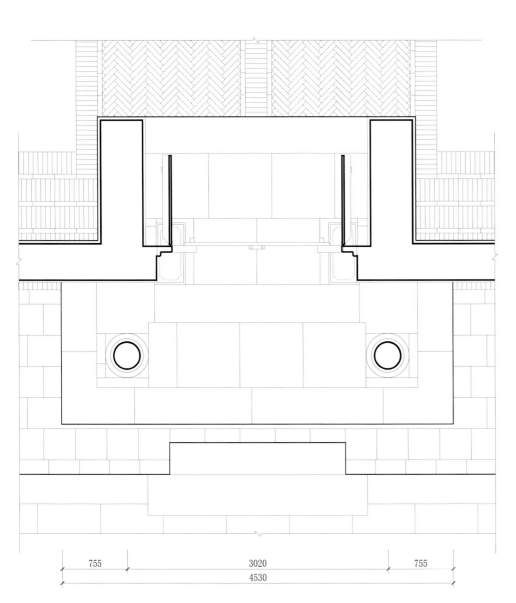

寝宫院门平面图
Plan of the Gate of the Chamber Yard

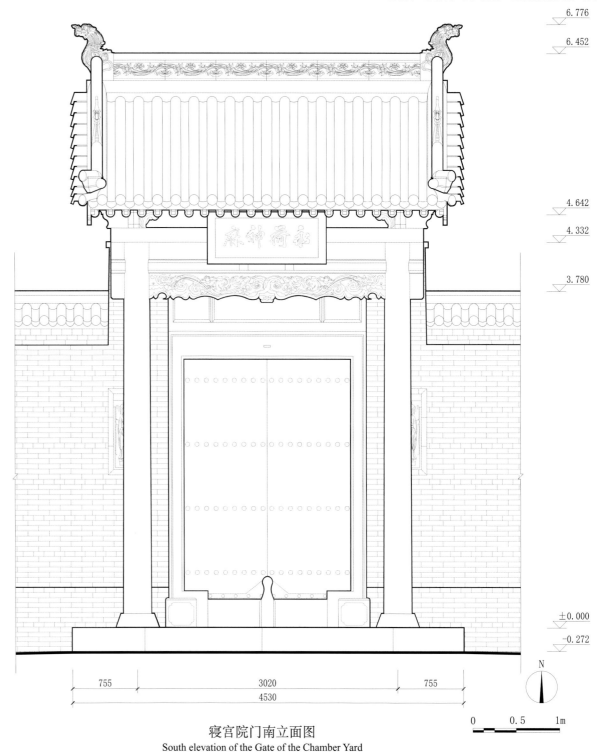

寝宫院门南立面图
South elevation of the Gate of the Chamber Yard

6.776
6.452
4.642
4.332
3.780
±0.000
-0.272

755　3020　755
4530

755　3020　755
4530

N

0　0.5　1m

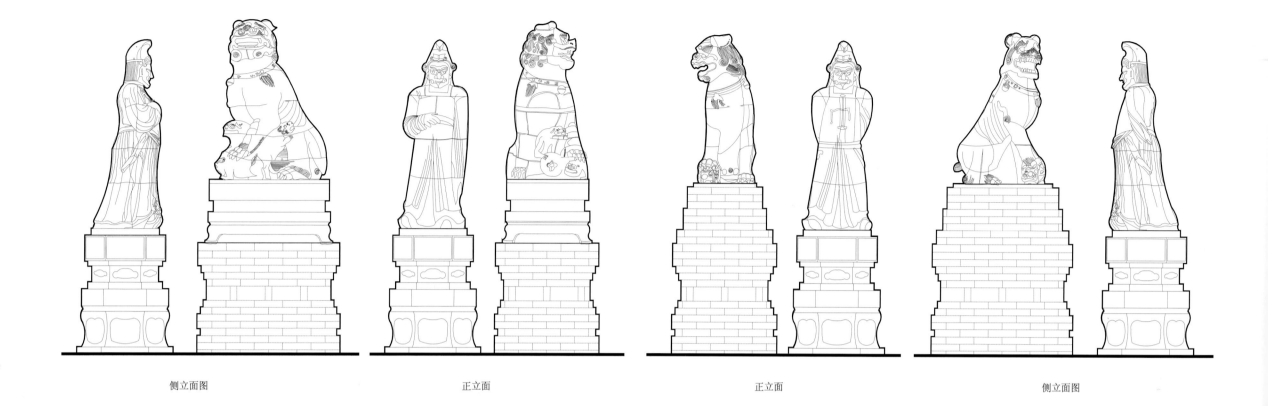

侧立面图 　　　　　　　　　　 正立面 　　　　　　　　　　 正立面 　　　　　　　　　　 侧立面图

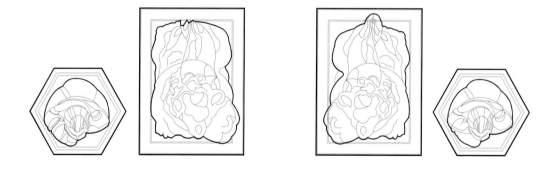

平面图 　　　　　　　　　　　　　　 平面图

寝宫院门西铁人铁狮 　　　　　　　　　 寝宫院门东铁人铁狮
Details of the Iron man on the east and west side of the Gate of the Chamber Yard

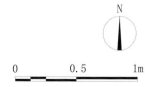

0　　　　　0.5　　　　　1m

气肃千秋坊

The Qisu qianqiu Memorial Archway

10.932
10.390
8.846
8.065
7.531
6.050
5.030
±0.000
-0.560
氣肅千秋
2585
2850
735
3280
735
2850
2585
15620
0 0.5 1m

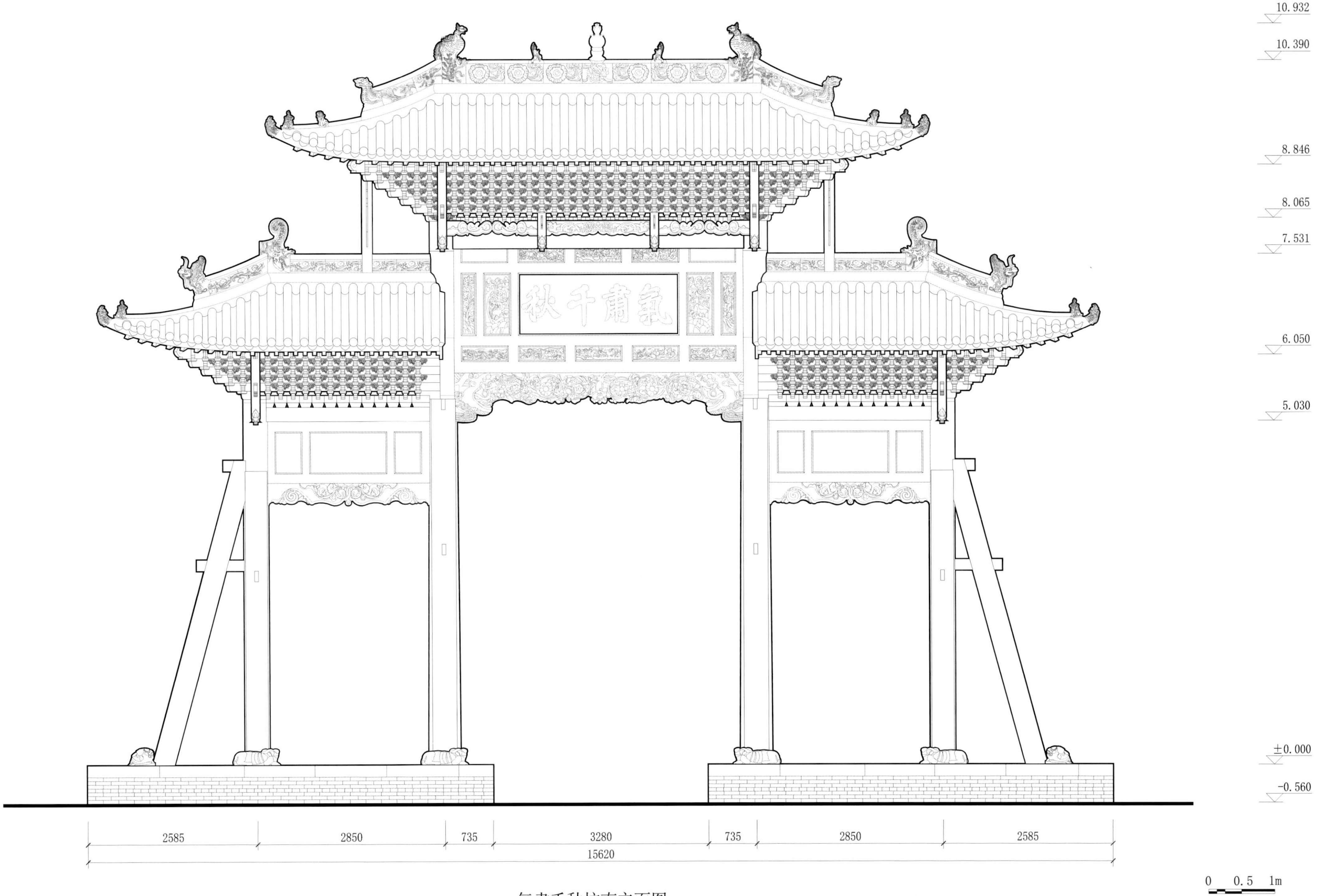

气肃千秋坊南立面图

South elevation of the Qisu qianqiu Memorial Archway

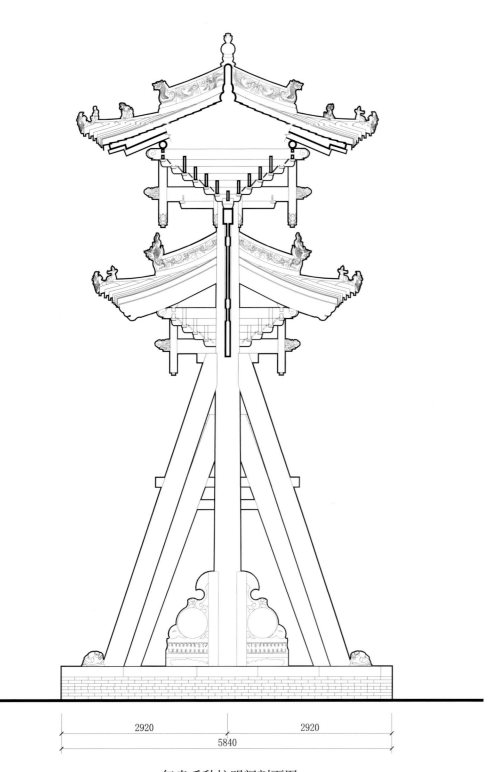

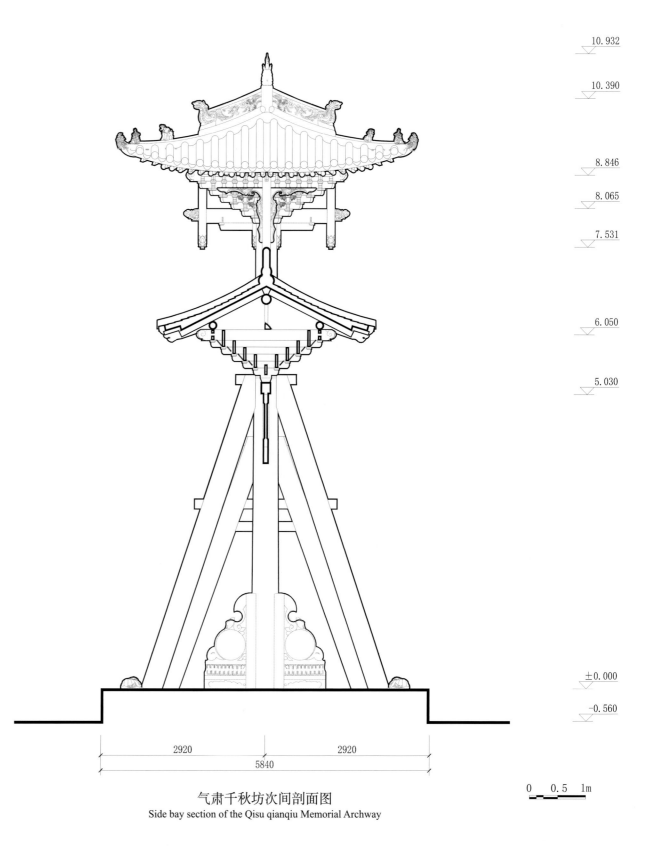

气肃千秋坊明间剖面图
Central bay section of the Qisu qianqiu Memorial Archway

气肃千秋坊次间剖面图
Side bay section of the Qisu qianqiu Memorial Archway

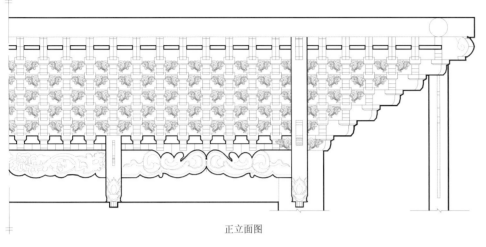

正立面图

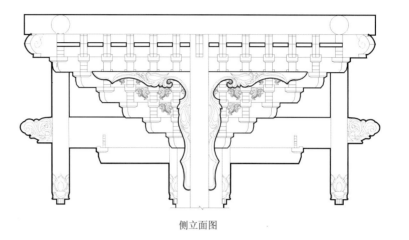

侧立面图

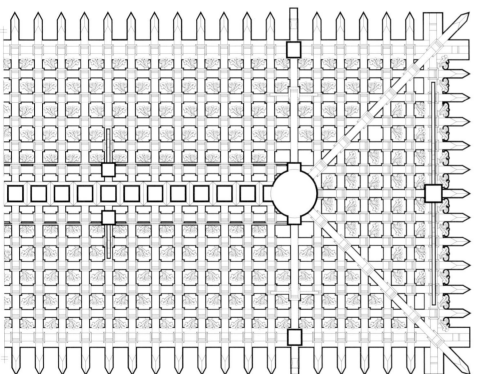

仰视平面图

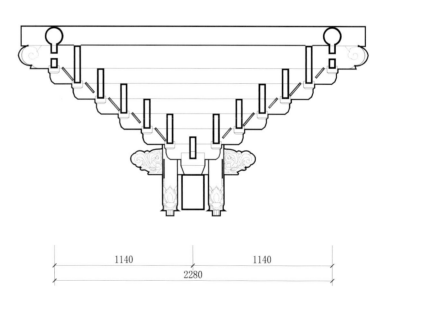

剖面图

气肃千秋坊明间斗栱大样图

Details of the bracket set of the central bay of the Qisu qianqiu Memorial Archway

0   0.25   0.5m

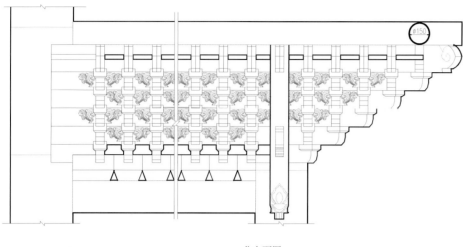

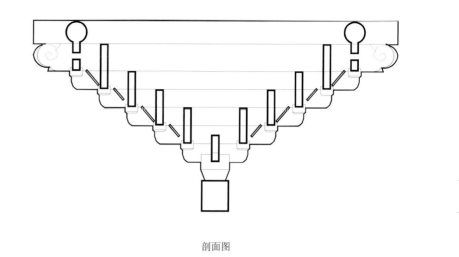

北立面图

剖面图

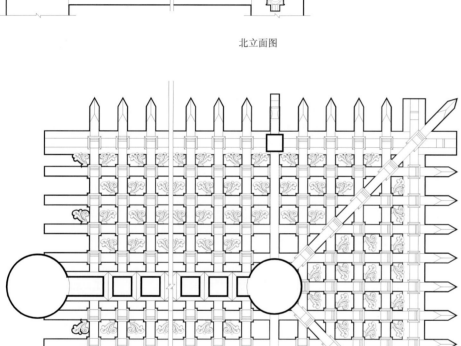

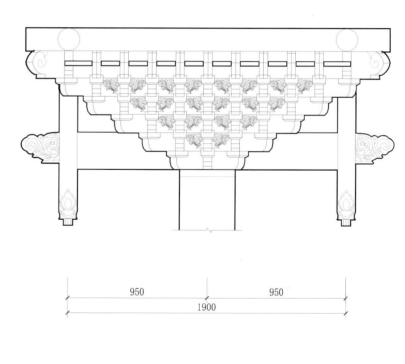

平面图

侧立面图

气肃千秋坊次间斗栱大样

Details of the bracket set of the side bay of the Qisu qianqiu Memorial Archway

0   0.25   0.5m

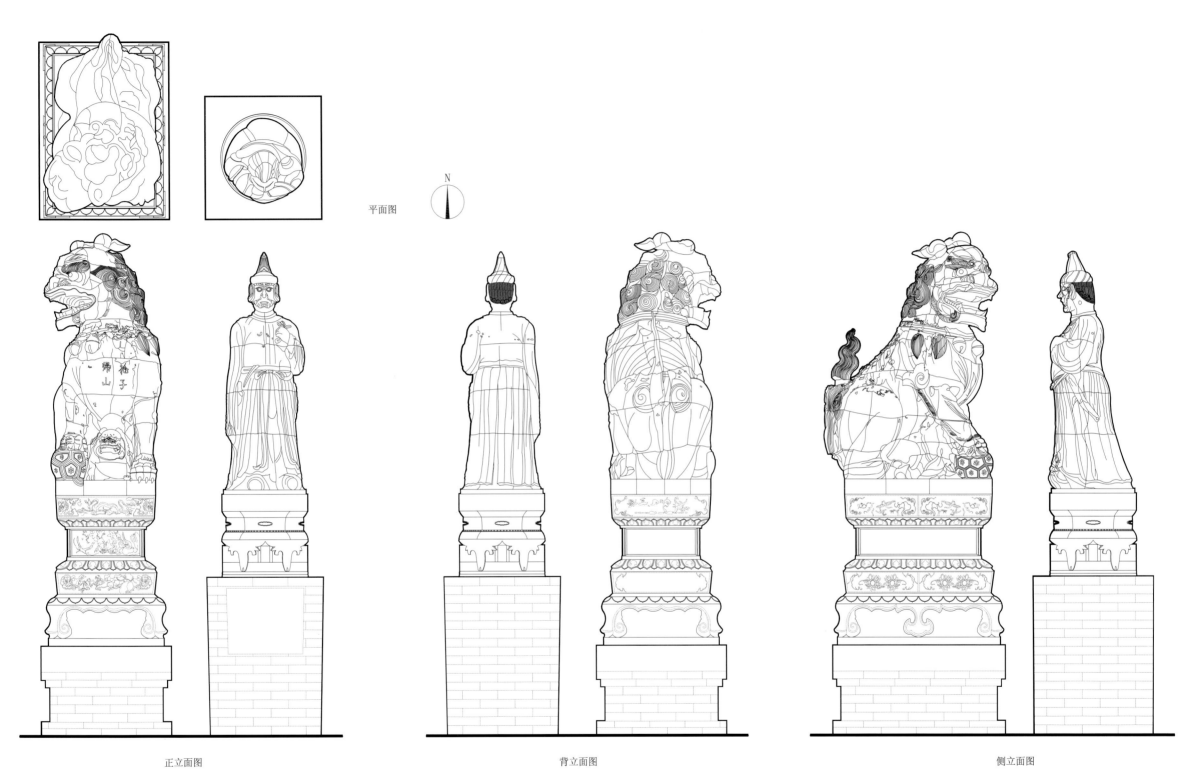

平面图

正立面图

背立面图

侧立面图

气肃千秋坊东铁人铁狮大样
Details of the Iron man on the east side of the Qisu qianqiu Memorial Archway

0   0.25   0.5m

143

平面图

N

正立面图

背立面图

侧立面图

气肃千秋西铁人铁狮大样图

Details of the Iron man on the west side of the Qisu qianqiu Memorial Archway

0   0.25   0.5m

印楼
The Yin Storied Building

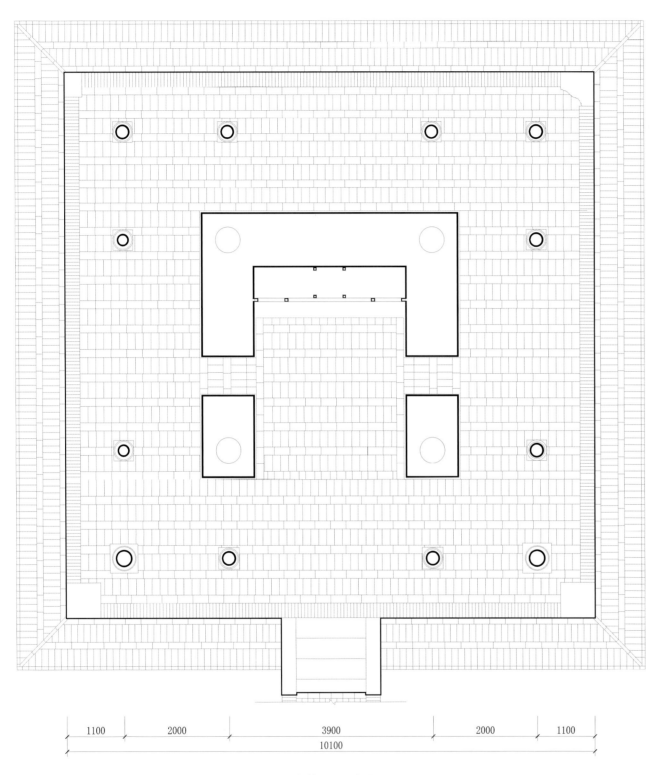

印楼一层平面图
First Floor Plan of the Yin Storied Building

N

0　0.5　1m

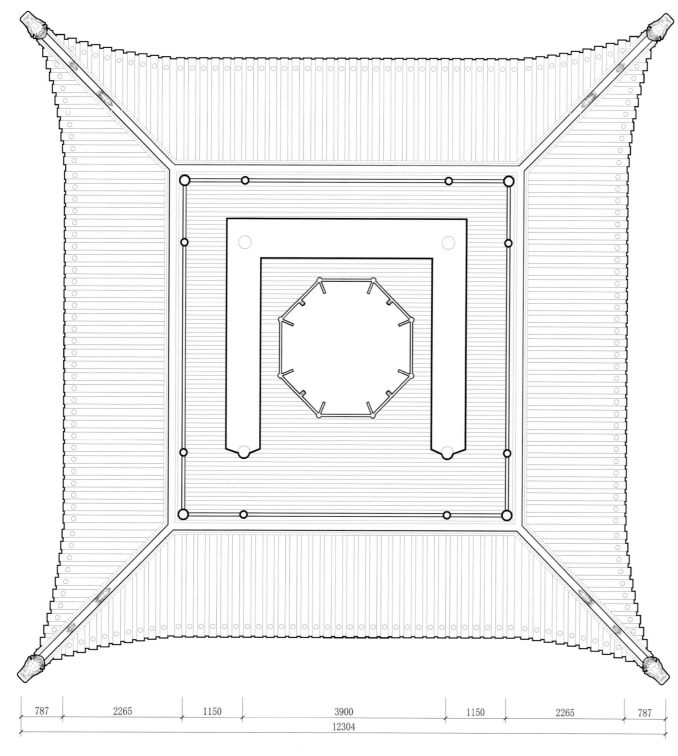

印楼二层平面图
Second Floor Plan of the Yin Storied Building

0   0.5   1m

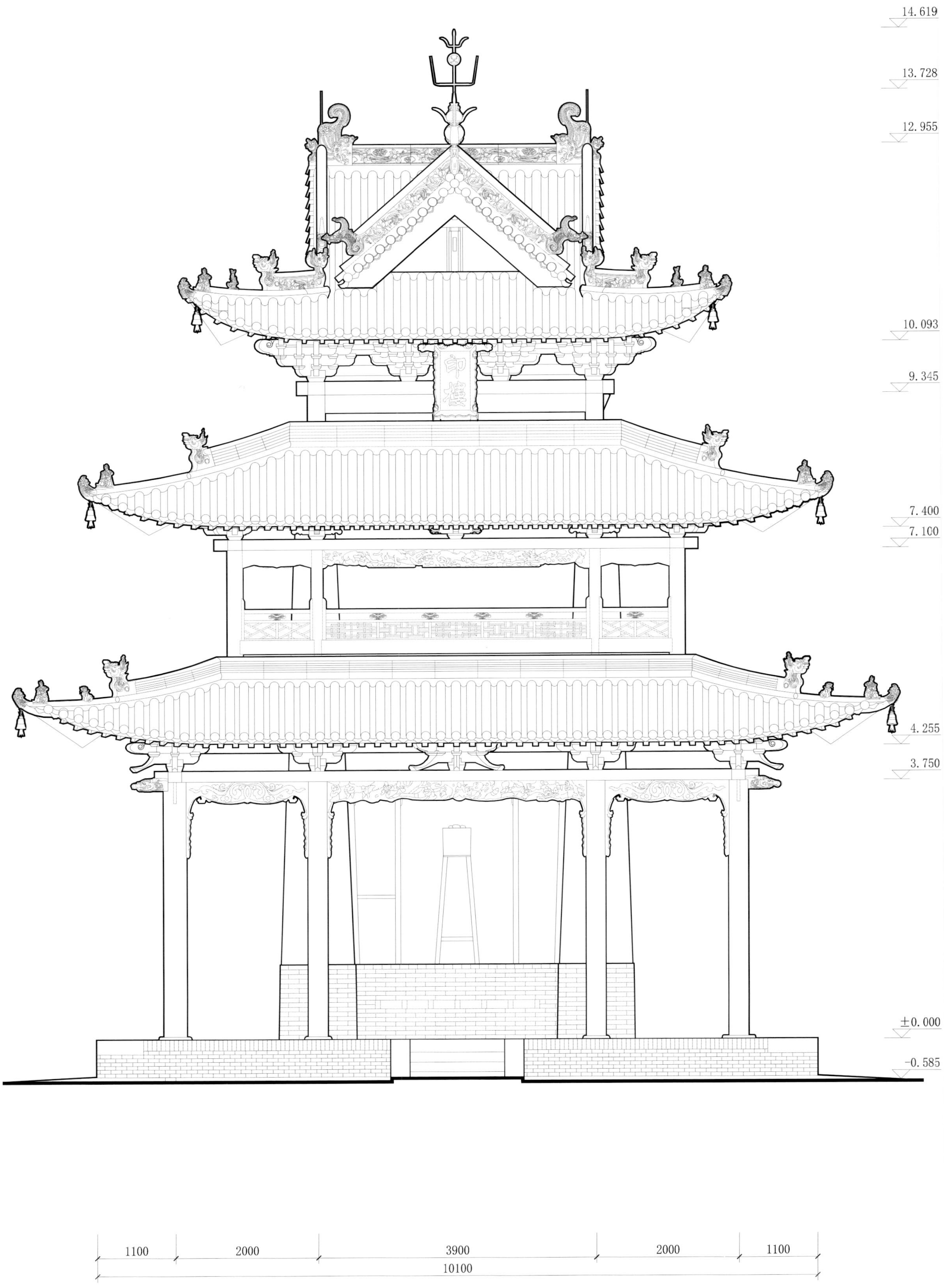

14.619
13.728
12.955

10.093
9.345

7.400
7.100

4.255
3.750

±0.000
-0.585

1100
2000
3900
10100
2000
1100

印楼西立面图
West elevation of the Yin Storied Building

0 0.5 1m

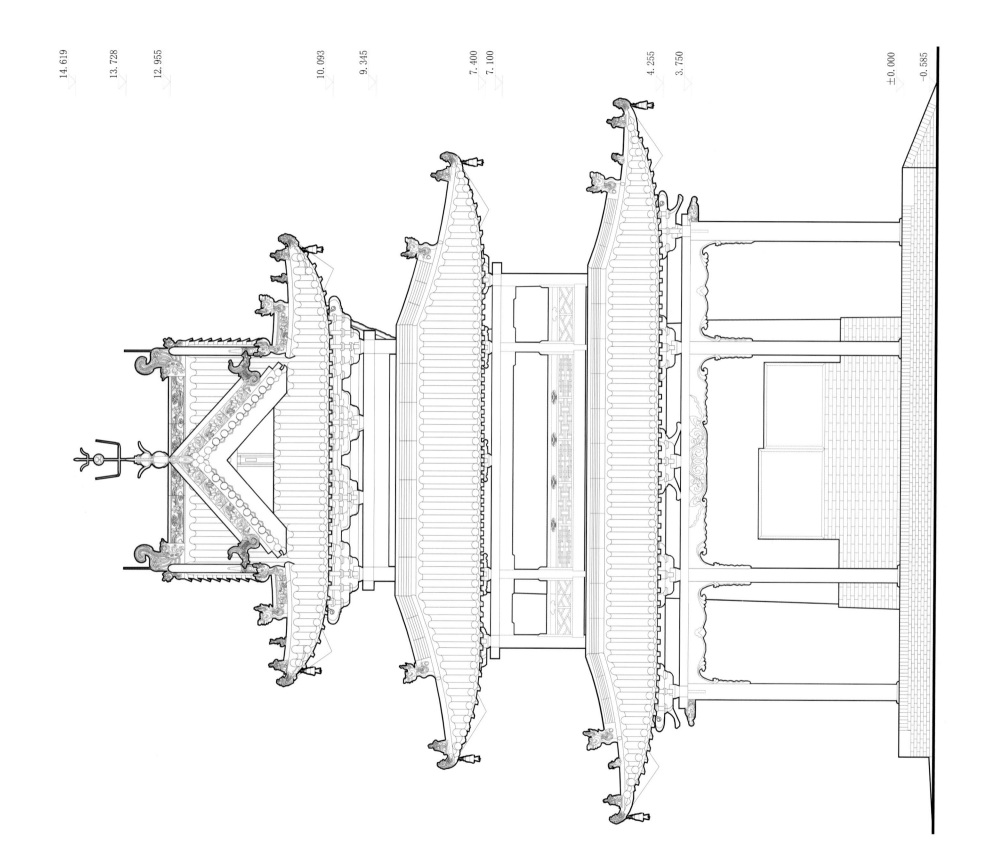

14.619
13.728
12.955
10.093
9.345
7.400
7.100
4.255
3.750
±0.000
-0.585

1423
1100
2000
11523
3900
2000
1100

0 0.5 1m

印楼北立面图
North elevation of the Yin Storied Building

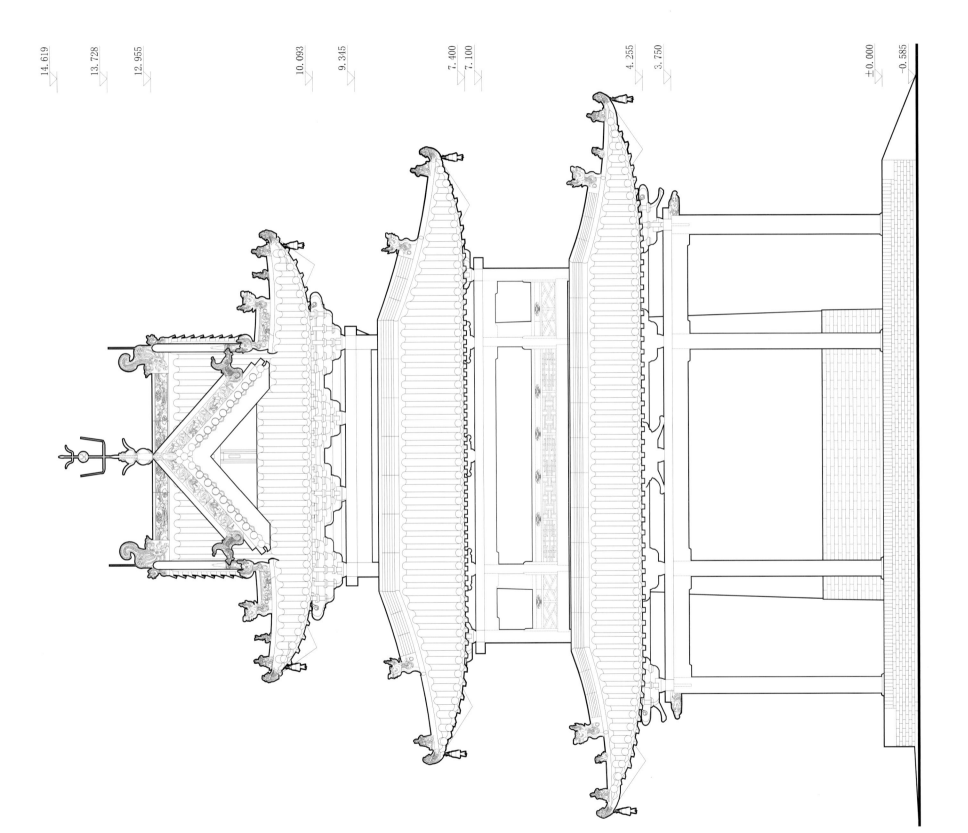

14.619
13.728
12.955

10.093
9.345

7.400
7.100

4.255
3.750

±0.000
-0.585

印楼东立面图
East elevation of the Yin Storied Building

 中国古建筑测绘大系 · 祠庙建筑 —— 解州关帝庙

1423
1100
2000
11523
3900
2000
1100

0 0.5 1m

148

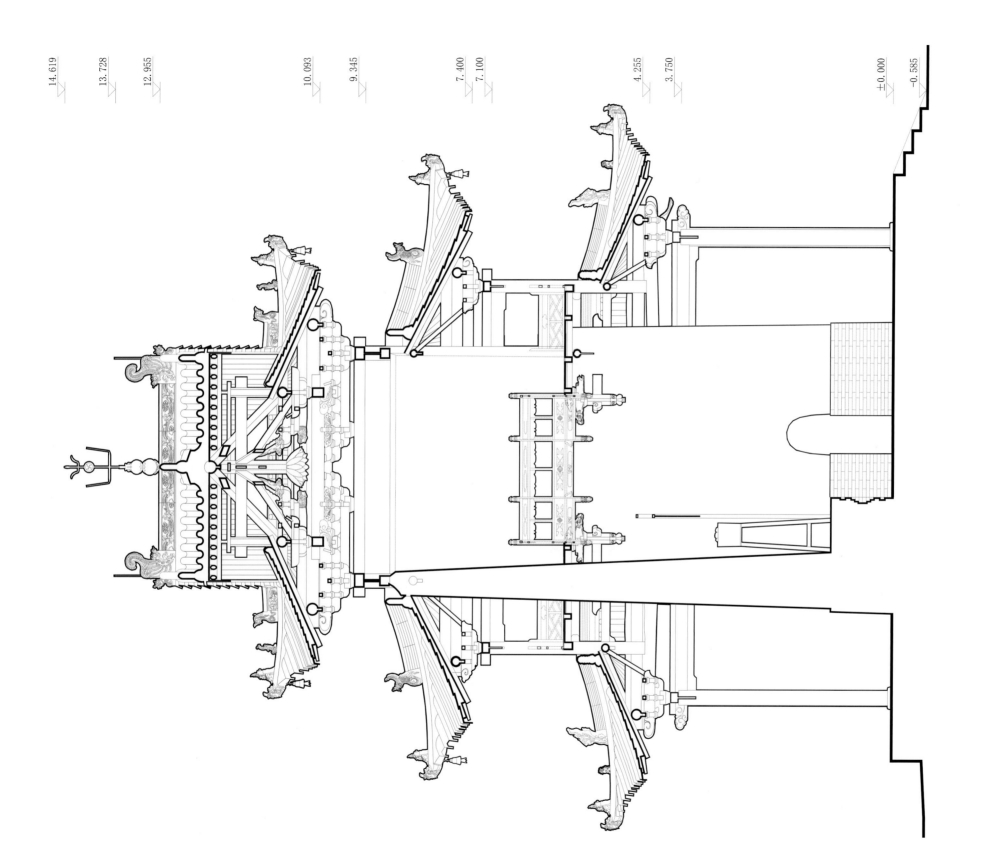

14.619
13.728
12.955
10.093
9.345
7.400
7.100
4.255
3.750
±0.000
-0.585

1423
1100
850
1150
11523
3900
1150
850
1100

0　0.5　1m

印楼横剖面图
Cross-section of the Yin Storied Building

14.619
13.728
12.955

10.093
9.345

7.400
7.100

4.255
3.750

±0.000
-0.585

0   0.5   1m

1100
850
1150
3900
10100
1150
850
1100

印楼纵剖面图
Longitudinal section of the Yin Storied Building

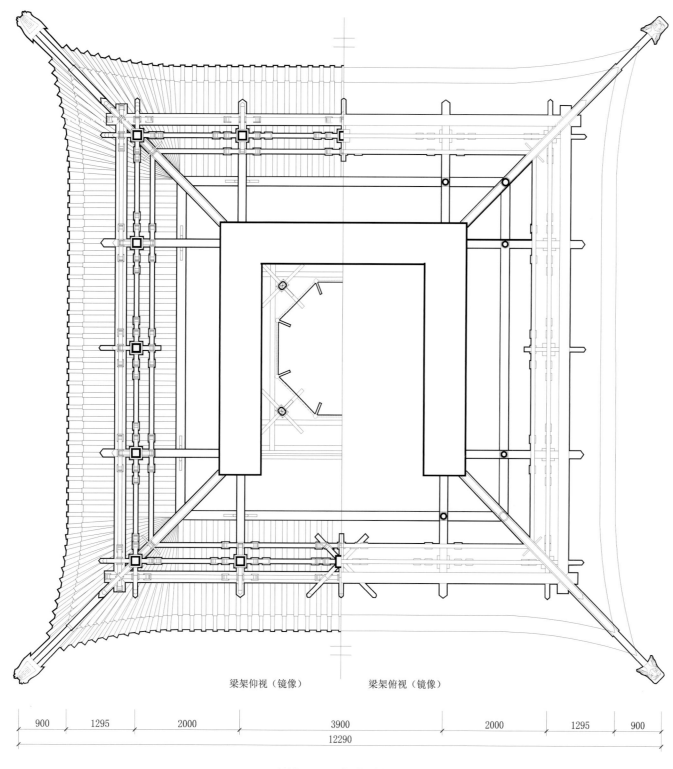

梁架仰视（镜像）　　梁架俯视（镜像）

| 900 | 1295 | 2000 | 3900 | 2000 | 1295 | 900 |

12290

印楼一层梁架仰俯视图
Top and bottom views of the beam frame of the first-floor of the Yin Storied Building

0　0.5　1m

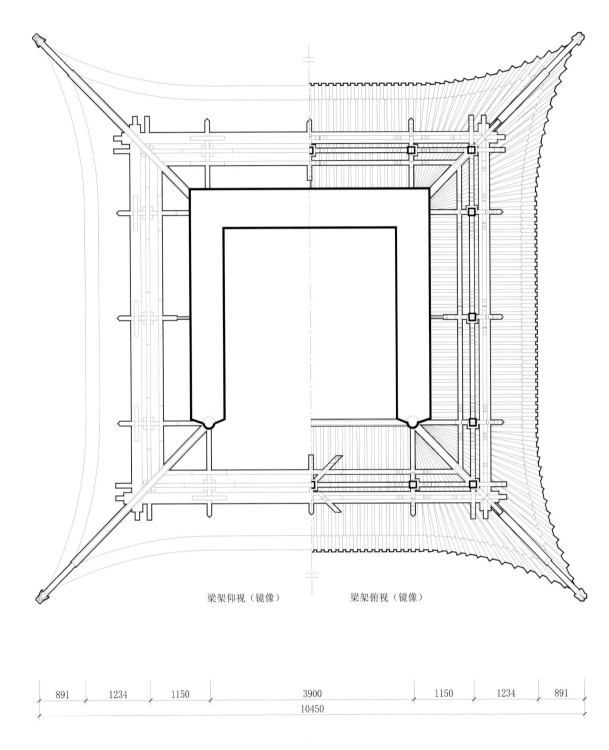

梁架仰视（镜像）　　　梁架俯视（镜像）

891　1234　1150　3900　1150　1234　891

10450

0　0.5　1m

印楼二层梁架仰俯视图
Top and bottom views of the beam frame of the second-floor of the Yin Storied Building

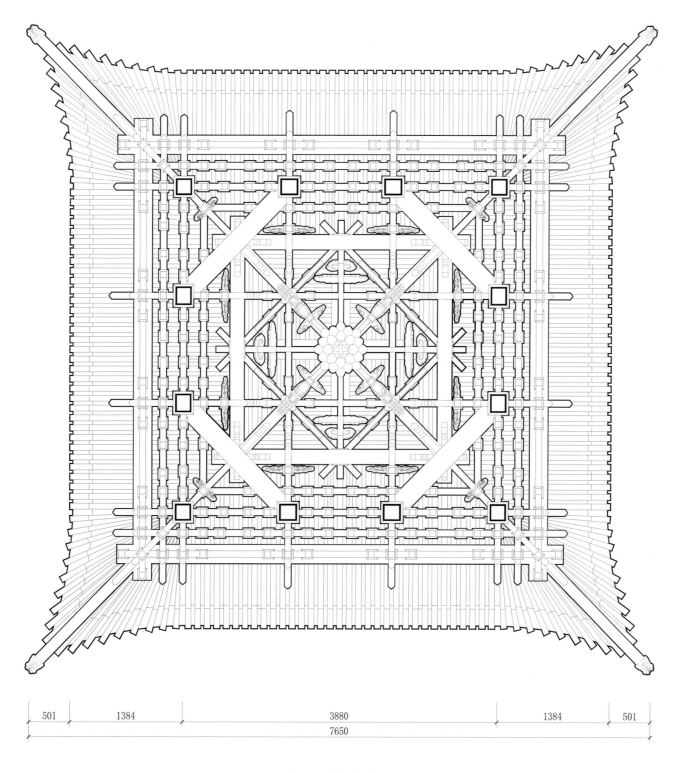

501 1384 3880 1384 501

7650

0 0.5 1m

印楼三层檐梁架仰视图

Bottom view of the beam frame and the eaves of the third-floor of the Yin Storied Building

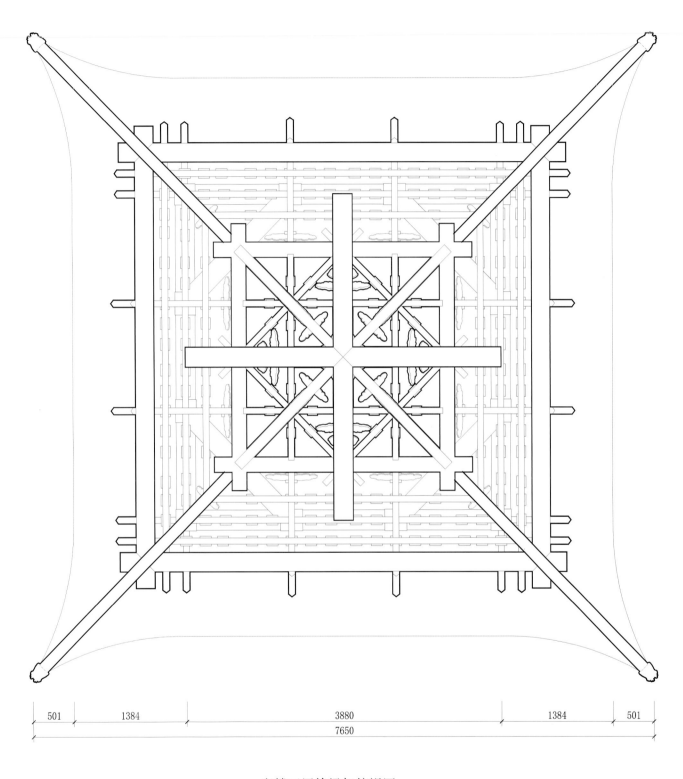

| 501 | 1384 | 3880 | 1384 | 501 |

7650

0   0.5   1m

印楼三层檐梁架俯视图
Top view of the beam frame and the eaves of the third-floor of the Yin Storied Building

刀楼
The Dao Storied Building

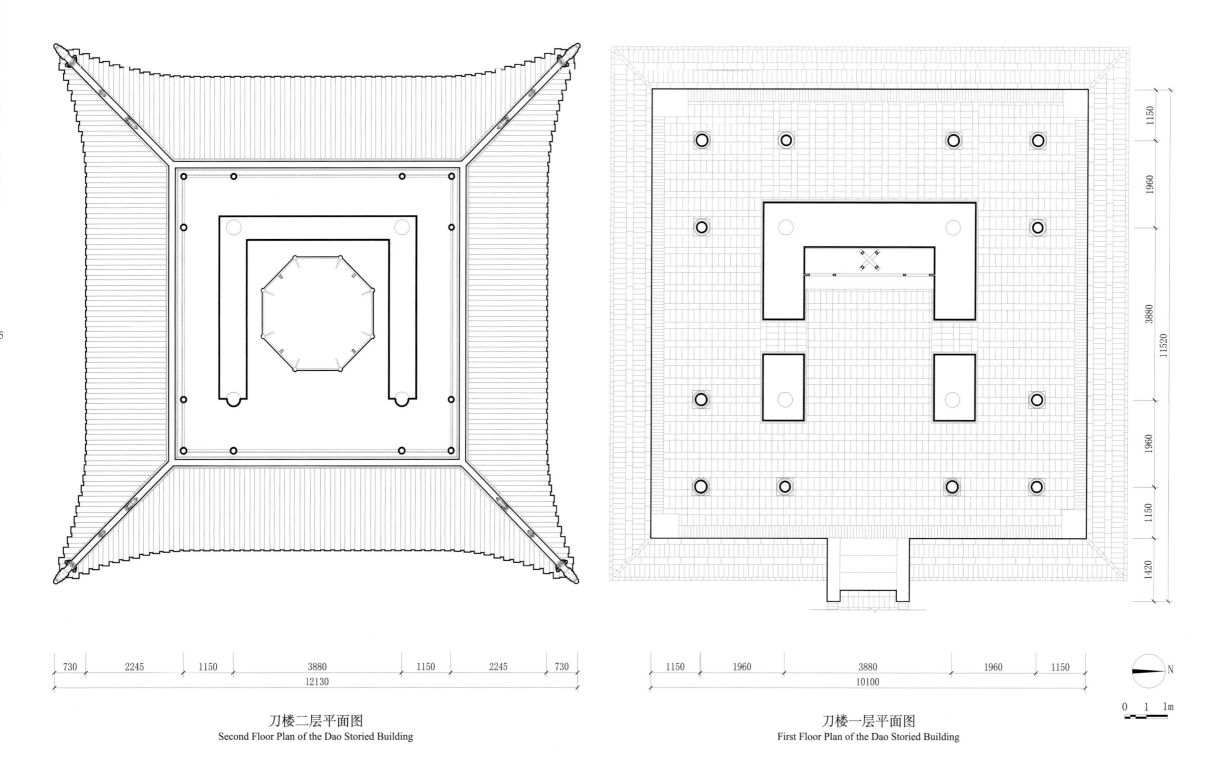

刀楼二层平面图
Second Floor Plan of the Dao Storied Building

刀楼一层平面图
First Floor Plan of the Dao Storied Building

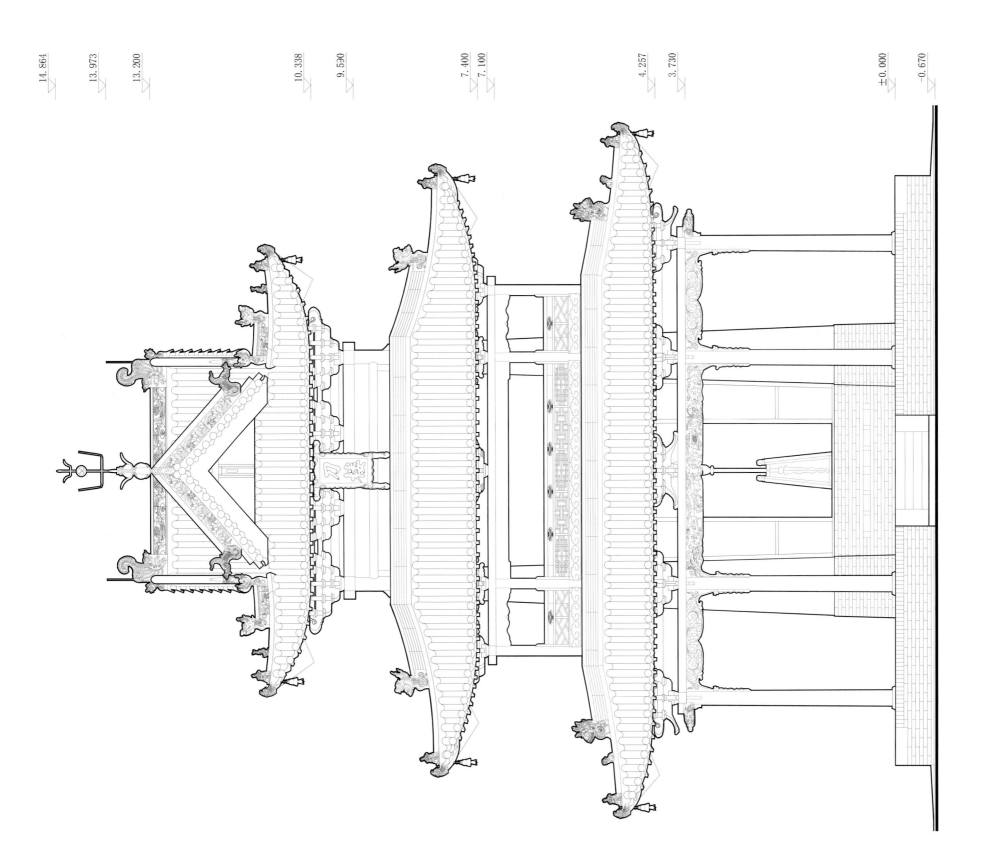

14.864
13.973
13.200
10.338
9.590
7.400
7.100
4.257
3.730
±0.000
-0.670

0 0.5 1m

刀楼东立面图

East elevation of the Dao Storied Building

1150
1960
3880
10100
1960
1150

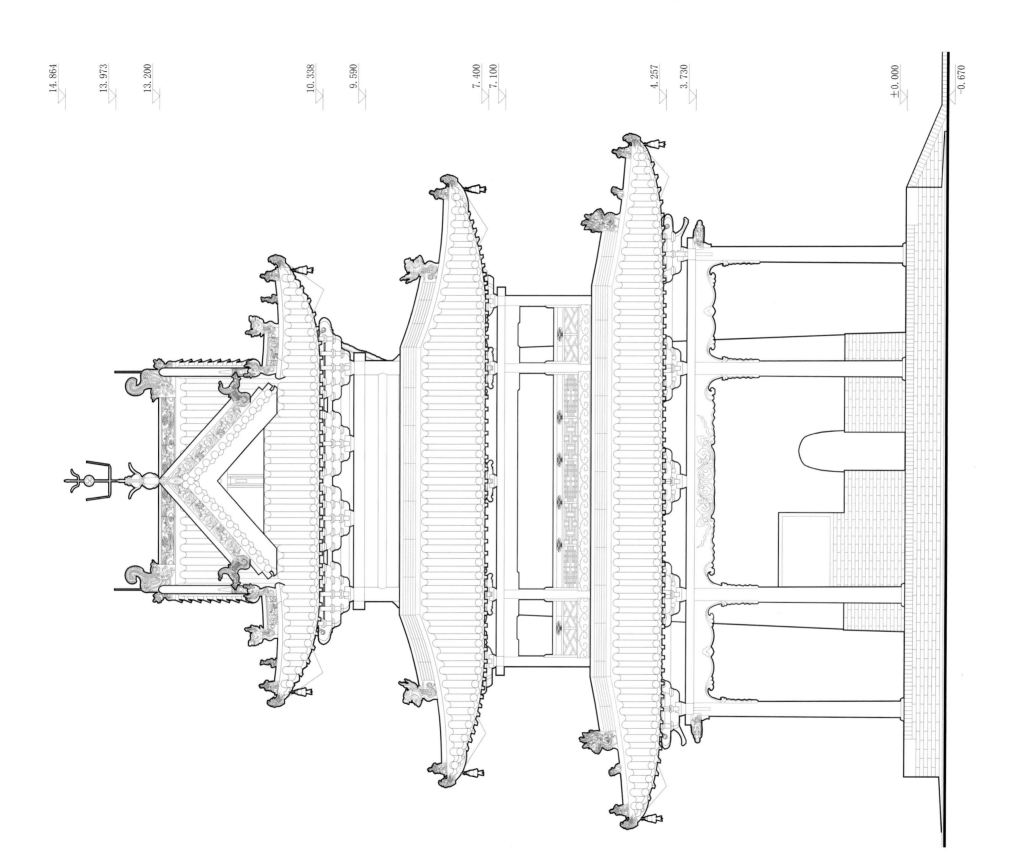

刀楼南立面图

South elevation of the Dao Storied Building

14.864 13.973 13.200 10.338 9.590 7.400 7.100 4.257 3.730 ±0.000 -0.670

1420 1150 1960 11520 3880 1960 1150

0 0.5 1m

中国古建筑测绘大系·祠庙建筑——解州关帝庙

157

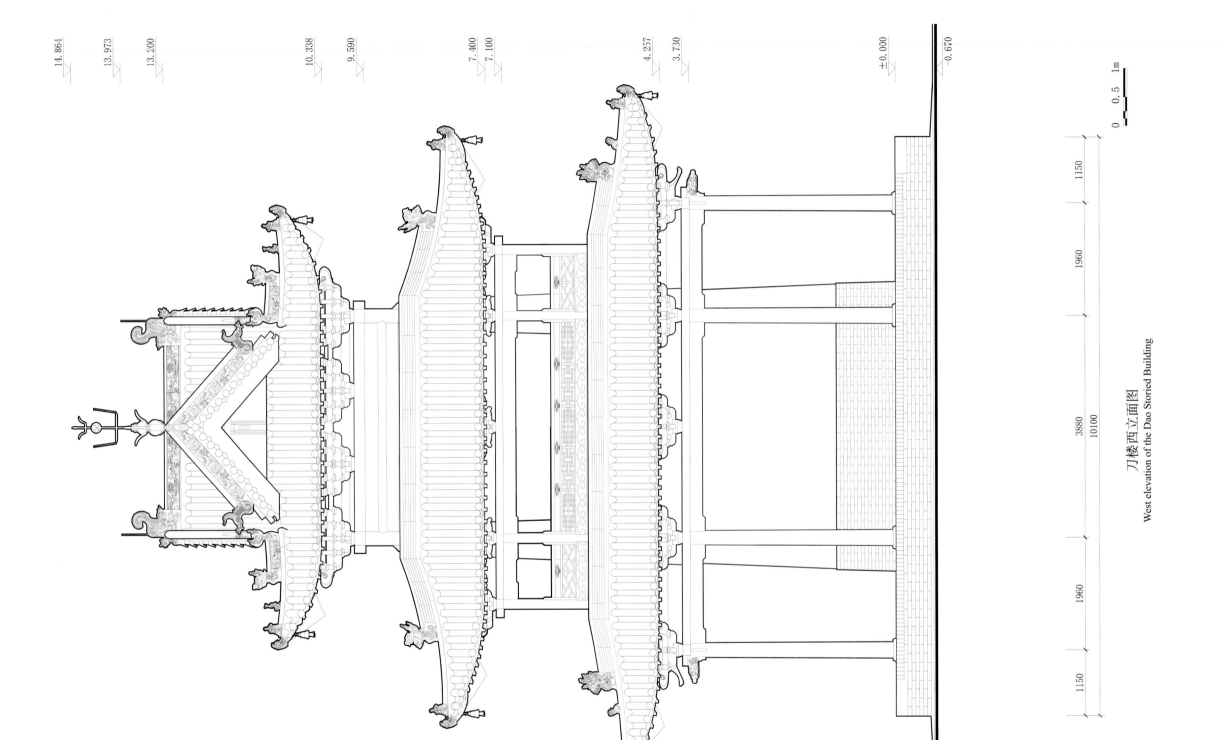

14 14.864

13.973
13.200

10.338
9.590

7.400
7.100

4.257
3.730

±0.000
-0.670

1150
1960
3880
10100
1960
1150

0 0.5 1m

刀楼西立面图
West elevation of the Dao Storied Building

中国古建筑测绘大系·祠庙建筑 —— 解州关帝庙

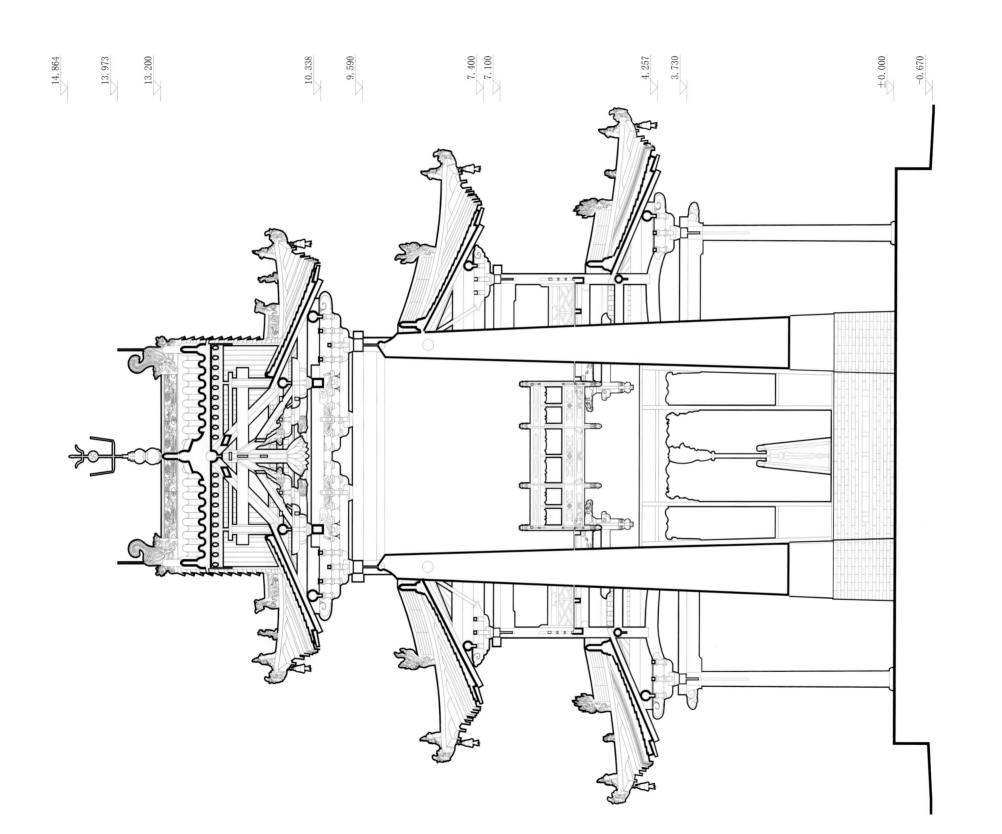

14.864
13.973
13.200
10.338
9.590
7.400
7.100
4.257
3.730
±0.000
-0.670

0 0.5 1m

1150
810
1150
1150
3880
10100
1150
810
1150

刀楼纵剖面图
Longitudinal section of the Dao Storied Building

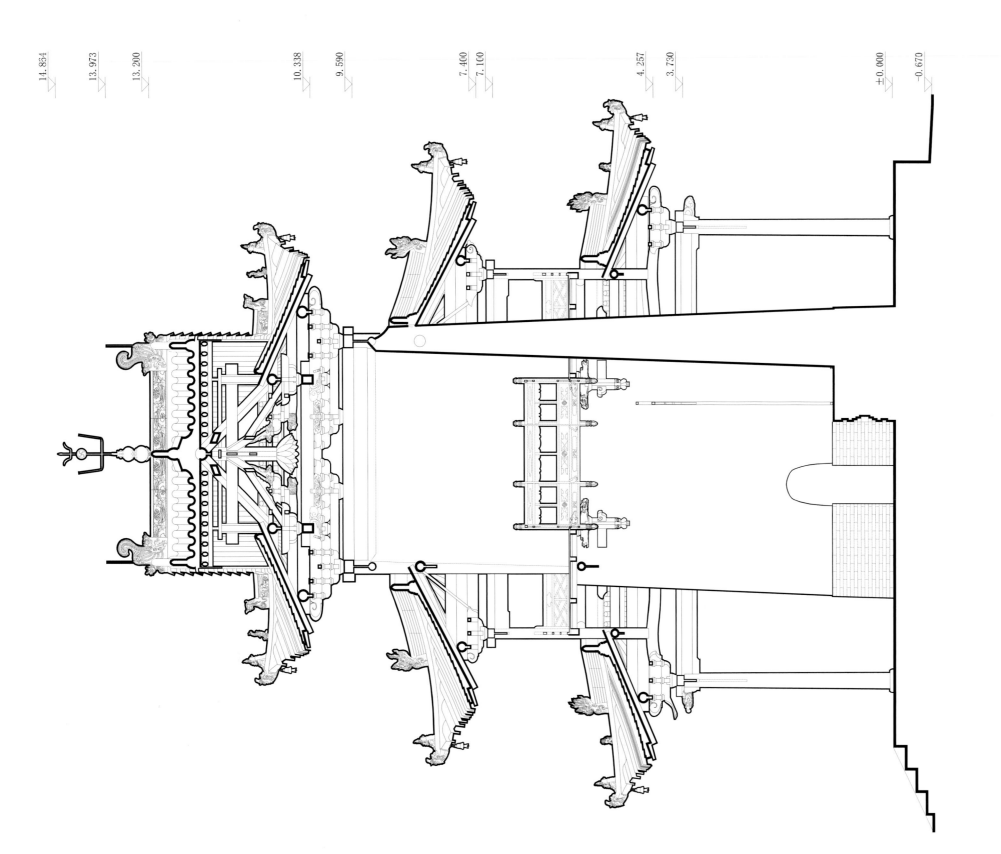

14.854
13.973
13.200
10.338
9.590
7.460
7.100
4.257
3.730
±0.000
-0.670

1150
810
1150
3880
11520
1150
810
1150
1420

0 0.5 1m

刀楼横剖面图
Cross-section of the Dao Storied Building

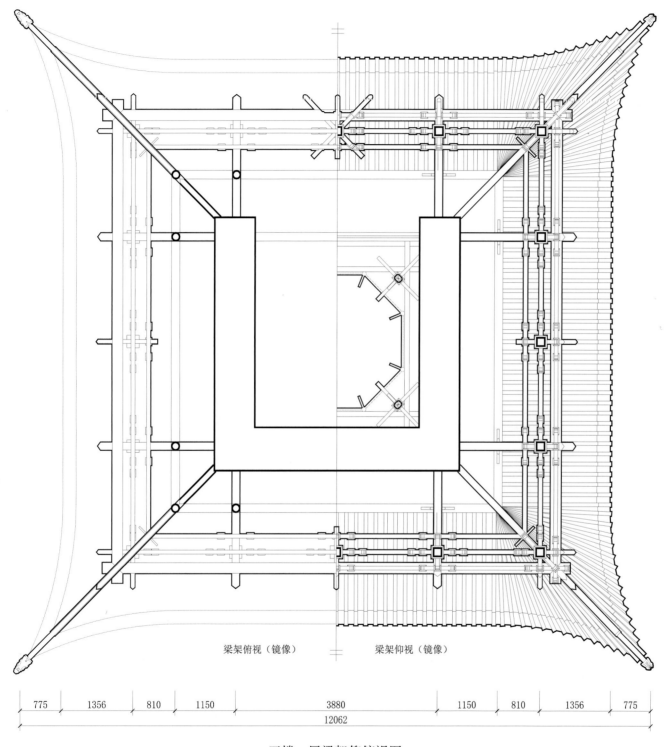

梁架俯视（镜像）　　梁架仰视（镜像）

| 775 | 1356 | 810 | 1150 | 3880 | 1150 | 810 | 1356 | 775 |

12062

刀楼一层梁架仰俯视图
Top and bottom views of the beam frame of the first-floor of the Dao Storied Building

0　0.5　1m

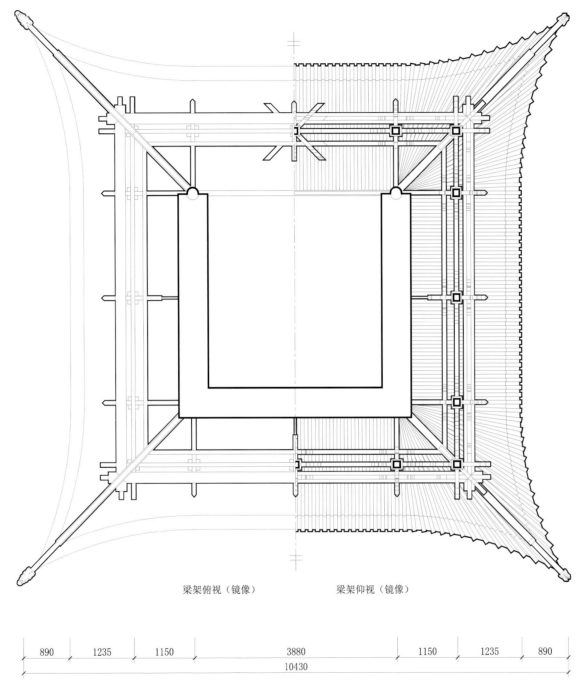

梁架俯视（镜像）　　　　梁架仰视（镜像）

890　1235　1150　3880　1150　1235　890
10430

0　0.5　1m

刀楼二层梁架仰俯视图
Top and bottom views of the beam frame of the second-floor of the Dao Storied Building

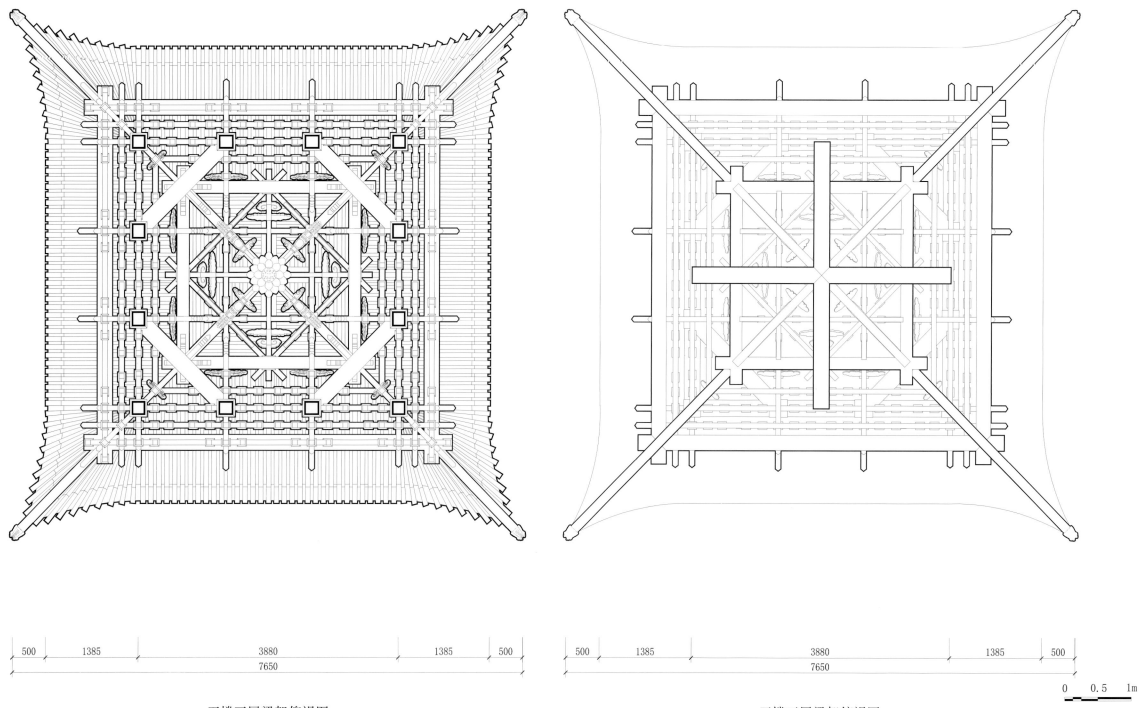

刀楼三层梁架仰视图
Bottom view of the beam frame of the third-floor of the Dao Storied Building

刀楼三层梁架俯视图
Top view of the beam frame of the third-floor of the Dao Storied Building

500 1385 3880 1385 500
7650

500 1385 3880 1385 500
7650

0　0.5　1m

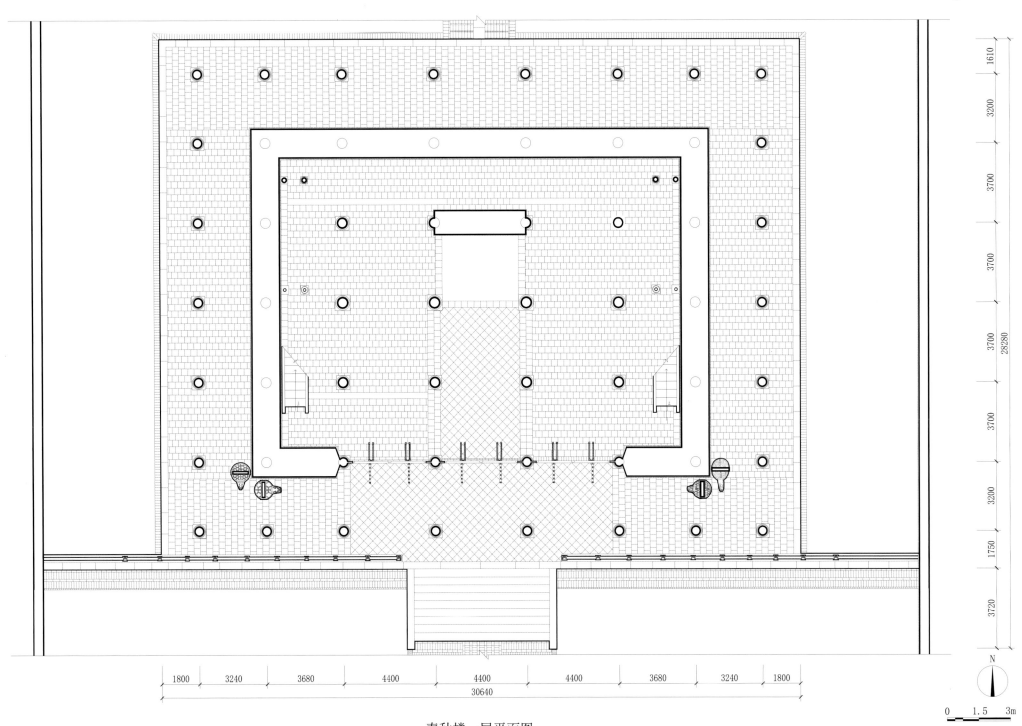

春秋楼一层平面图
First Floor Plan of the Chunqiu Storied Building

N

0   1.5   3m

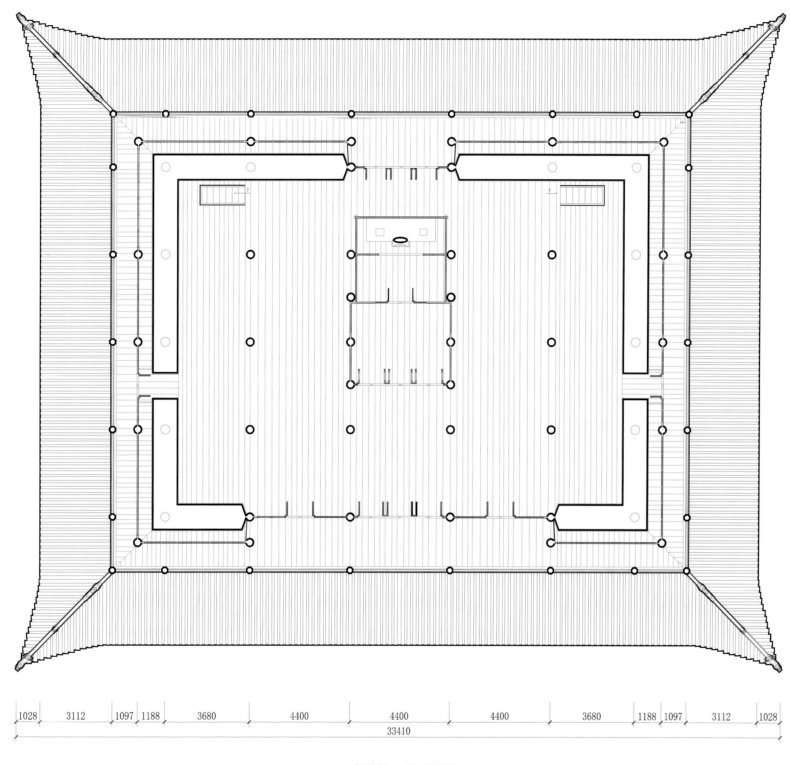
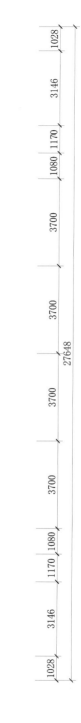

春秋楼二层平面图
Second Floor Plan of the Chunqiu Storied Building

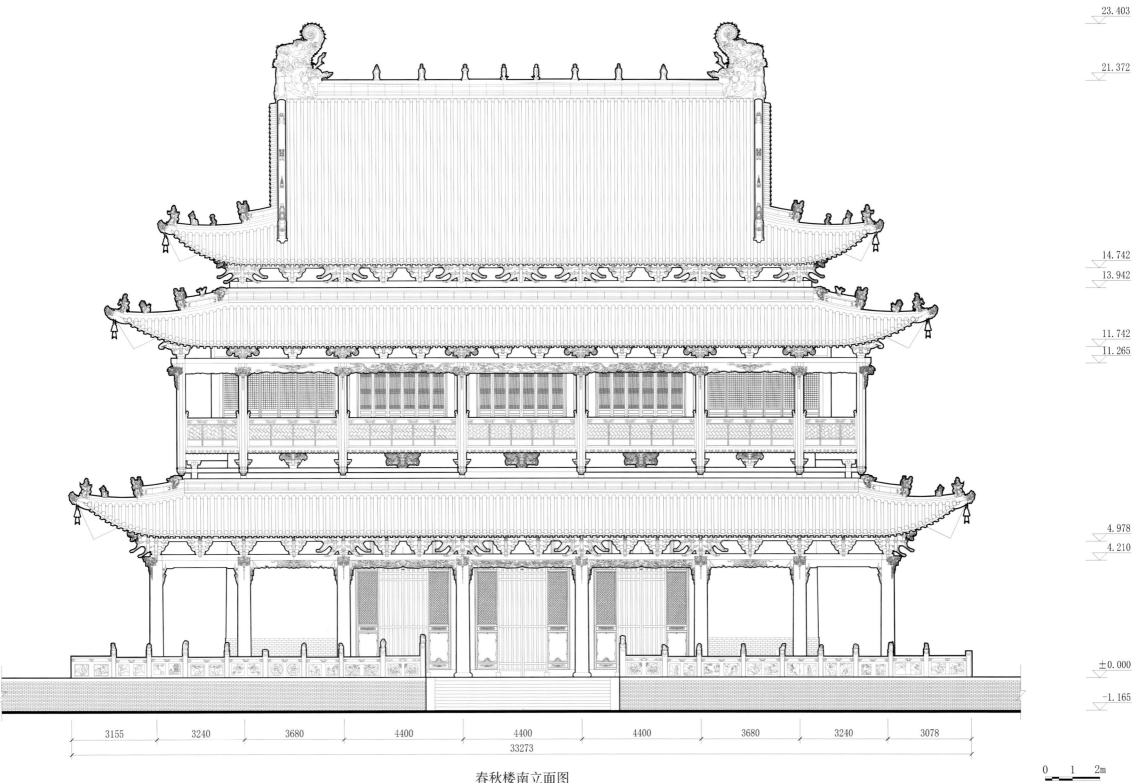

23.403

21.372

14.742

13.942

11.742

11.265

4.978

4.210

±0.000

-1.165

3155　3240　3680　4400　4400　4400　3680　3240　3078

33273

春秋楼南立面图
South elevation of the Chunqiu Storied Building

0　1　2m

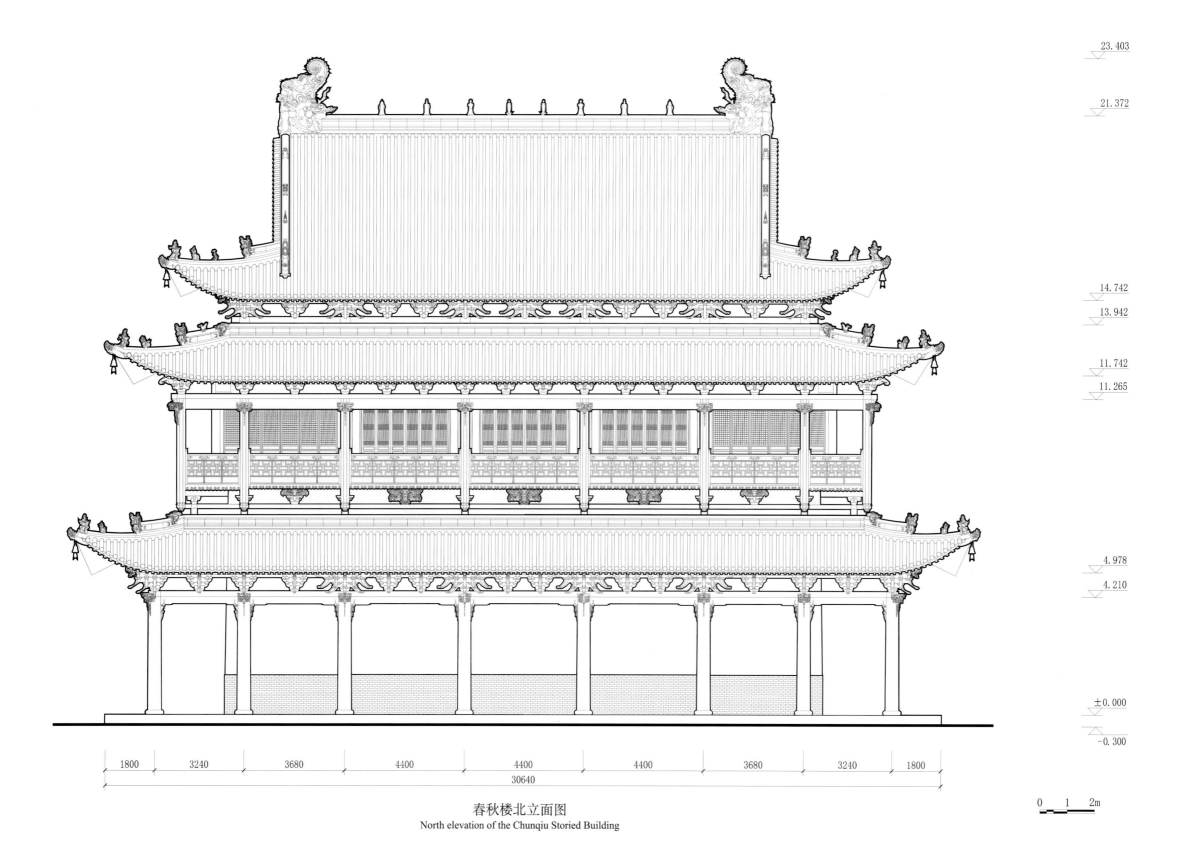

23.403

21.372

14.742

13.942

11.742

11.265

4.978

4.210

±0.000

-0.300

| 1800 | 3240 | 3680 | 4400 | 4400 | 4400 | 3680 | 3240 | 1800 |

30640

春秋楼北立面图
North elevation of the Chunqiu Storied Building

0   1   2m

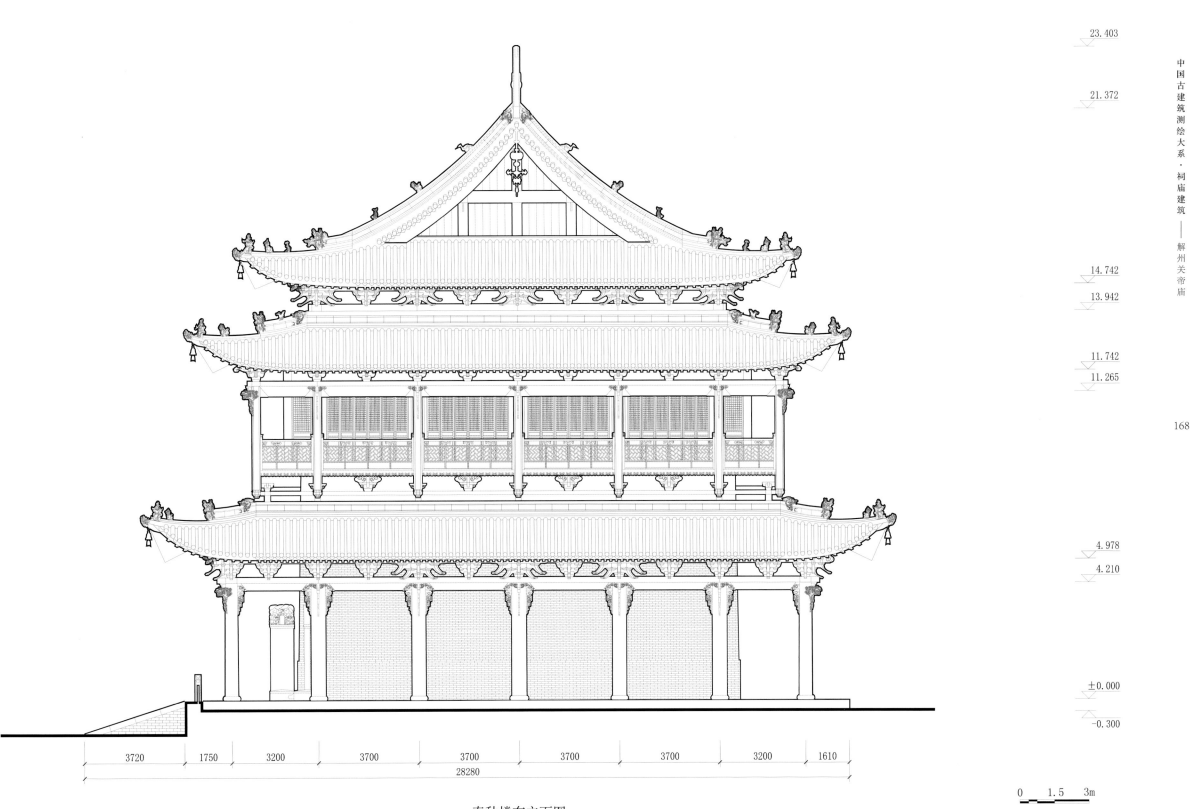

23.403

21.372

14.742

13.942

11.742

11.265

4.978

4.210

±0.000

-0.300

3720 1750 3200 3700 3700 3700 3700 3200 1610

28280

0　1.5　3m

春秋楼东立面图

East elevation of the Chunqiu Storied Building

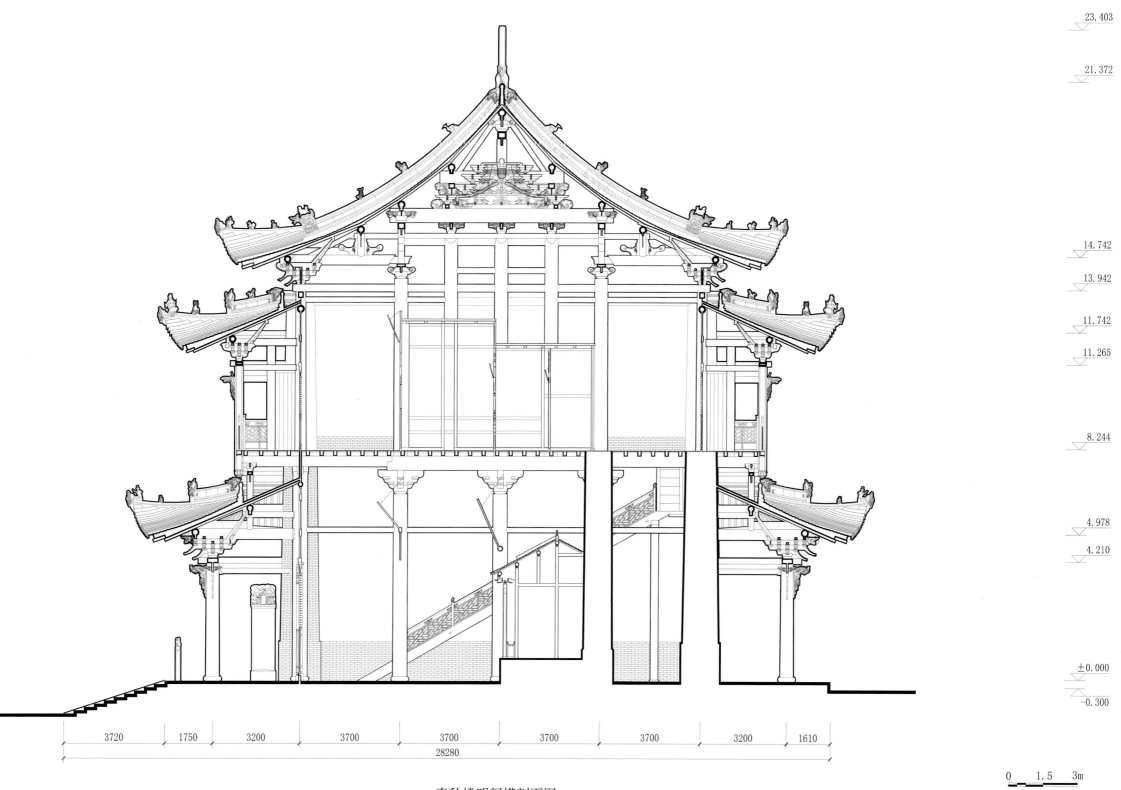

23.403

21.372

14.742

13.942

11.742

11.265

8.244

4.978

4.210

±0.000

-0.300

| 3720 | 1750 | 3200 | 3700 | 3700 | 3700 | 3700 | 3200 | 1610 |

28280

0   1.5   3m

春秋楼明间横剖面图
Central bay section of the Chunqiu Storied Building

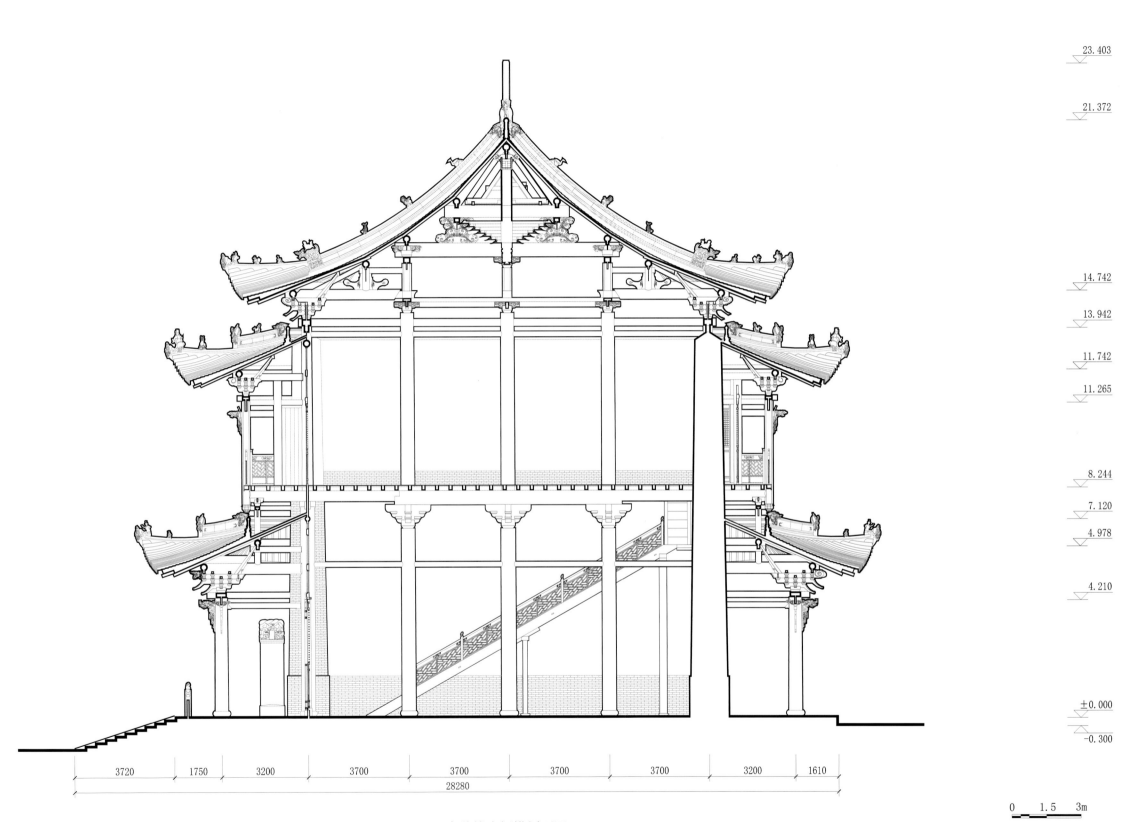

23.403

21.372

14.742

13.942

11.742

11.265

8.244

7.120

4.978

4.210

±0.000

-0.300

| 3720 | 1750 | 3200 | 3700 | 3700 | 3700 | 3700 | 3200 | 1610 |

28280

0   1.5   3m

春秋楼次间横剖面图
Second bay section of the Chunqiu Storied Building

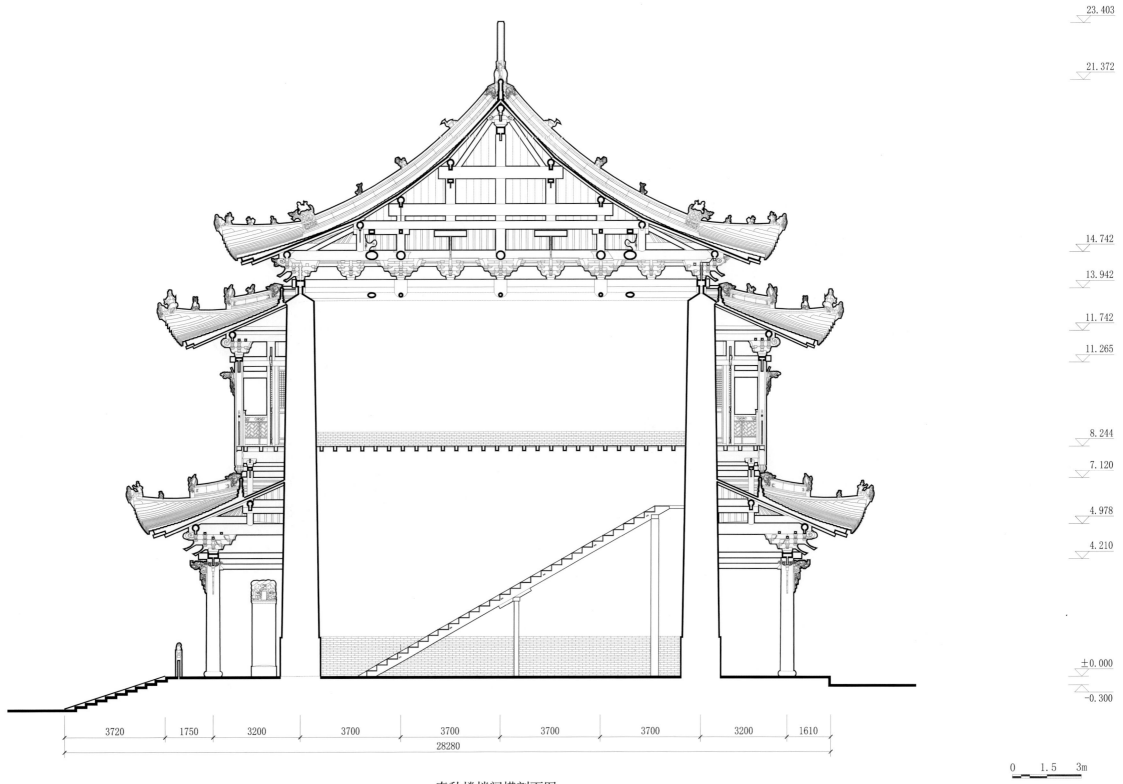

春秋楼梢间横剖面图
Side bay section of the Chunqiu Storied Building

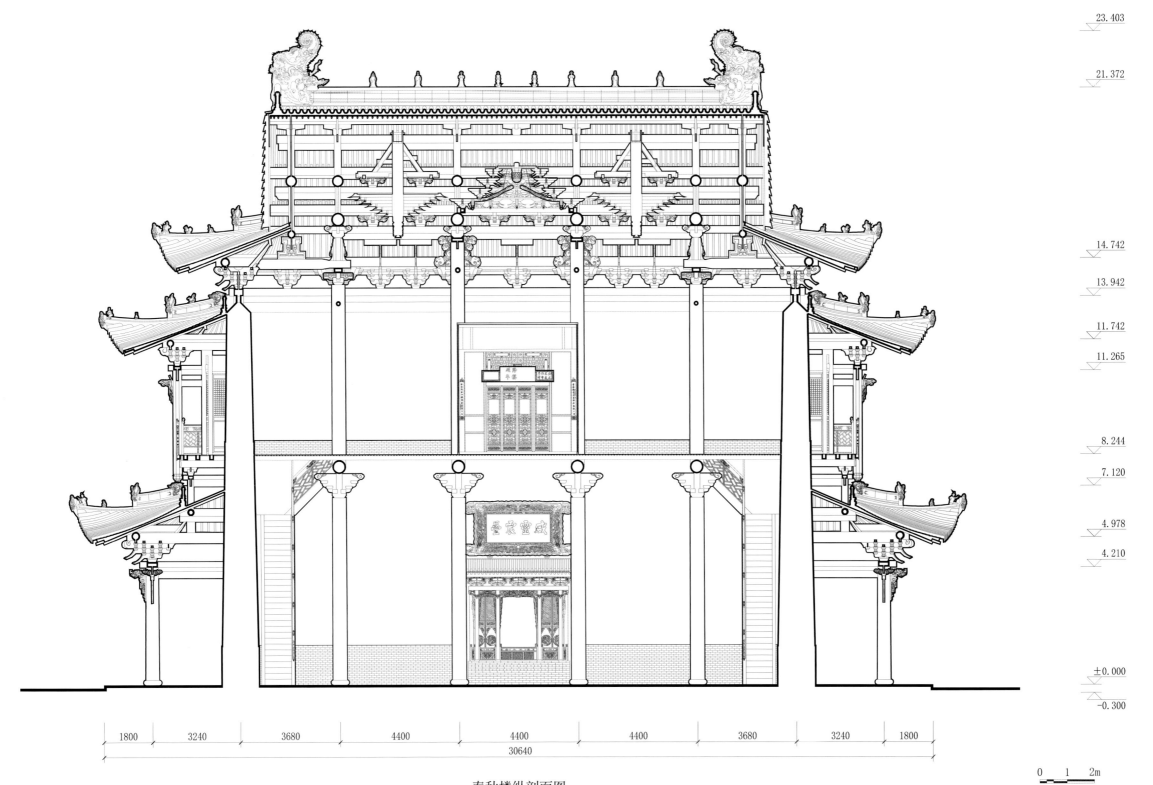

23.403

21.372

14.742

13.942

11.742

11.265

8.244

7.120

4.978

4.210

±0.000

-0.300

| 1800 | 3240 | 3680 | 4400 | 4400 | 4400 | 3680 | 3240 | 1800 |

30640

0 1 2m

春秋楼纵剖面图
Longitudinal section of the Chunqiu Storied Building

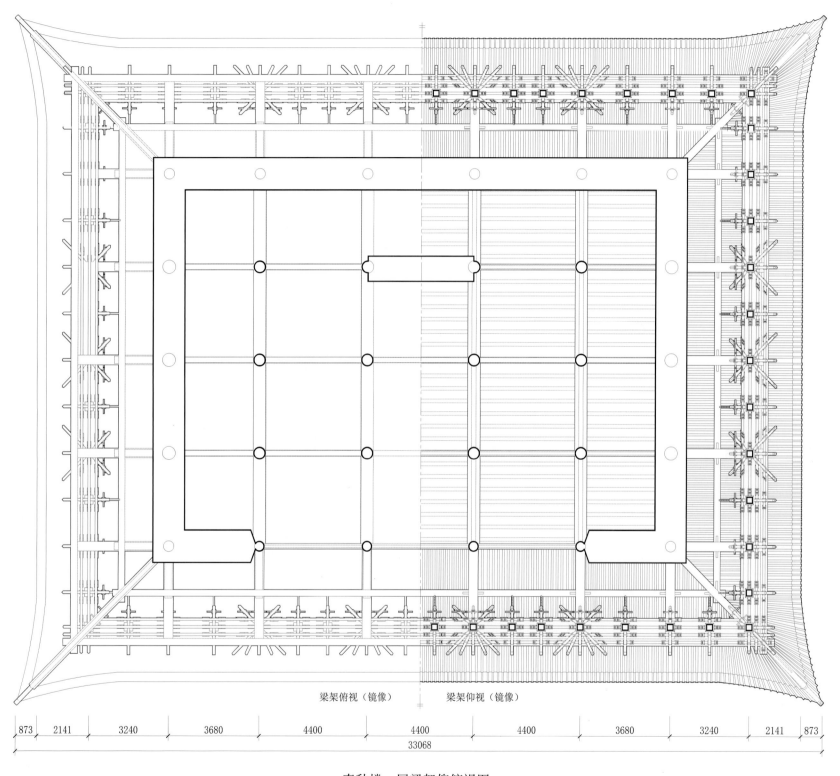
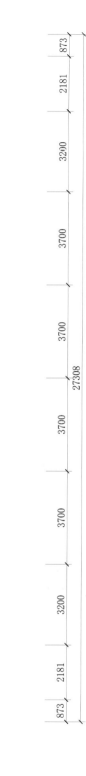

873
2181
3200
3700
3700
27308
3700
3700
3200
2181
873

梁架俯视（镜像）　　梁架仰视（镜像）

873　2141　3240　3680　4400　4400　4400　3680　3240　2141　873
33068

0　1.5　3m

春秋楼一层梁架仰俯视图
Top and bottom views of the beam frame of the first-floor of the Chunqiu Storied Building

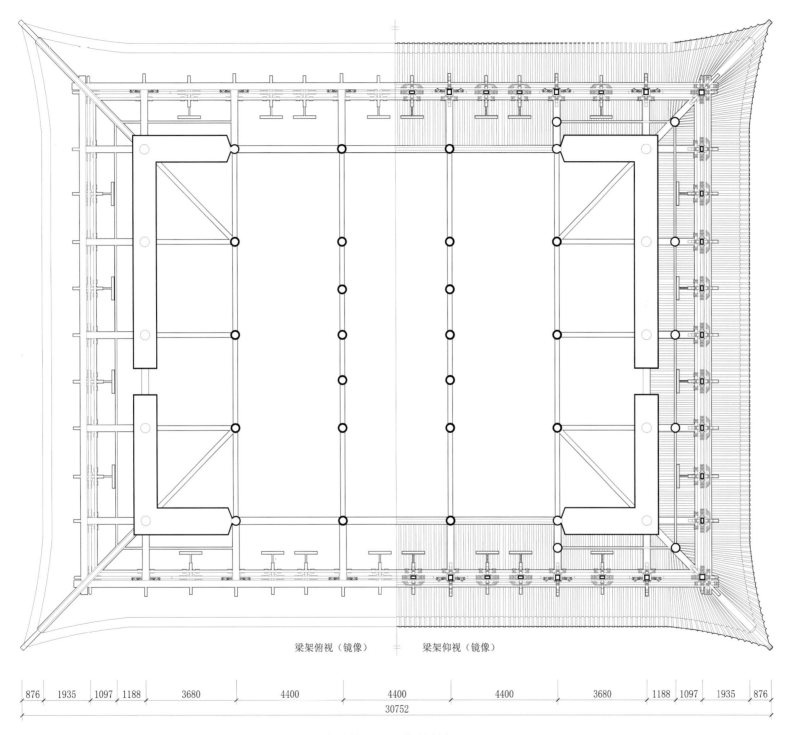

梁架俯视（镜像） ＝ 梁架仰视（镜像）

春秋楼二层梁架仰俯视图
Top and bottom views of the beam frame of the second-floor of the Chunqiu Storied Building

0　　1.5　　3m

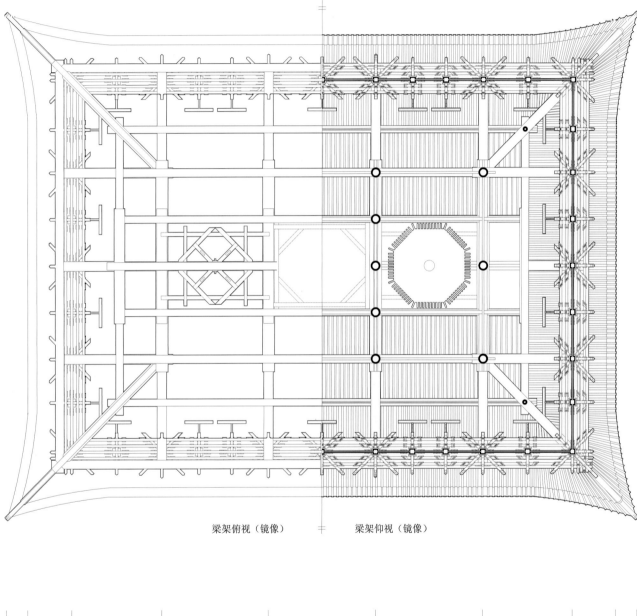
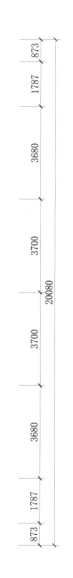

梁架俯视（镜像）　　　　梁架仰视（镜像）

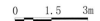

873 | 1787 | 3680 | 4400 | 4400 | 4400 | 3680 | 1787 | 873

25880

春秋楼上檐梁架仰俯视图

Top and bottom views of the beam frame of the third-floor of the Chunqiu Storied Building

0　1.5　3m

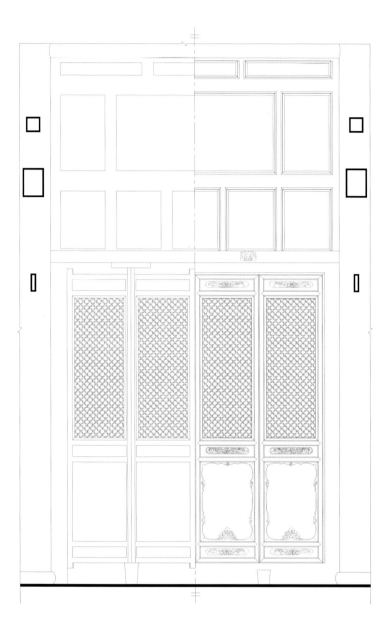

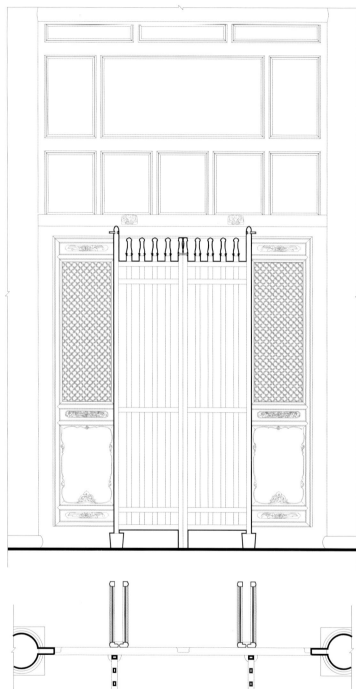

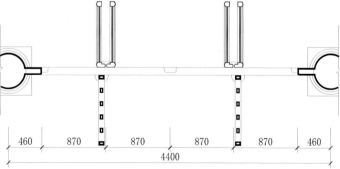

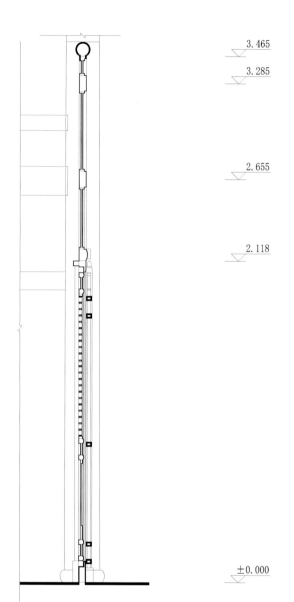

3.465

3.285

2.655

2.118

±0.000

460　870　870　870　870　460

4400

春秋楼一层格扇门大样图
Details of the partition door of the first-floor of the Chunqiu Storied Building

0　0.5　1m

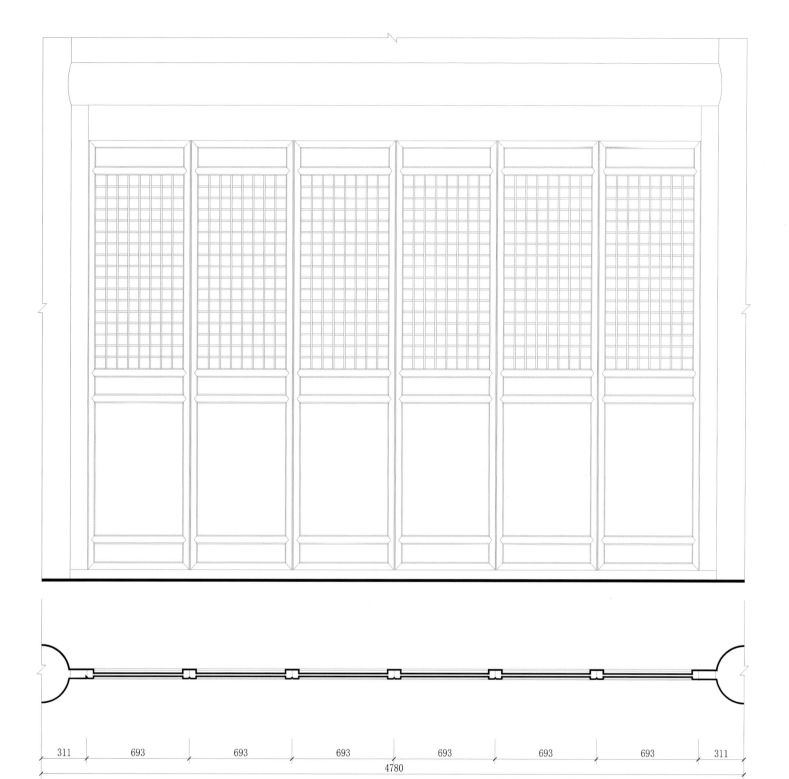
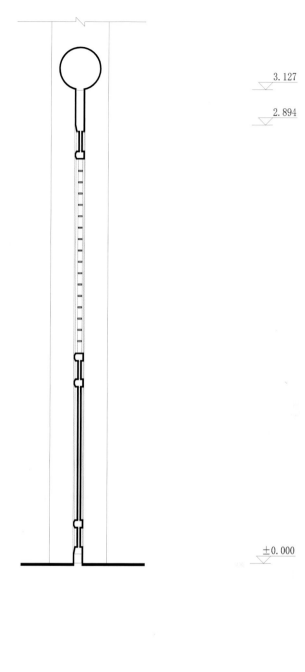

3.127

2.894

±0.000

311   693   693   693   693   693   693   311
4780

春秋楼二层格扇门大样图
Details of the partition door of the second-floor of the Chunqiu Storied Building

0   0.25   0.5m

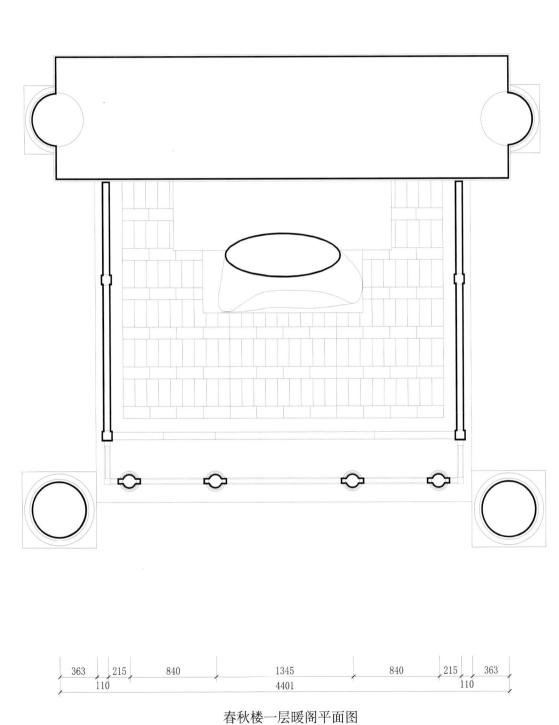

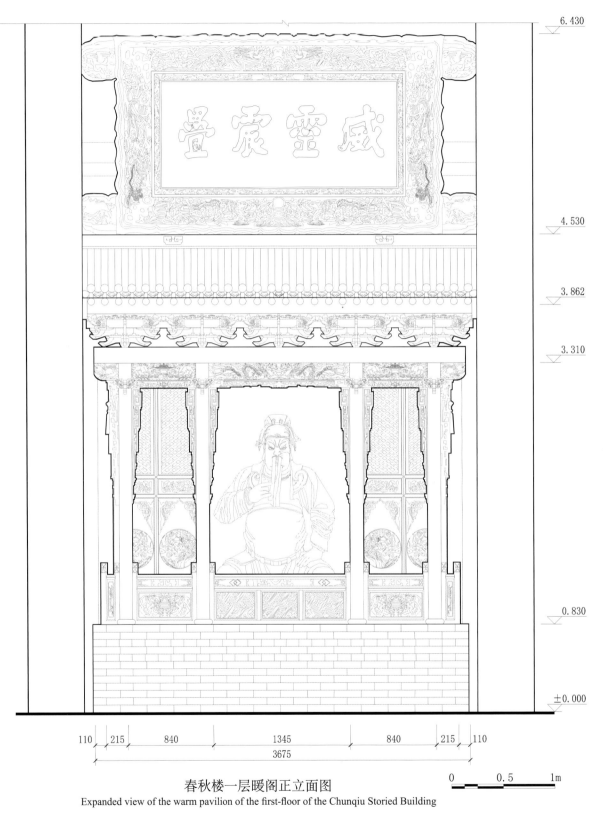

春秋楼一层暖阁平面图
Plan of the warm pavilion of the first-floor of the Chunqiu Storied Building

春秋楼一层暖阁正立面图
Expanded view of the warm pavilion of the first-floor of the Chunqiu Storied Building

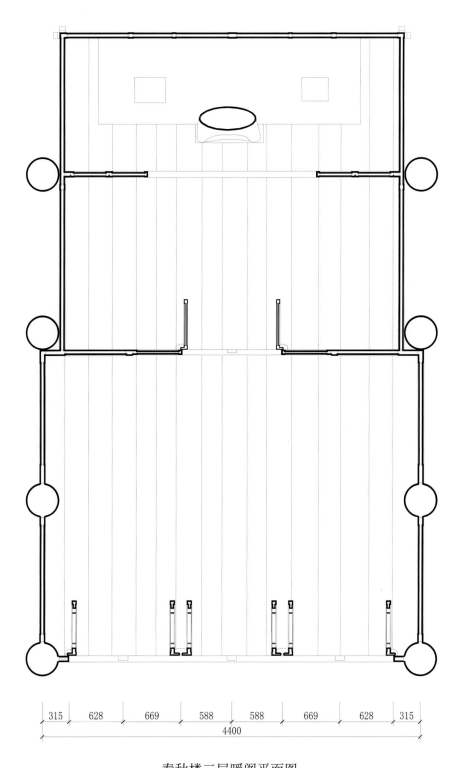

315 628 669 588 588 669 628 315
4400

春秋楼二层暖阁平面图
Plan of the warm pavilion of the second-floor of the Chunqiu Storied Building

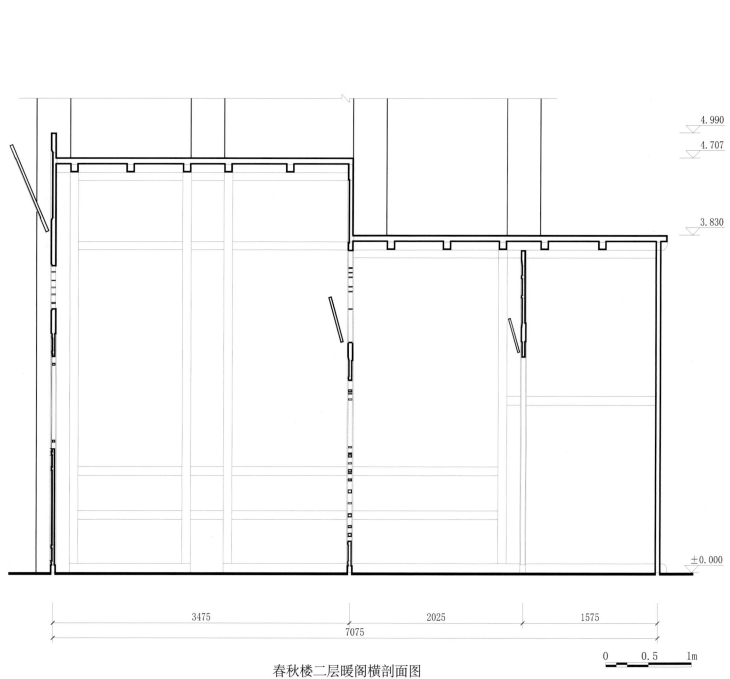

4.990
4.707

3.830

±0.000

3475 2025 1575
7075

0 0.5 1m

春秋楼二层暖阁横剖面图
Cross section of the warm pavilion of the second-floor of the Chunqiu Storied Building

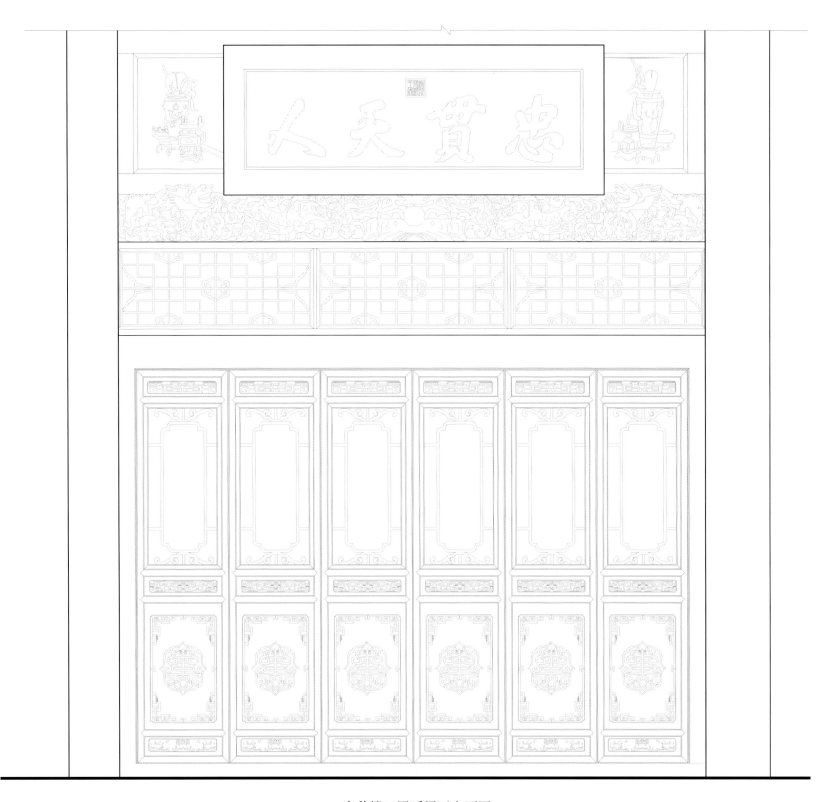

春秋楼二层暖阁正立面图
Front elevation of the warm pavilion of the second-floor of the Chunqiu Storied Building

0   0.25   0.5m

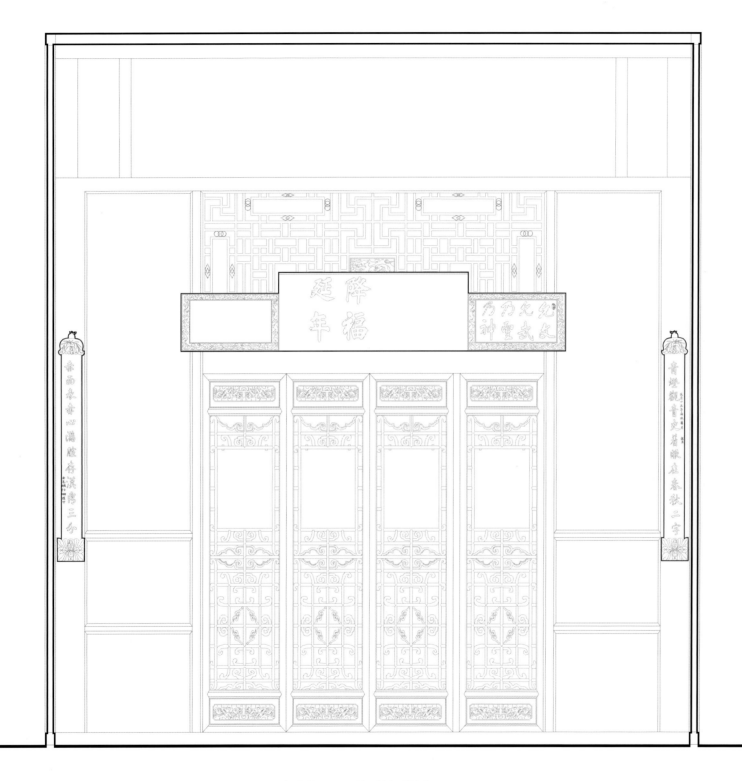

春秋楼二层暖阁外间纵剖面图

Longitudinal section of the warm pavilion's outer bay of the second-floor of the Chunqiu Storied Building

0   0.25   0.5m

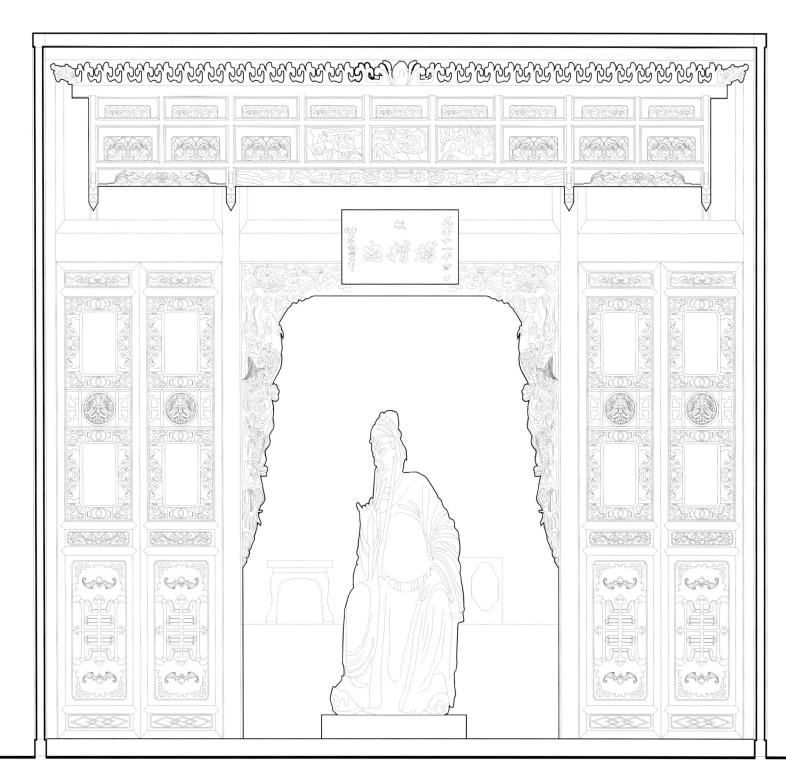

春秋楼二层暖阁里间纵剖面图
Longitudinal section of the warm pavilion's inner bay of the second-floor of the Chunqiu Storied Building

0   0.25   0.5m

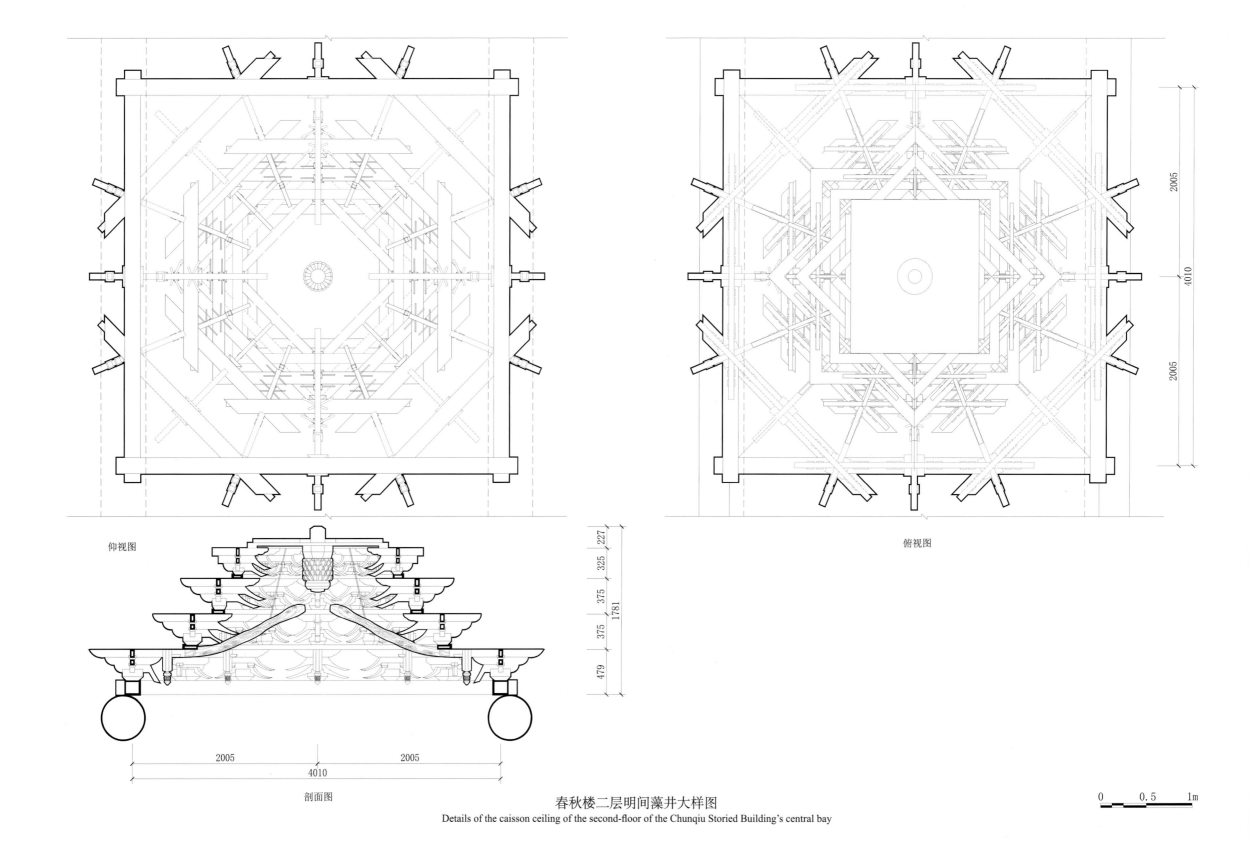

仰视图

俯视图

剖面图

227
325
375
375
1781
479

2005
4010
2005

2005
4010
2005

2005
4010

春秋楼二层明间藻井大样图
Details of the caisson ceiling of the second-floor of the Chunqiu Storied Building's central bay

0   0.5   1m

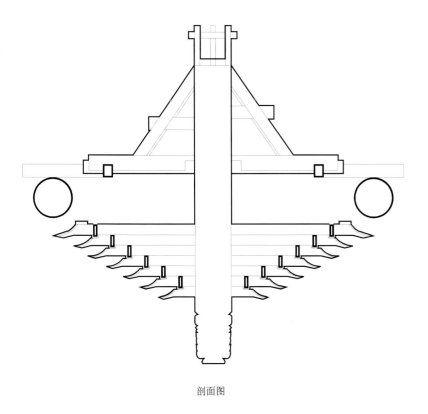

剖面图

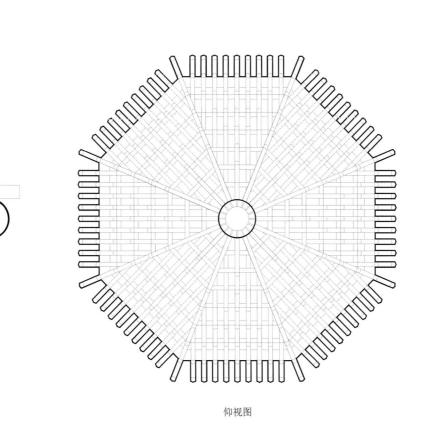

仰视图

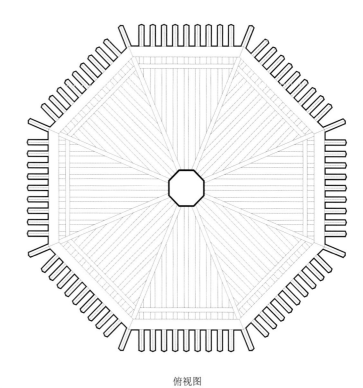

俯视图

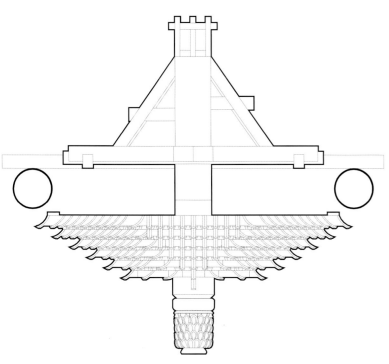

立面图

春秋楼二层次间藻井大样图

Details of the caisson ceiling of the second-floor of the Chunqiu Storied Building's side bay

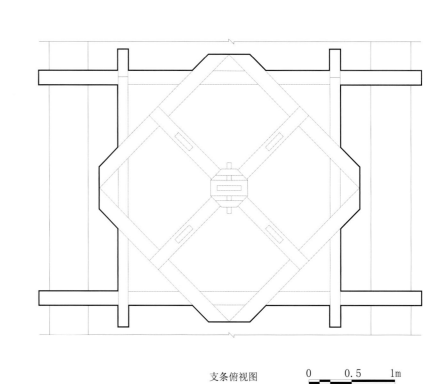

支条俯视图

0   0.5   1m

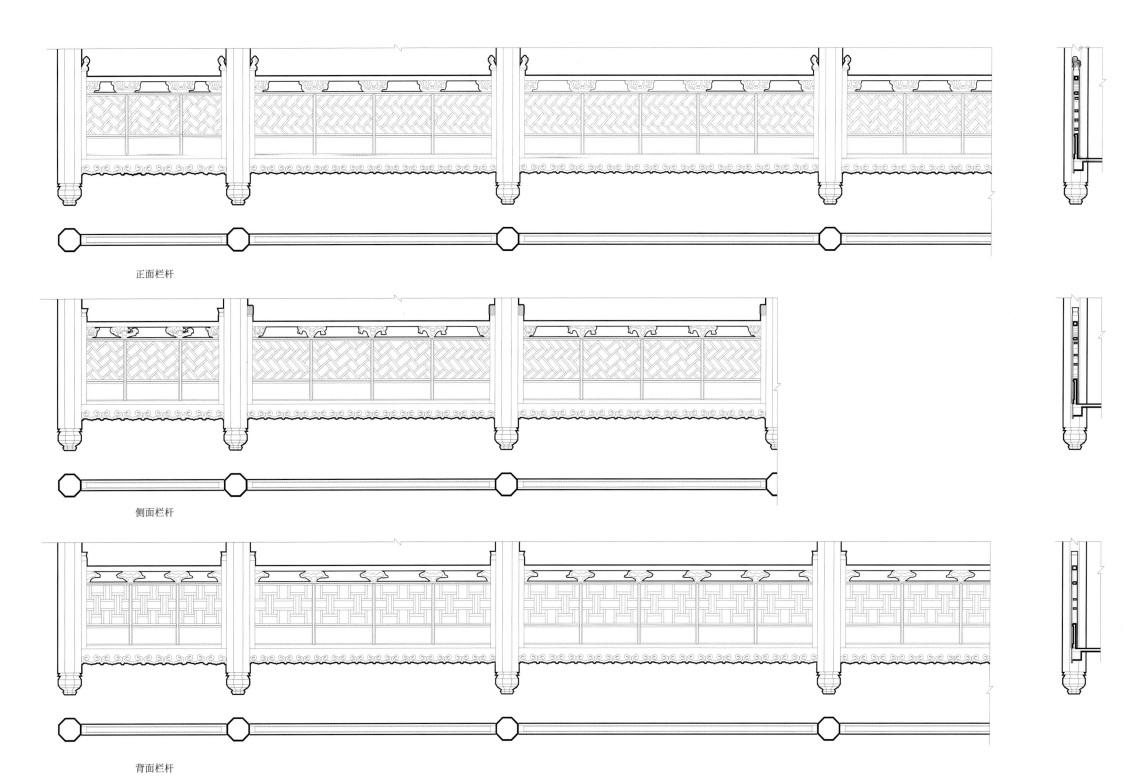

正面栏杆

侧面栏杆

背面栏杆

春秋楼二层栏杆大样图
Details of the balustrade of the second-floor of the Chunqiu Storied Building

0   0.5   2m

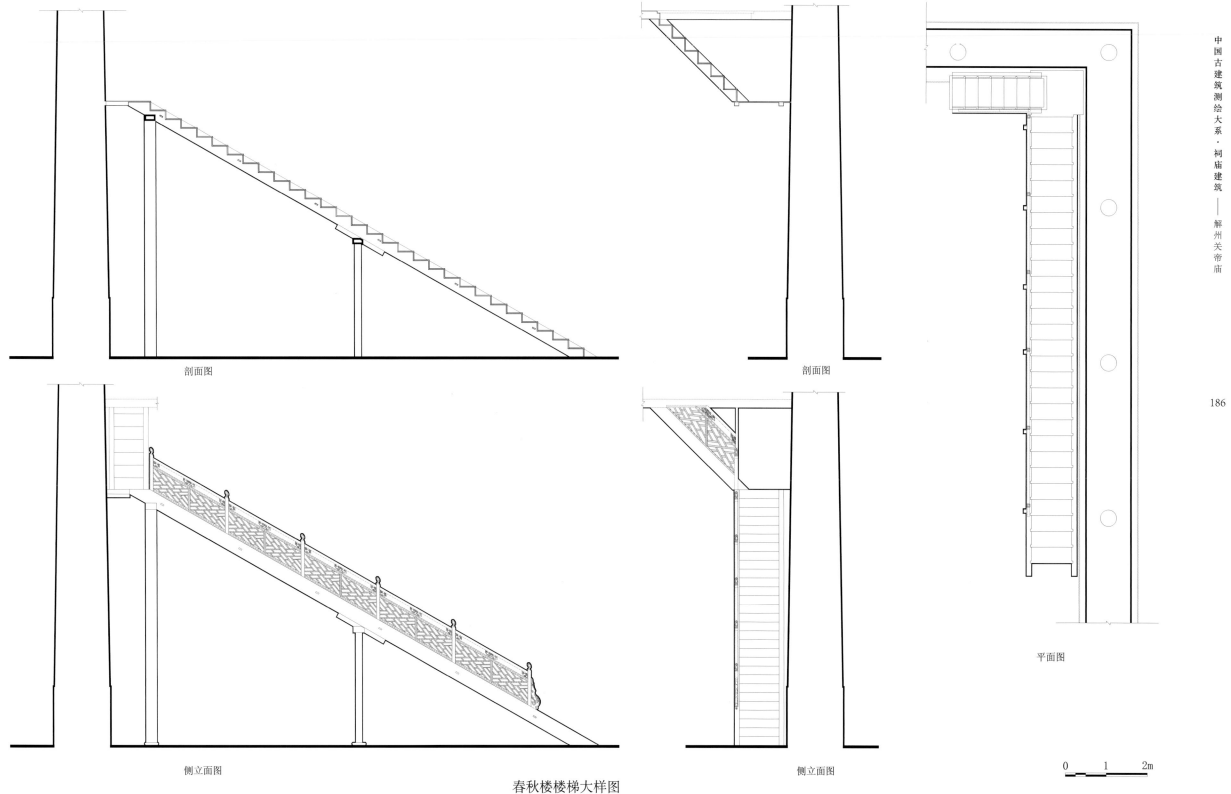

剖面图

剖面图

平面图

侧立面图

侧立面图

0　1　2m

春秋楼楼梯大样图
Details of the staircase of the Chunqiu Storied Building

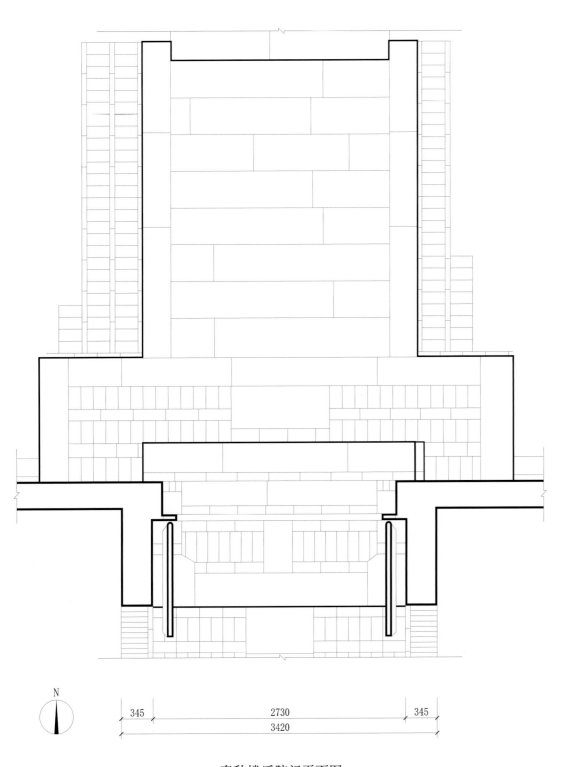

春秋楼后院门平面图
Plan of the gate of the Chunqiu Storied Building's yard

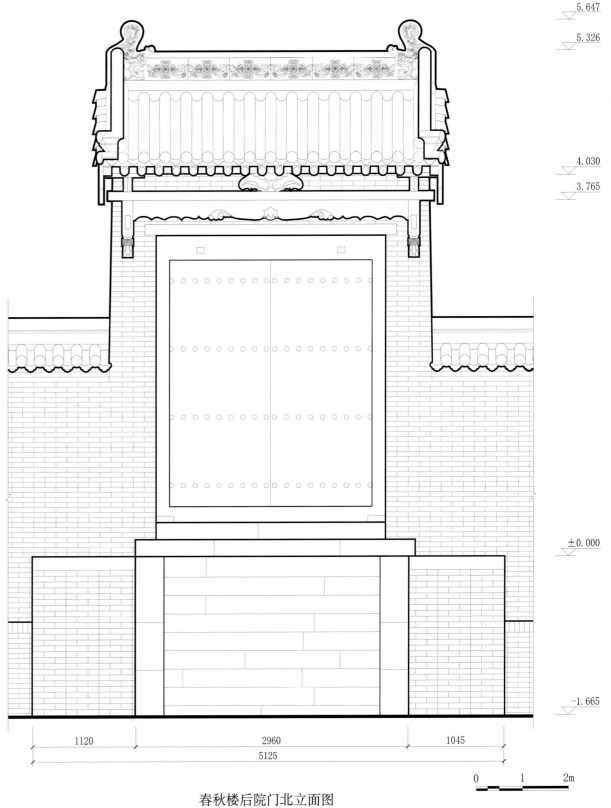

春秋楼后院门北立面图
North elevation of the gate of the Chunqiu Storied Building's yard

厚载门
The Houzai Gate

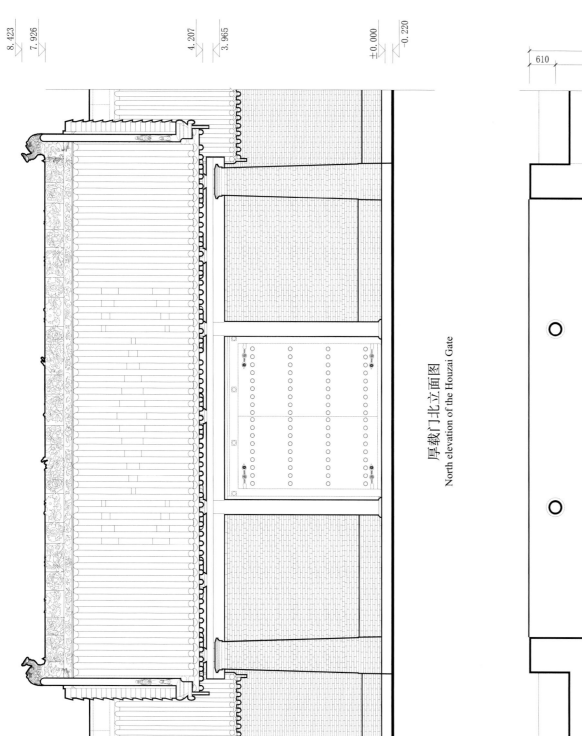

厚载门北立面图
North elevation of the Houzai Gate

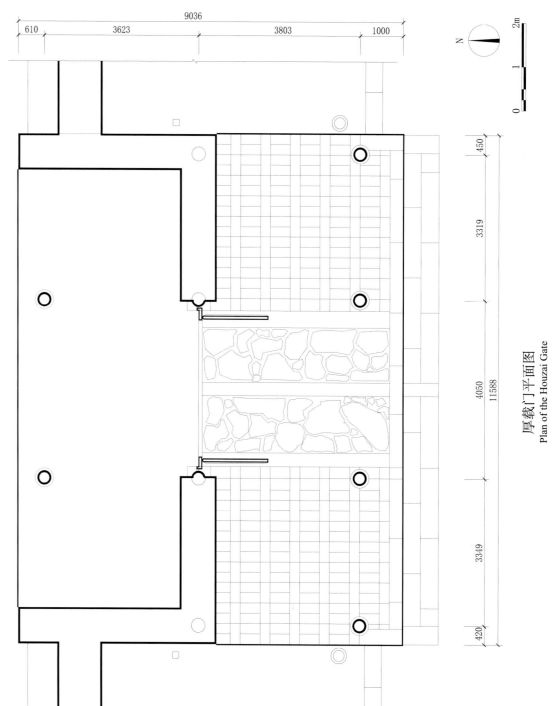

厚载门平面图
Plan of the Houzai Gate

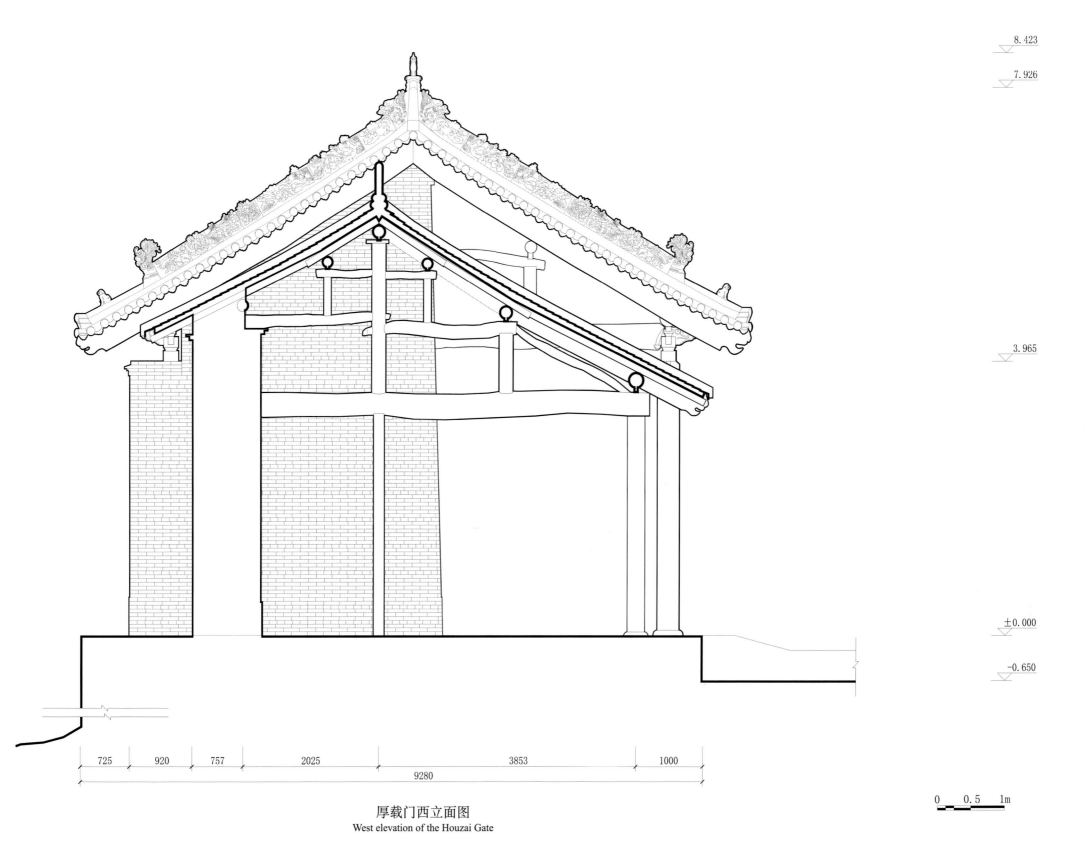

8.423

7.926

3.965

±0.000

-0.650

725 920 757 2025 3853 1000

9280

0 0.5 1m

厚载门西立面图
West elevation of the Houzai Gate

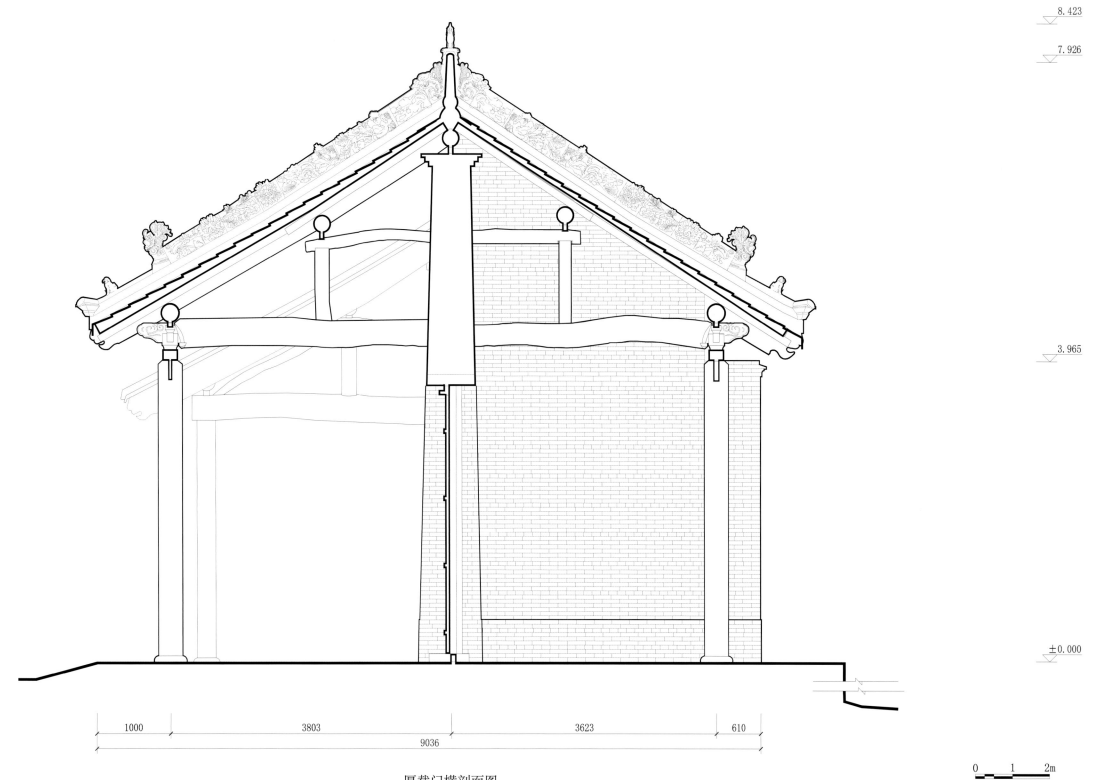

8.423

7.926

3.965

±0.000

1000　　　3803　　　3623　　610

9036

厚载门横剖面图
Cross-section of the Houzai Gate

0　　1　　2m

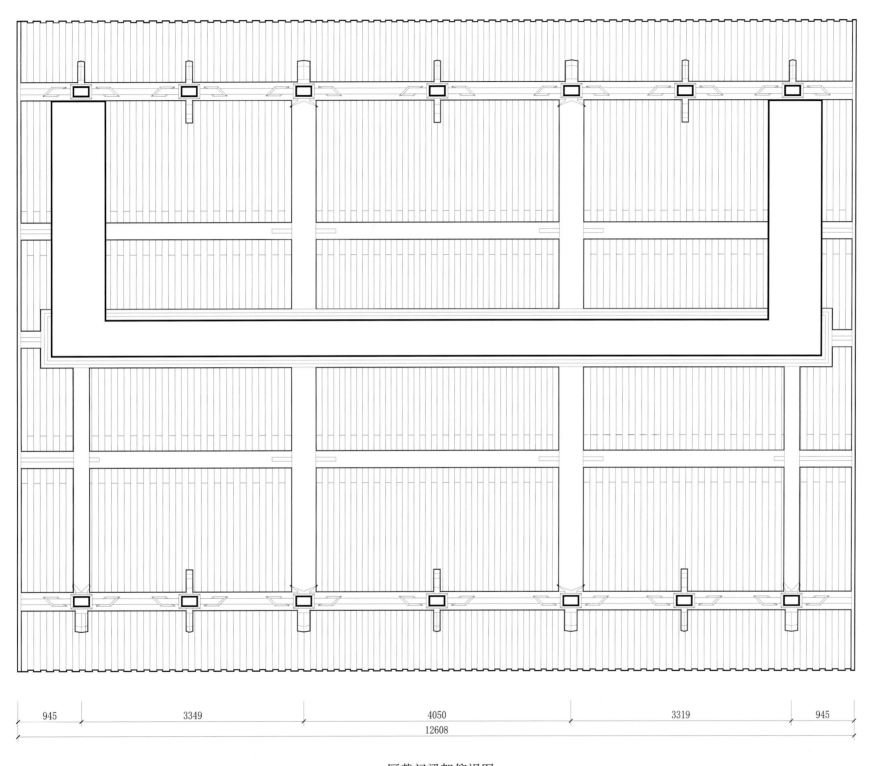

厚载门梁架仰视图
Bottom view of the beam frame of the Houzai Gate

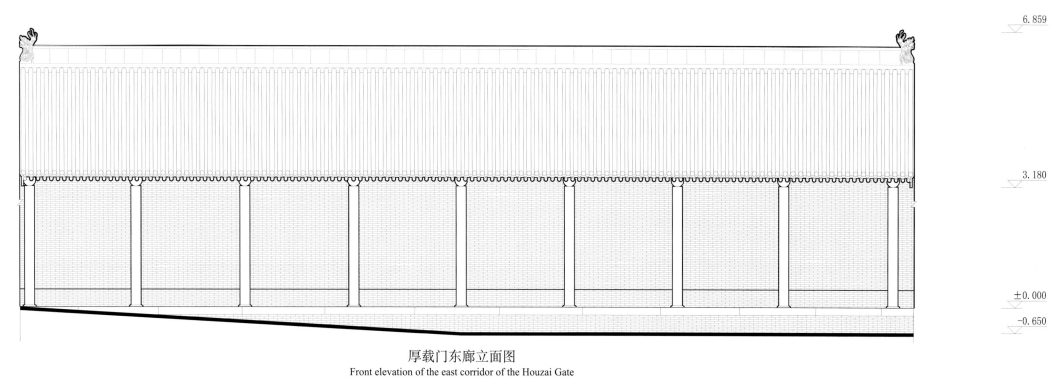

6.859

3.180

±0.000

-0.650

厚载门东廊立面图
Front elevation of the east corridor of the Houzai Gate

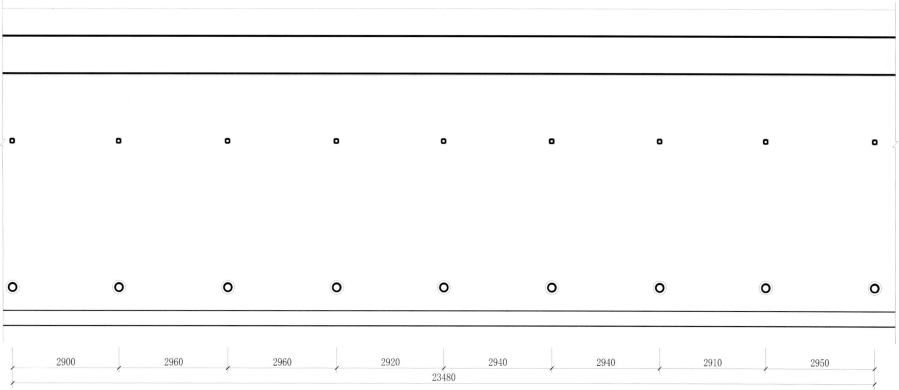

700
1035
1750
8335
3850
1000

2900　2960　2960　2920　2940　2940　2910　2950

23480

厚载门东廊平面图
Plan of the east corridor of the Houzai Gate

N

0　1　2m

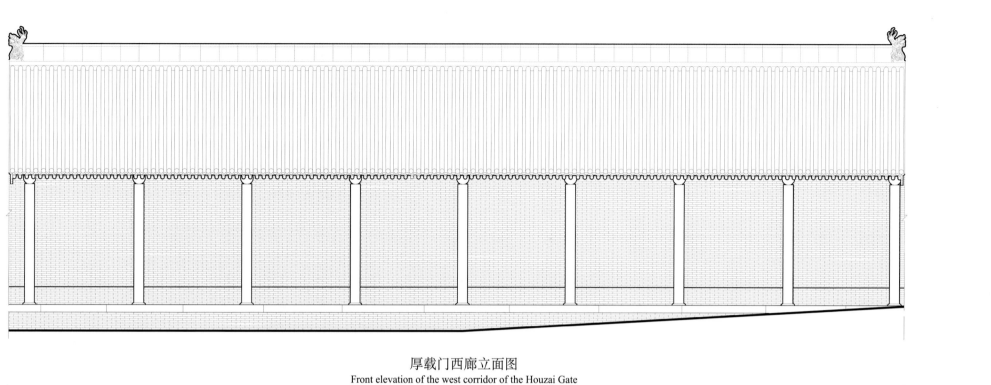

厚载门西廊立面图
Front elevation of the west corridor of the Houzai Gate

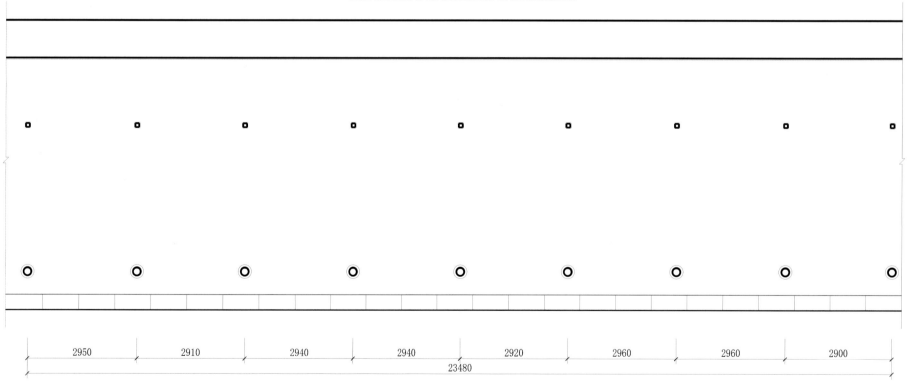

厚载门西廊平面图
Plan of the west corridor of the Houzai Gate

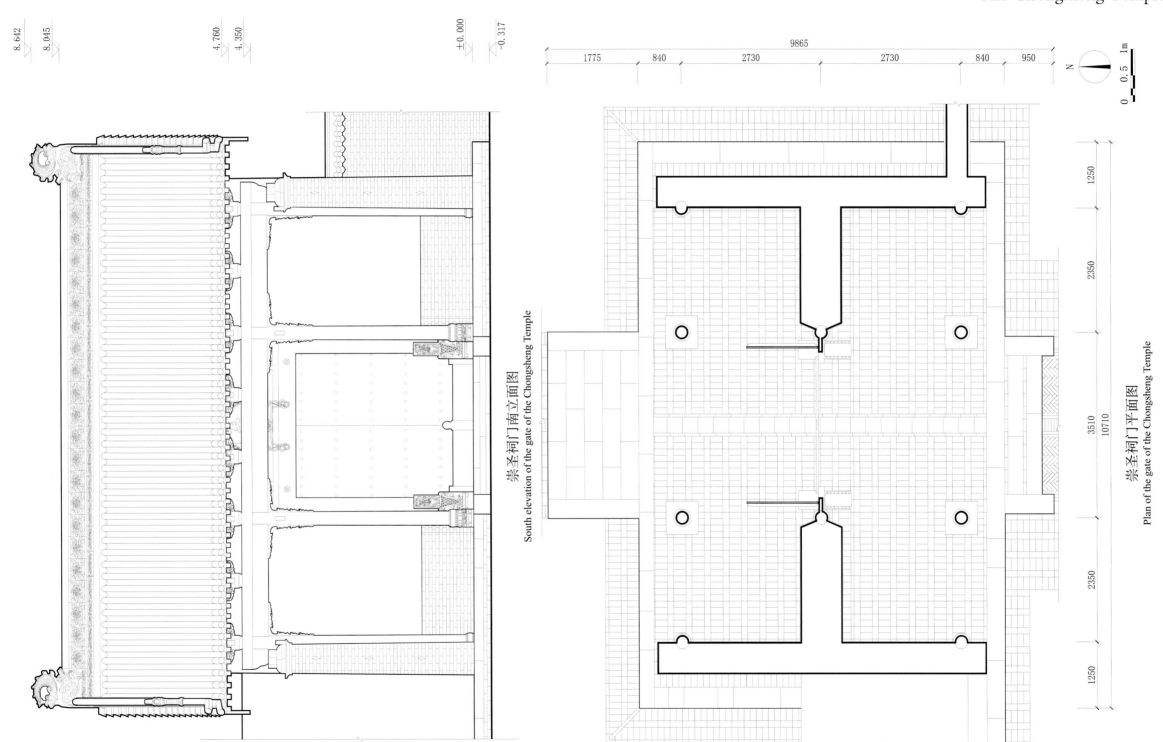

8.642
8.045
4.760
4.350
±0.000
-0.317

崇圣祠门南立面图
South elevation of the gate of the Chongsheng Temple

9865
1775 840 2730 2730 840 950

1250
2350
3510
10710
2350
1250

崇圣祠门平面图
Plan of the gate of the Chongsheng Temple

N

0 0.5 1m

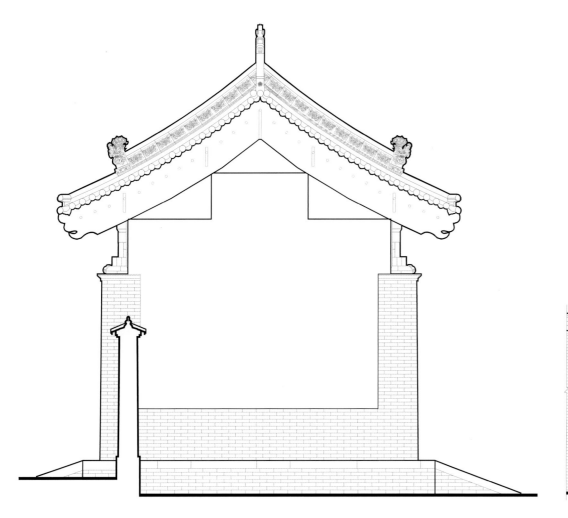

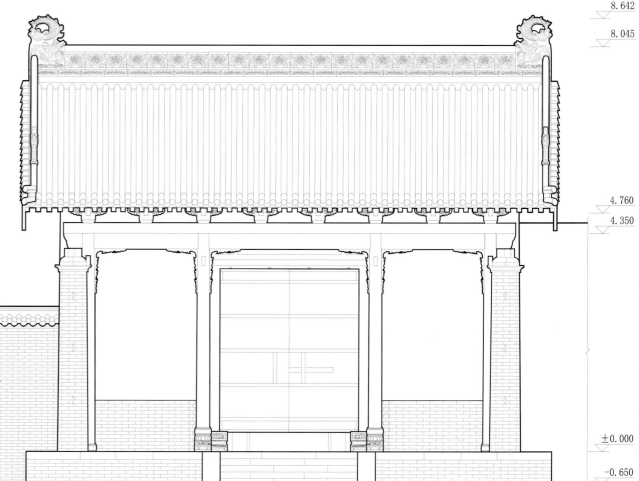

8.642

8.045

4.760

4.350

±0.000

-0.650

950　840　2730　2730　840　1775

9865

崇圣祠门东立面图

East elevation of the gate of the Chongsheng Temple

1250　2350　3510　2350　1250

10710

0　1　2m

崇圣祠门北立面图

North elevation of the gate of the Chongsheng Temple

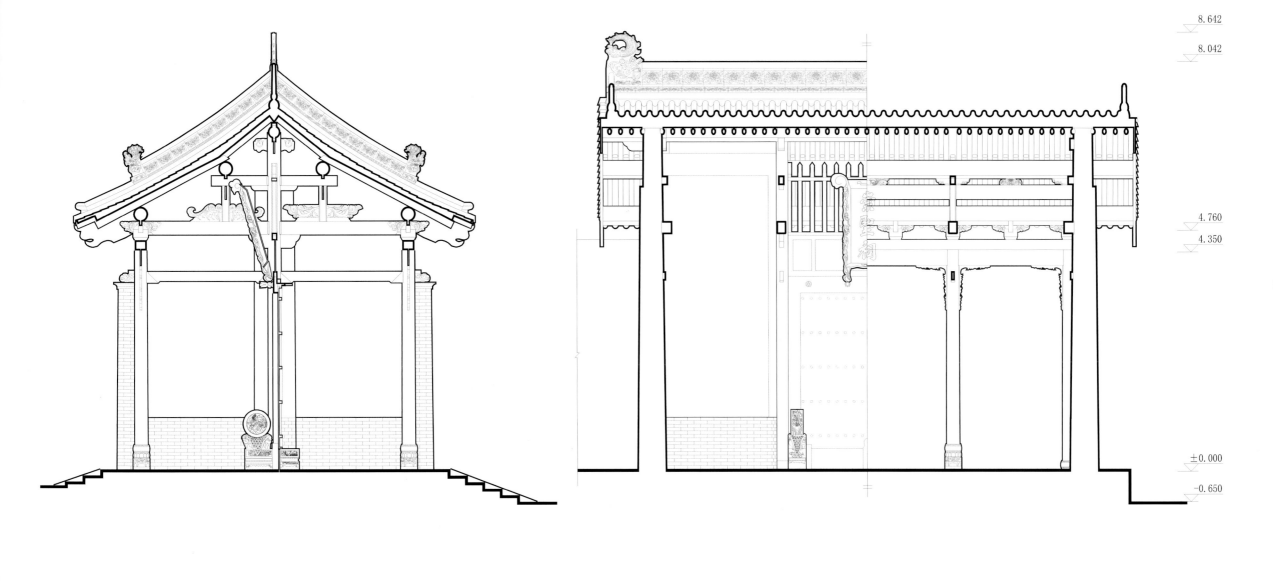

8.642
8.042
4.760
4.350
±0.000
-0.650

950 840 2730 2730 840 1775
9865

1250 2350 3510 2350 1250
10710

0   1   2m

崇圣祠门明间横剖面图
Central bay section of the gate of the Chongsheng Temple

崇圣祠门纵剖面图
Longitudinal section of the gate of the Chongsheng Temple

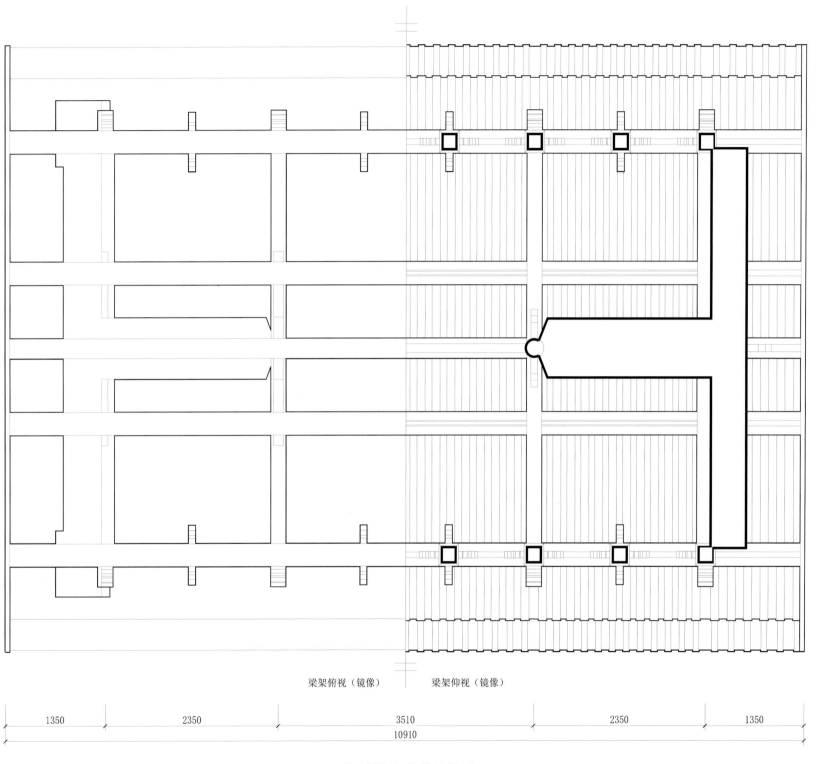

梁架俯视（镜像）　　梁架仰视（镜像）

崇圣祠门梁架仰俯视图
Top and bottom view of the beam frame of the gate of the Chongsheng Temple

1275　2730　8010　2730　1275

1350　2350　3510　2350　1350
10910

0　0.5　1m

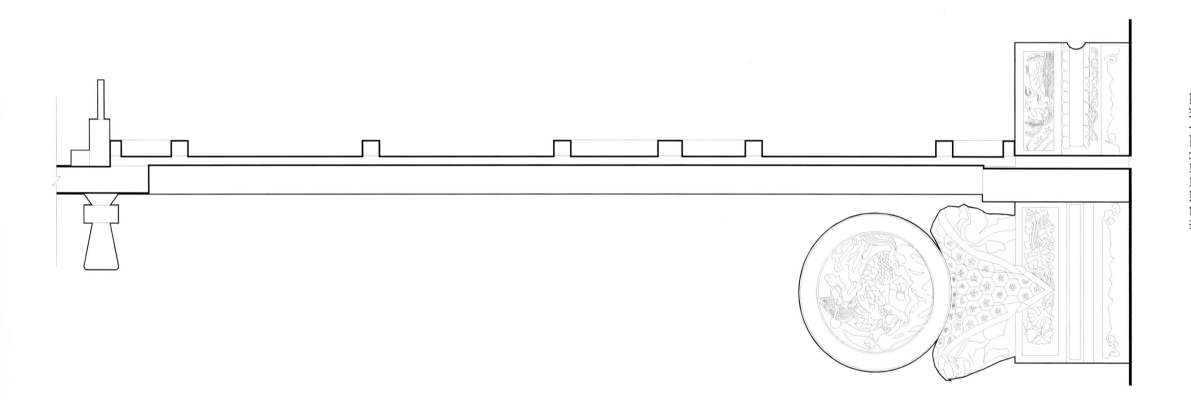

0 0.1 0.2m

崇圣祠门门枕石大样图

Details of the bearing stone of the gate of the Chongsheng Temple

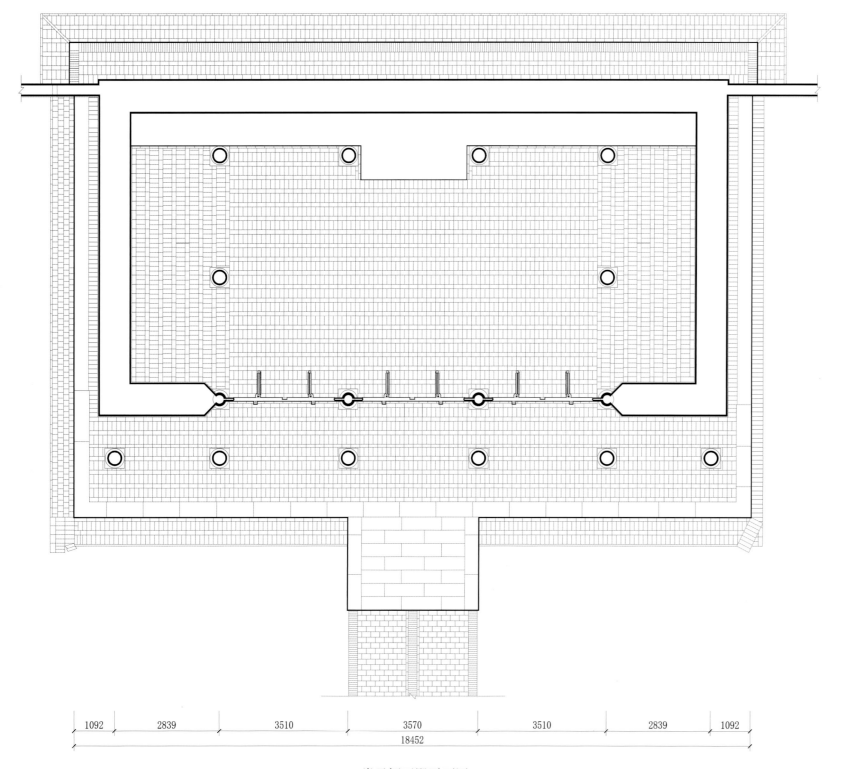

崇圣祠正殿平面图
Plan of the main hall of the Chongsheng Temple

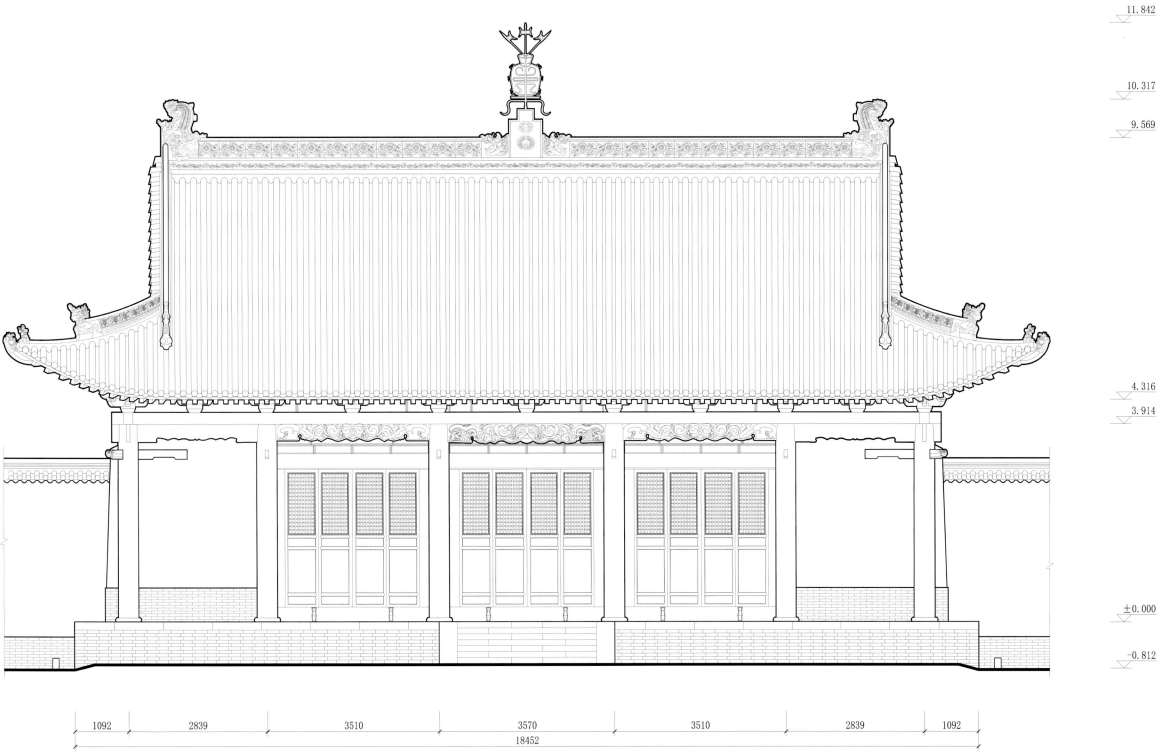

11.842

10.317

9.569

4.316

3.914

200

±0.000

-0.812

| 1092 | 2839 | 3510 | 3570 | 3510 | 2839 | 1092 |

18452

0 0.5 1m

崇圣祠正殿南立面图
South elevation of the main hall of the Chongsheng Temple

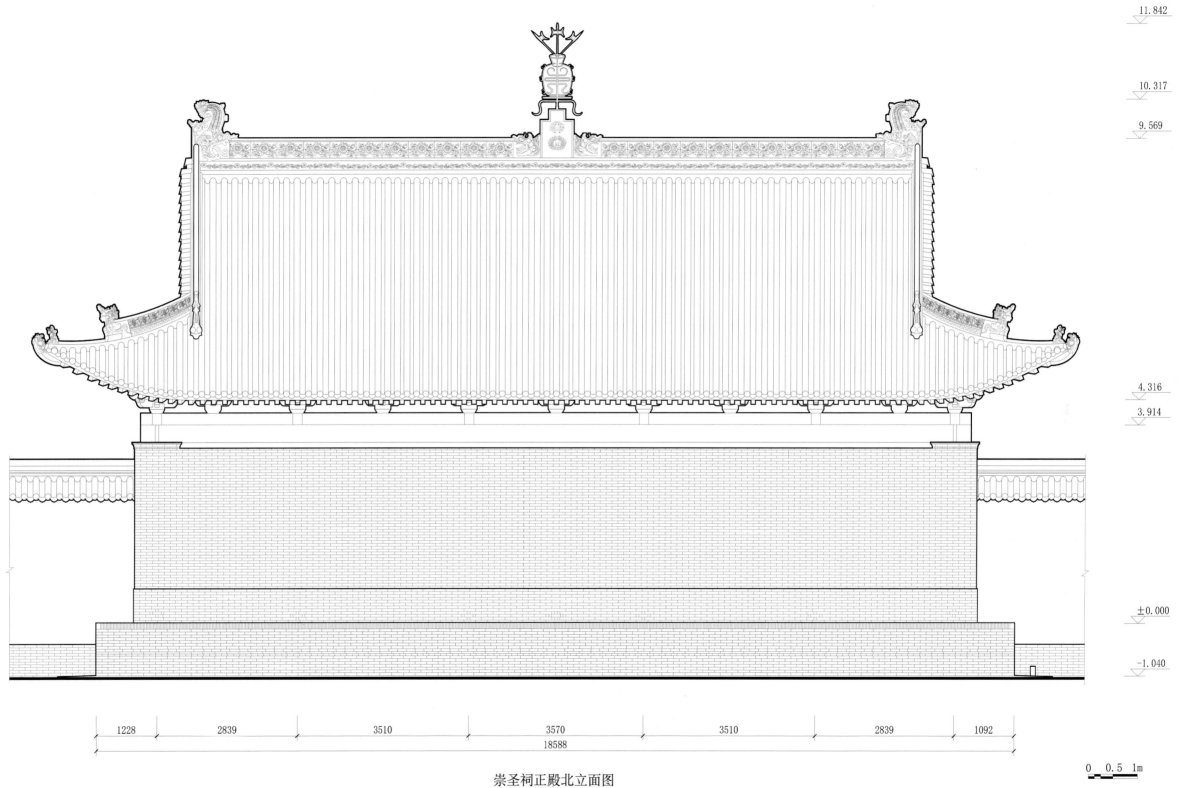

11.842

10.317

9.569

4.316

3.914

±0.000

-1.040

1228　　2839　　3510　　3570　　3510　　2839　　1092

18588

0　0.5　1m

崇圣祠正殿北立面图

North elevation of the main hall of the Chongsheng Temple

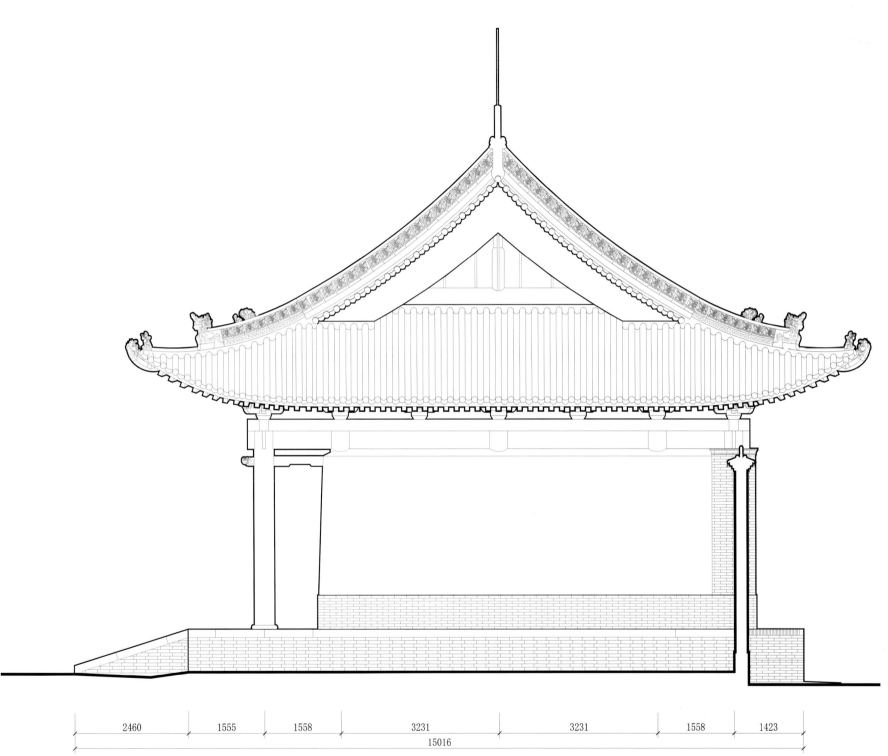

11.842

10.317

9.569

4.316

3.914

±0.000

-0.812

-1.040

| 2460 | 1555 | 1558 | 3231 | 3231 | 1558 | 1423 |

15016

崇圣祠正殿东立面图

East elevation of the main hall of the Chongsheng Temple

0   0.5   1m

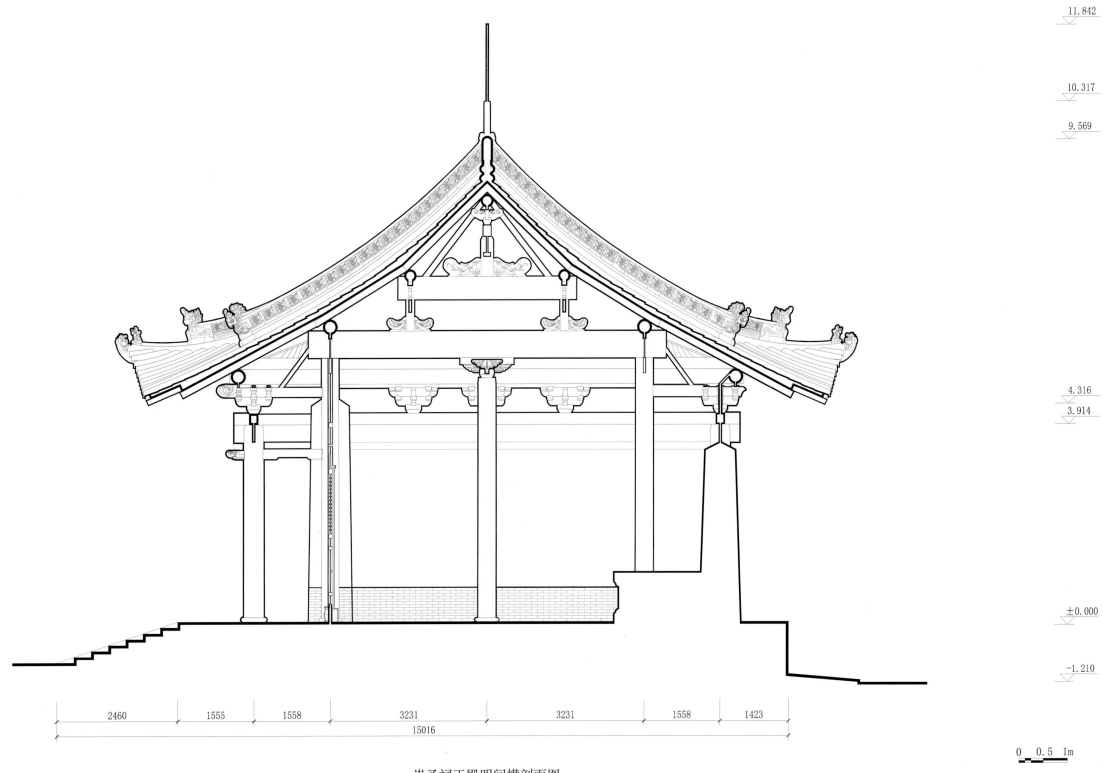

11.842

10.317

9.569

4.316

3.914

±0.000

−1.210

| 2460 | 1555 | 1558 | 3231 | 3231 | 1558 | 1423 |

15016

0 0.5 1m

崇圣祠正殿明间横剖面图

Central bay section of the main hall of the Chongsheng Temple

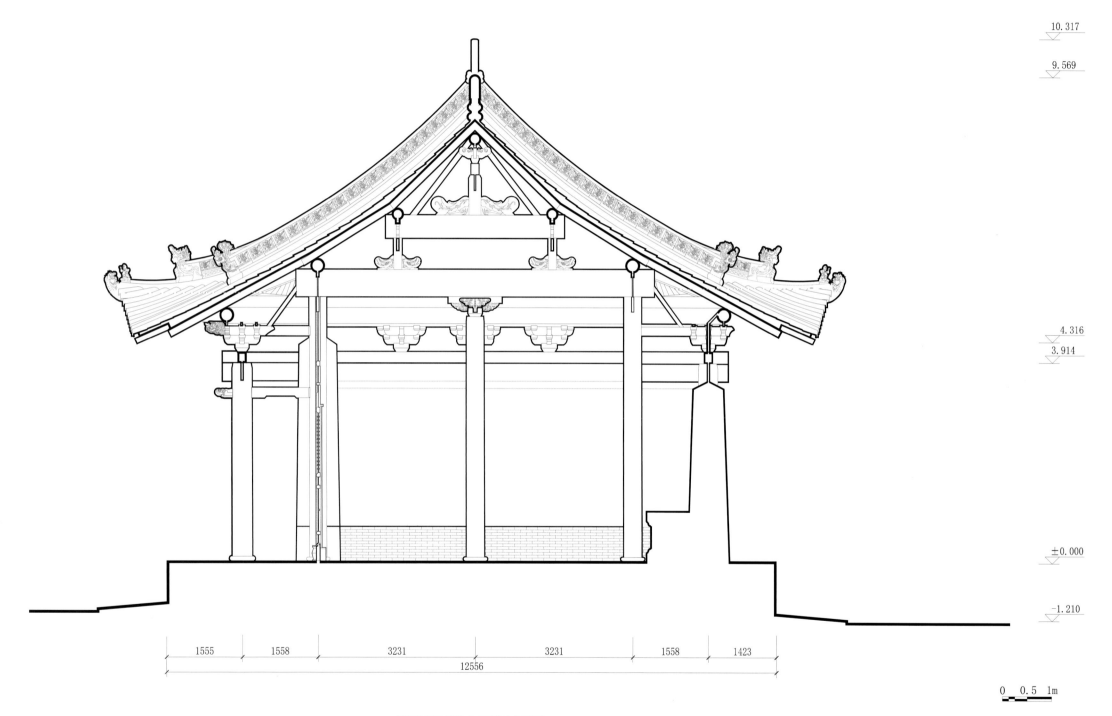

10.317

9.569

4.316

3.914

±0.000

-1.210

| 1555 | 1558 | 3231 | 3231 | 1558 | 1423 |

12556

0   0.5   1m

崇圣祠正殿次间横剖面图

Second bay section of the main hall of the Chongsheng Temple

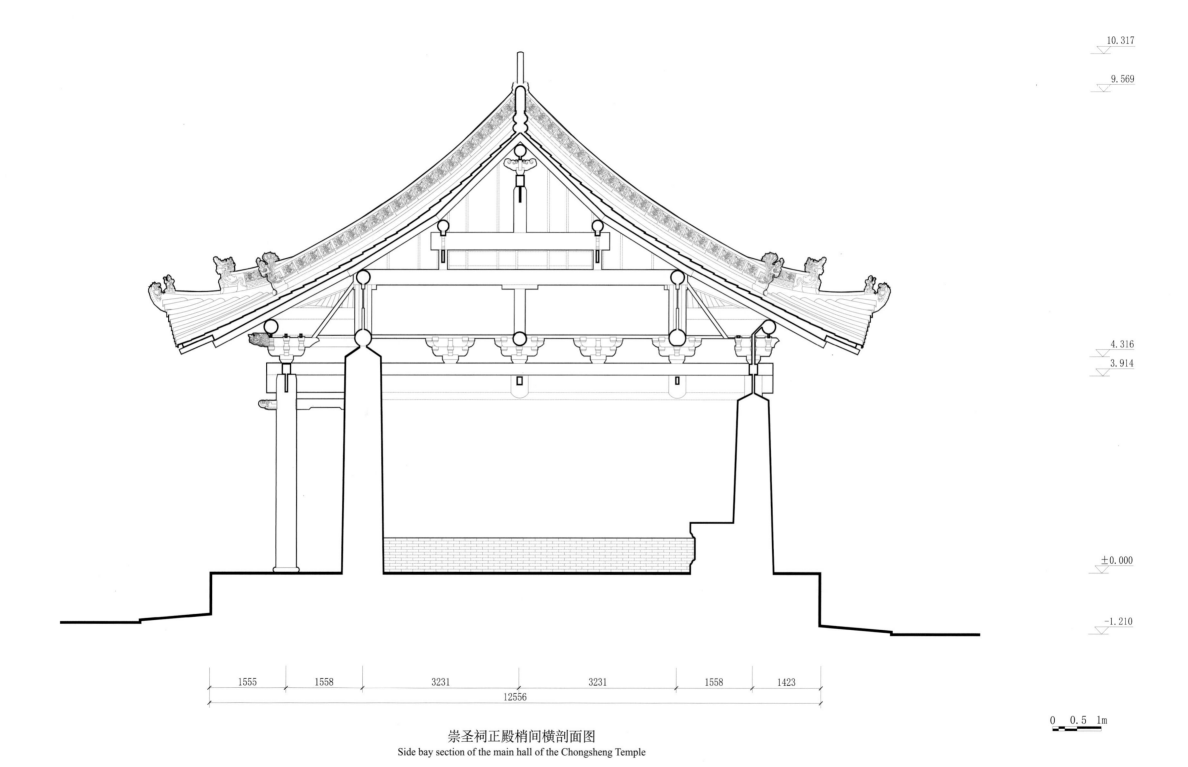

10.317

9.569

4.316

3.914

±0.000

-1.210

| 1555 | 1558 | 3231 | 3231 | 1558 | 1423 |

12556

0   0.5   1m

崇圣祠正殿梢间横剖面图
Side bay section of the main hall of the Chongsheng Temple

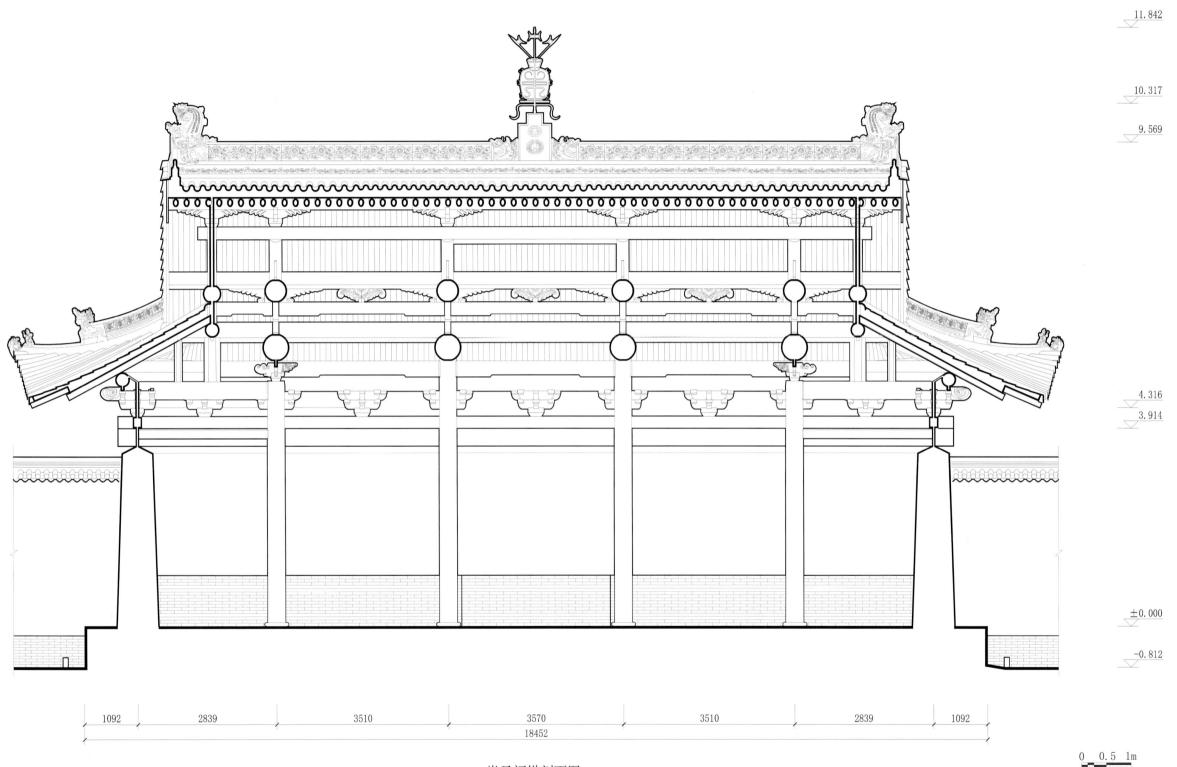

11.842

10.317

9.569

4.316

3.914

±0.000

−0.812

| 1092 | 2839 | 3510 | 3570 | 3510 | 2839 | 1092 |

18452

0 0.5 1m

崇圣祠纵剖面图
Longitudinal section of the main hall of the Chongsheng Temple

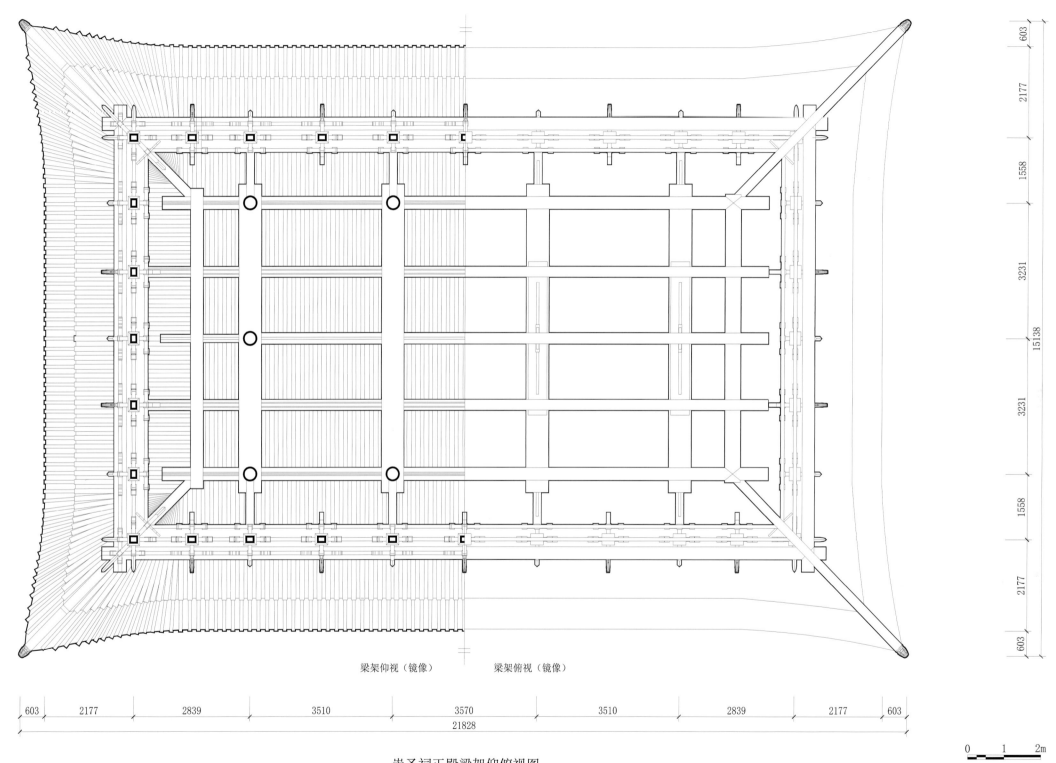

梁架仰视（镜像）　　　梁架俯视（镜像）

603　2177　2839　3510　3570　3510　2839　2177　603
21828

崇圣祠正殿梁架仰俯视图
Top and bottom of the beam frame of the main hall of the Chongsheng Temple

0　1　2m

中国古建筑测绘大系·祠庙建筑——解州关帝庙

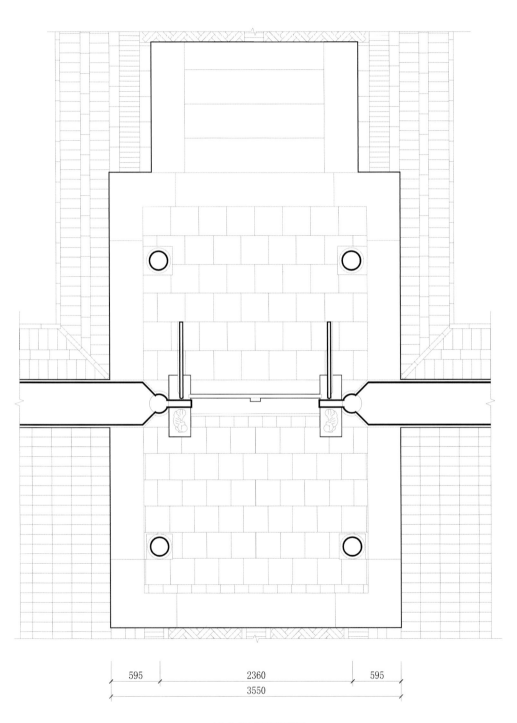

胡公祠门平面图
Plan of the gate of the Hugong Temple

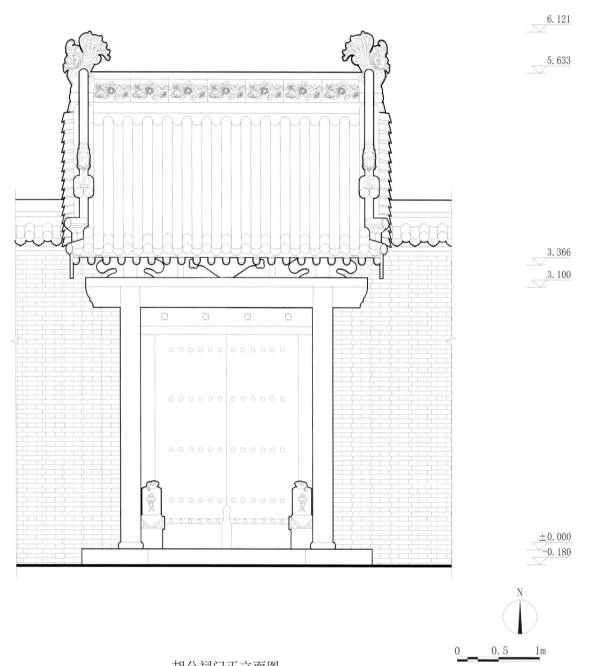

胡公祠门正立面图
Front elevation of the gate of the Hugong Temple

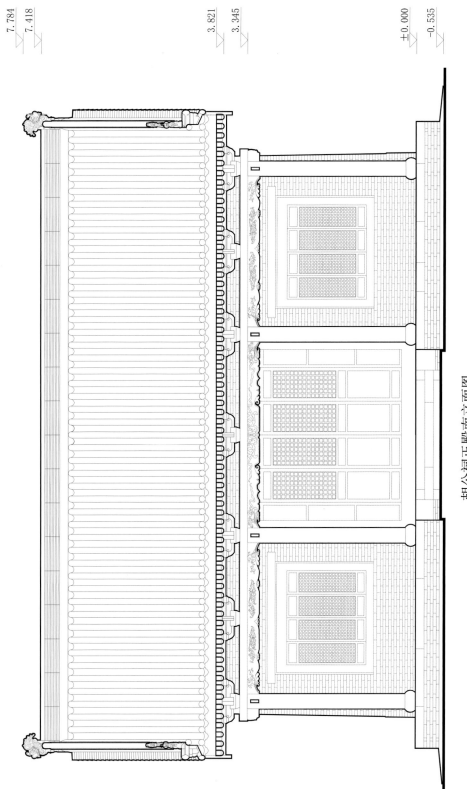

7.784
7.418
3.821
3.345
±0.000
−0.535

胡公祠正殿南立面图
South elevation of the main hall of the Hugong Temple

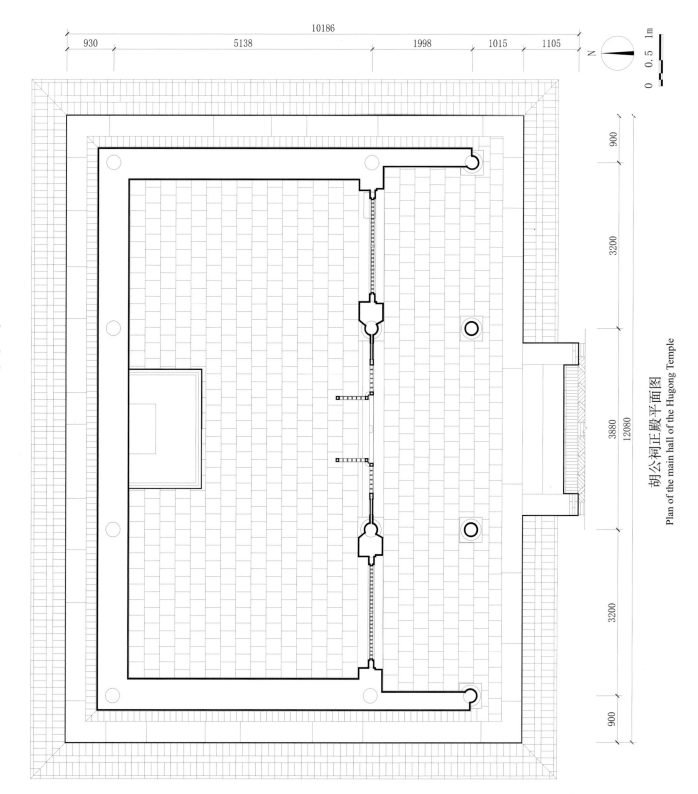

10186
930   5138   1998   1015   1105
900
3200
3880   12080
3200
900

N
0   0.5   1m

胡公祠正殿平面图
Plan of the main hall of the Hugong Temple

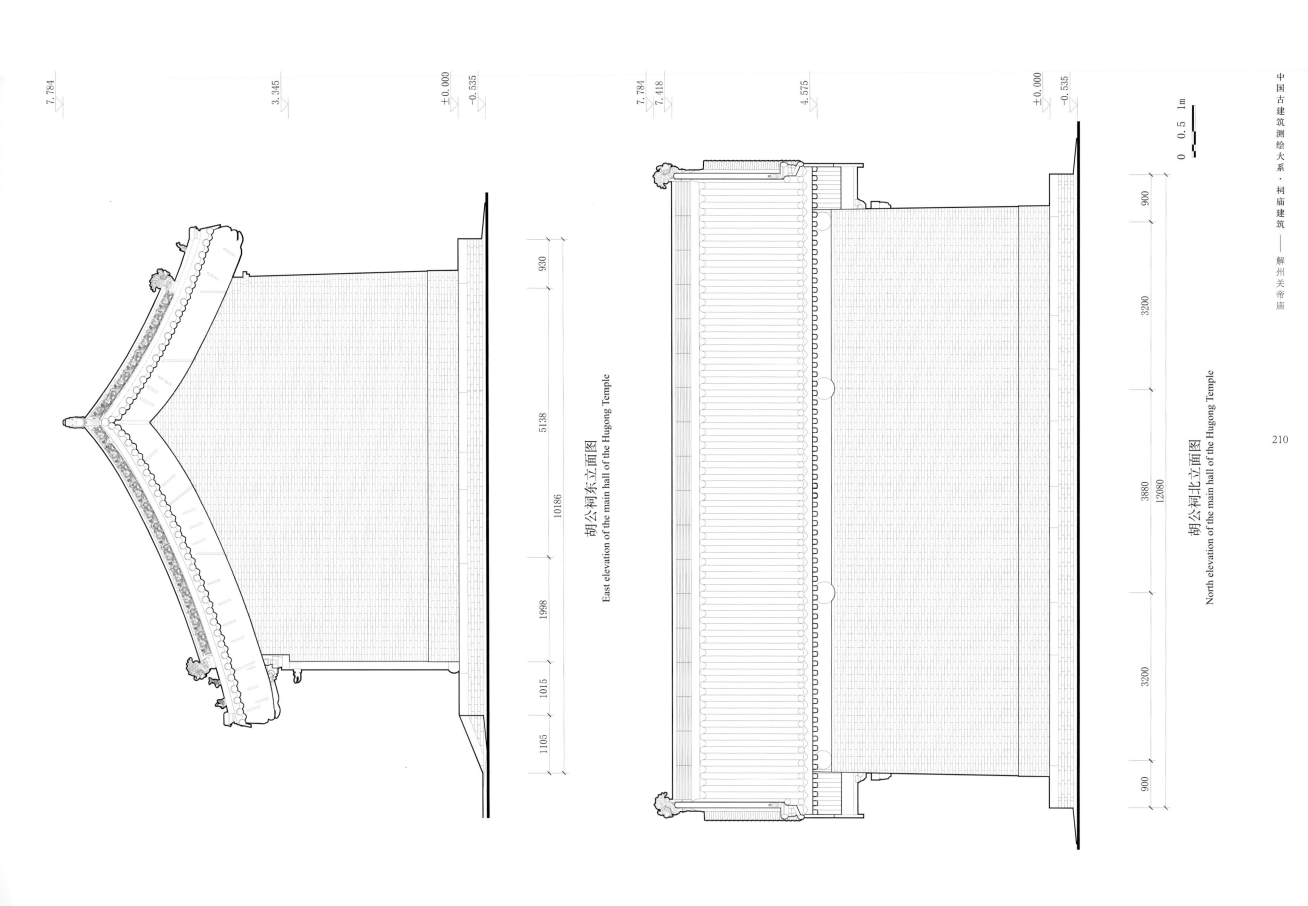

7.784

3.345

±0.000
−0.535

930

5138

10186

1998

1015

1105

胡公祠东立面图
East elevation of the main hall of the Hugong Temple

7.784
7.418

4.575

±0.000
−0.535

0 0.5 1m

900

3200

3880

12080

3200

900

胡公祠北立面图
North elevation of the main hall of the Hugong Temple

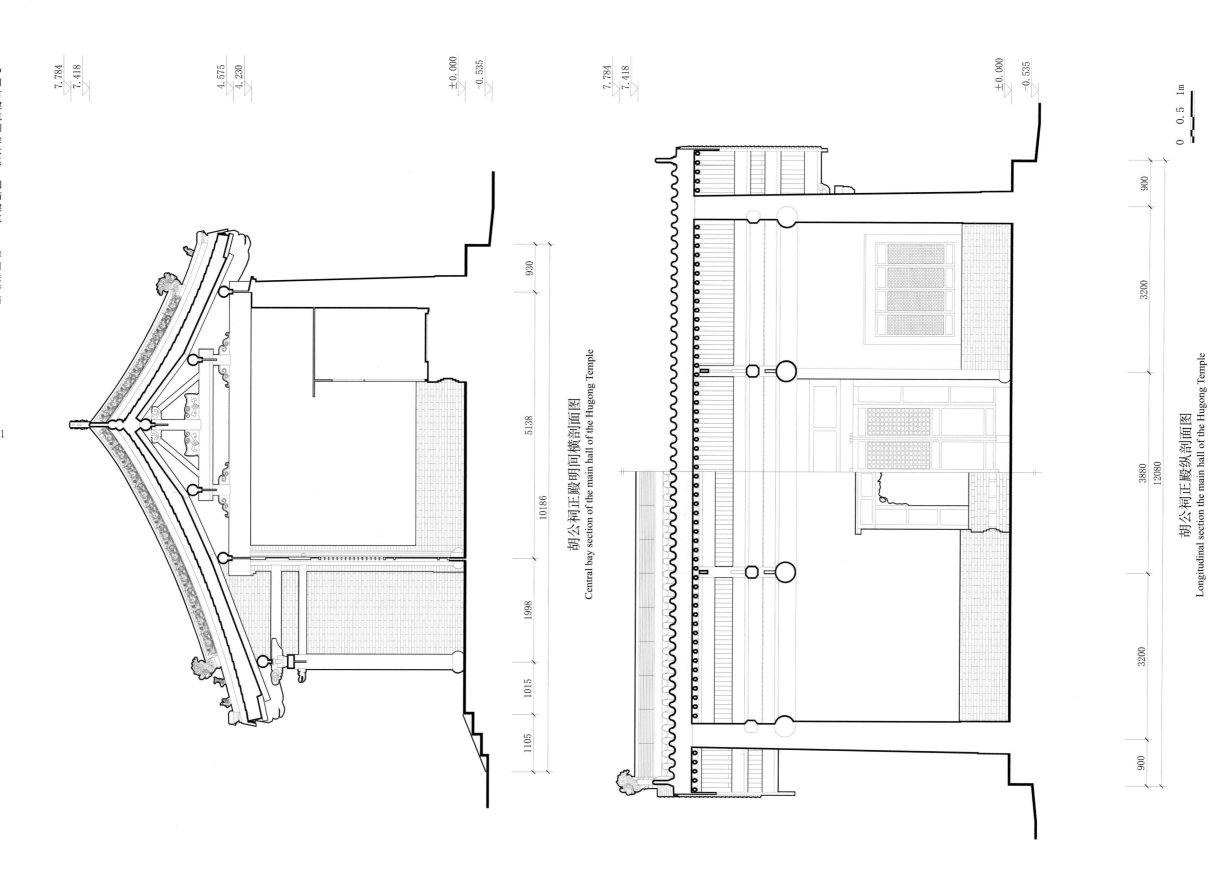

7.784
7.418

4.575
4.230

±0.000
-0.535

胡公祠正殿明间横剖面图
Central bay section of the main hall of the Hugong Temple

930

5138

10186

1998

1015

1105

7.784
7.418

±0.000
-0.535

胡公祠正殿纵剖面图
Longitudinal section the main hall of the Hugong Temple

900

3200

3880
12080

3200

900

0　0.5　1m

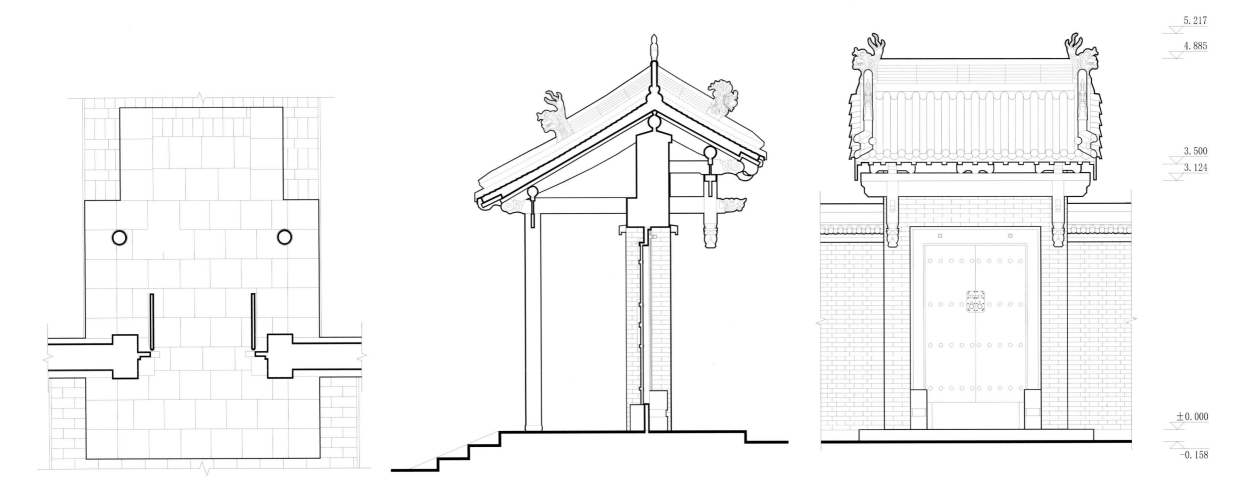

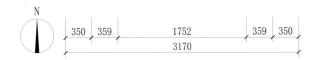

道正司垂花门平面图
Plan of the floral pendant gate of the Daozheng House

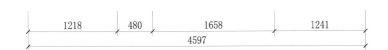

道正司垂花门横剖面图
Cross-section of the floral pendant gate of the Daozheng House

道正司垂花门正立面图
Front elevation of the floral pendant gate of the Daozheng House

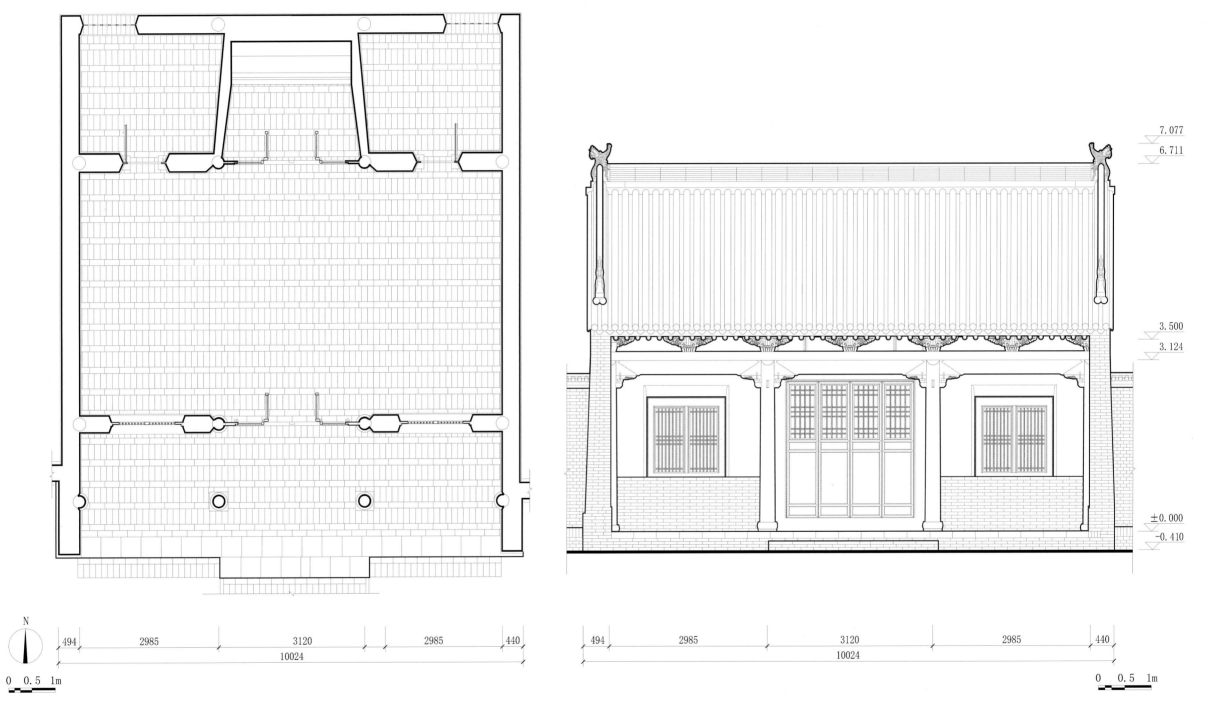

N

494 | 2985 | 3120 | 2985 | 440
10024

0 0.5 1m

494 | 2985 | 3120 | 2985 | 440
10024

0 0.5 1m

7.077
6.711

3.500
3.124

±0.000
-0.410

道正司平面图
Plan of the Daozheng House

道正司南立面图
South elevation of the Daozheng House

道正司北立面图
North elevation of the Daozheng House

道正司西立面图
West elevation of the Daozheng House

7.077
6.711

2.930

±0.000

410
2985
3120
9910
2985
410

7.077

±0.000
-0.410
-0.864

428
1160
603
11764
9348
225

0 0.5 1m

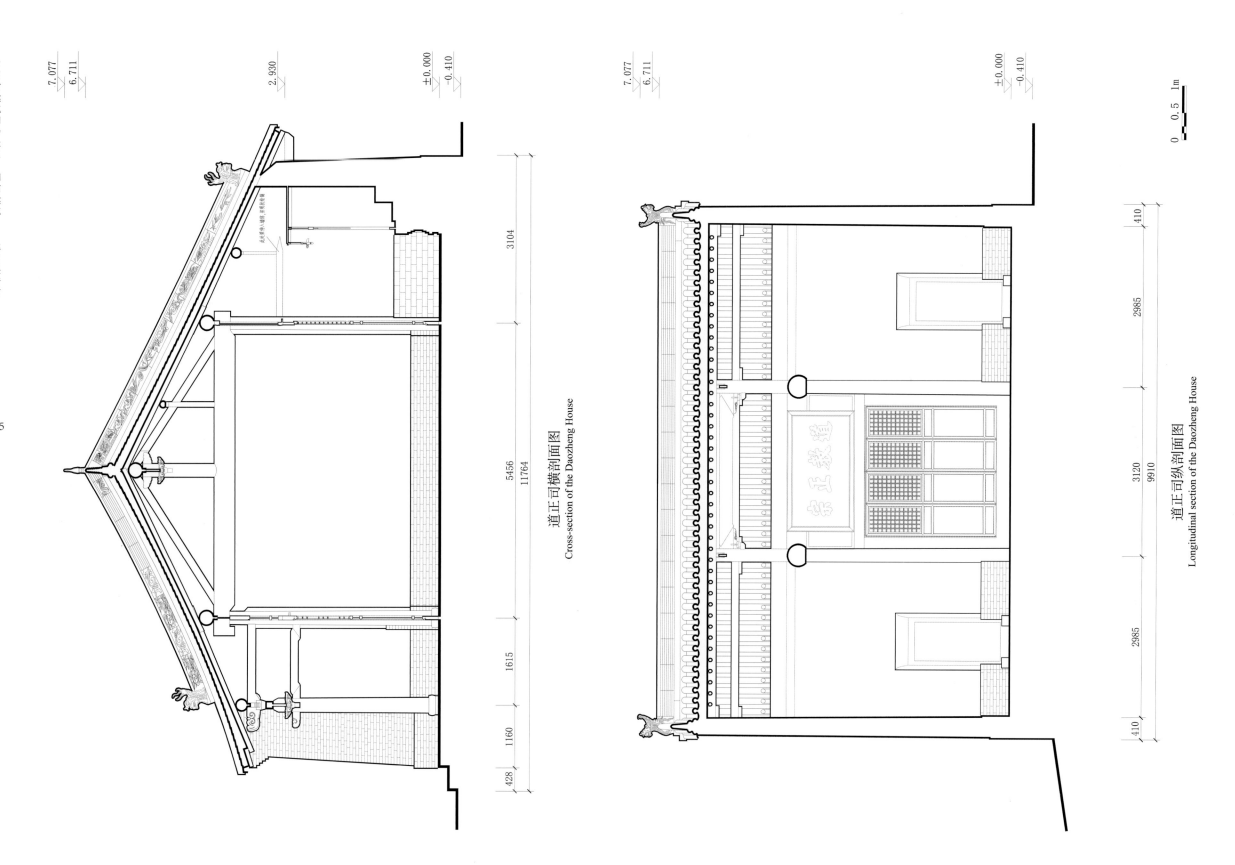

7.077
6.711

2.930

±0.000
-0.410

3104

5456
11764

1615

1160

428

道正司横剖面图
Cross-section of the Daozheng House

7.077
6.711

±0.000
-0.410

0 0.5 1m

410

2985

3120
9910

2985

410

道正司纵剖面图
Longitudinal section of the Daozheng House

结义园

Jieyi Garden

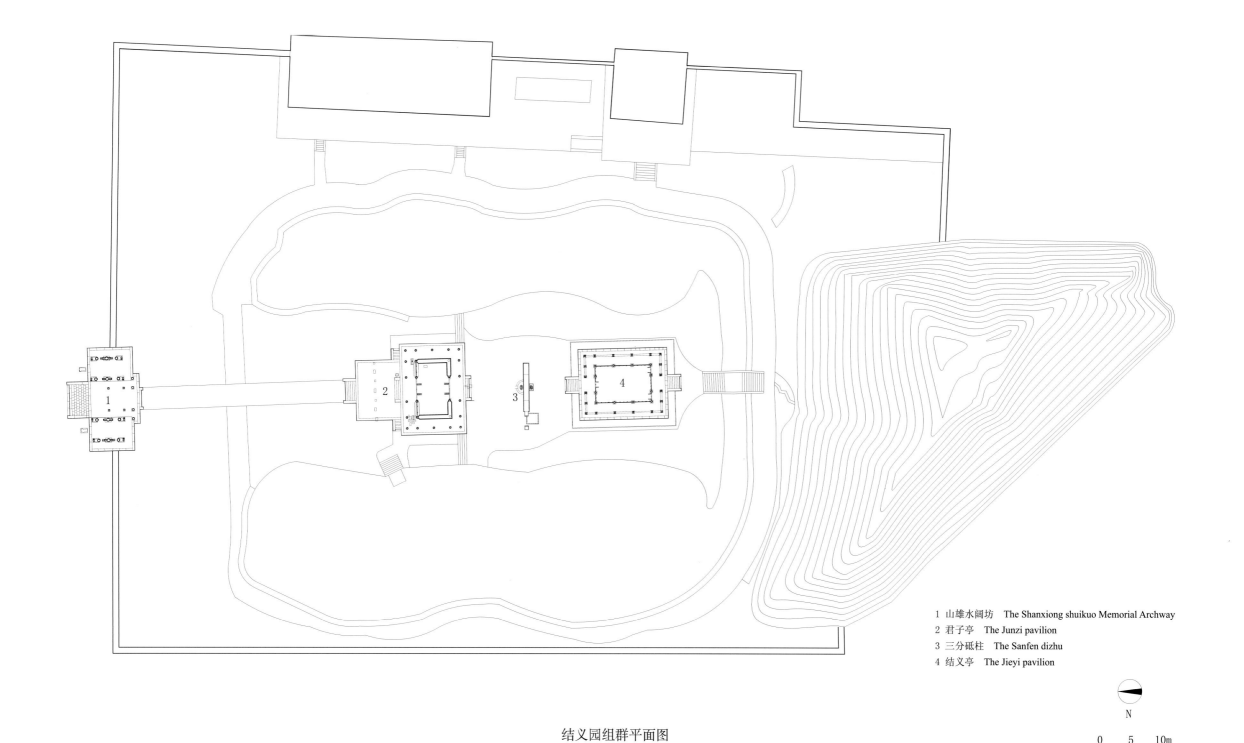

1 山雄水阔坊　The Shanxiong shuikuo Memorial Archway
2 君子亭　The Junzi pavilion
3 三分砥柱　The Sanfen dizhu
4 结义亭　The Jieyi pavilion

结义园组群平面图
Plan of the groups of the Jieyi Garden

N

0　5　10m

山雄水阔坊
The Shanxiong shuikuo Memorial Archway

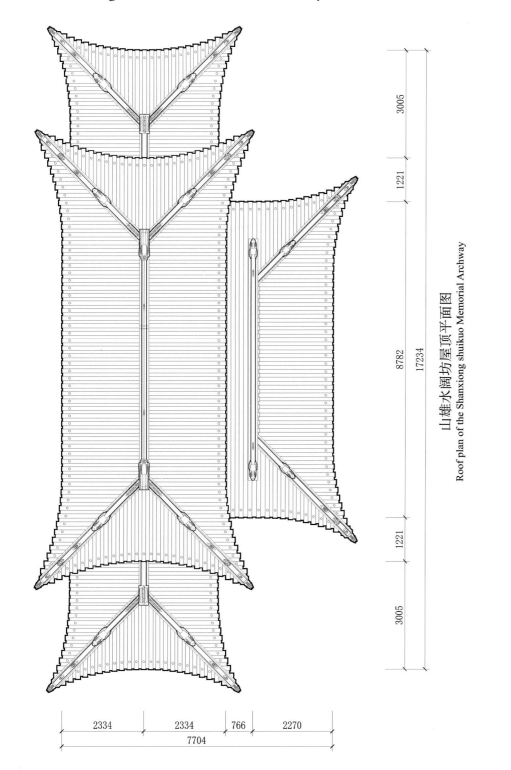

山雄水阔坊屋顶平面图
Roof plan of the Shanxiong shuikuo Memorial Archway

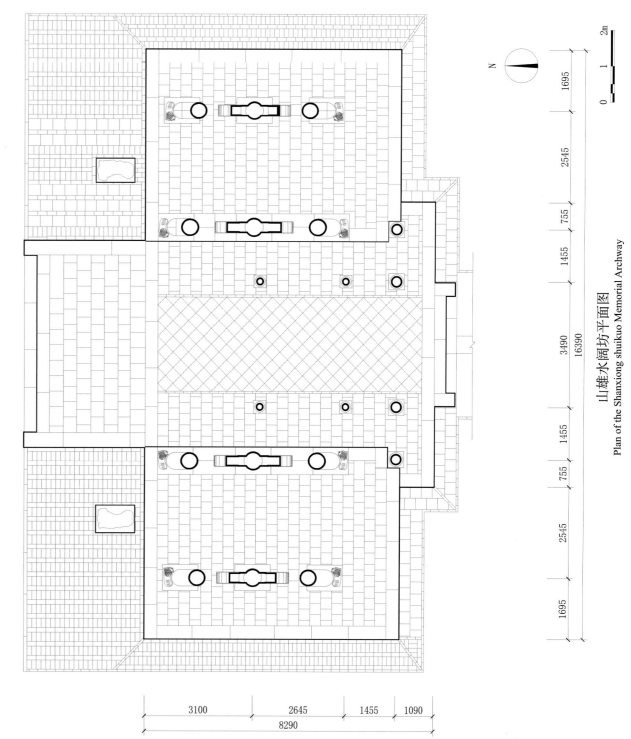

山雄水阔坊平面图
Plan of the Shanxiong shuikuo Memorial Archway

11.632

8.830

8.295

6.164

结义园

±0.000

-1.006

| 1695 | 3250 | 1505 | 3490 | 1505 | 3250 | 1695 |

16390

山雄水阔坊北立面图
North elevation of the Shanxiong shuikuo Memorial Archway

0   0.5   1m

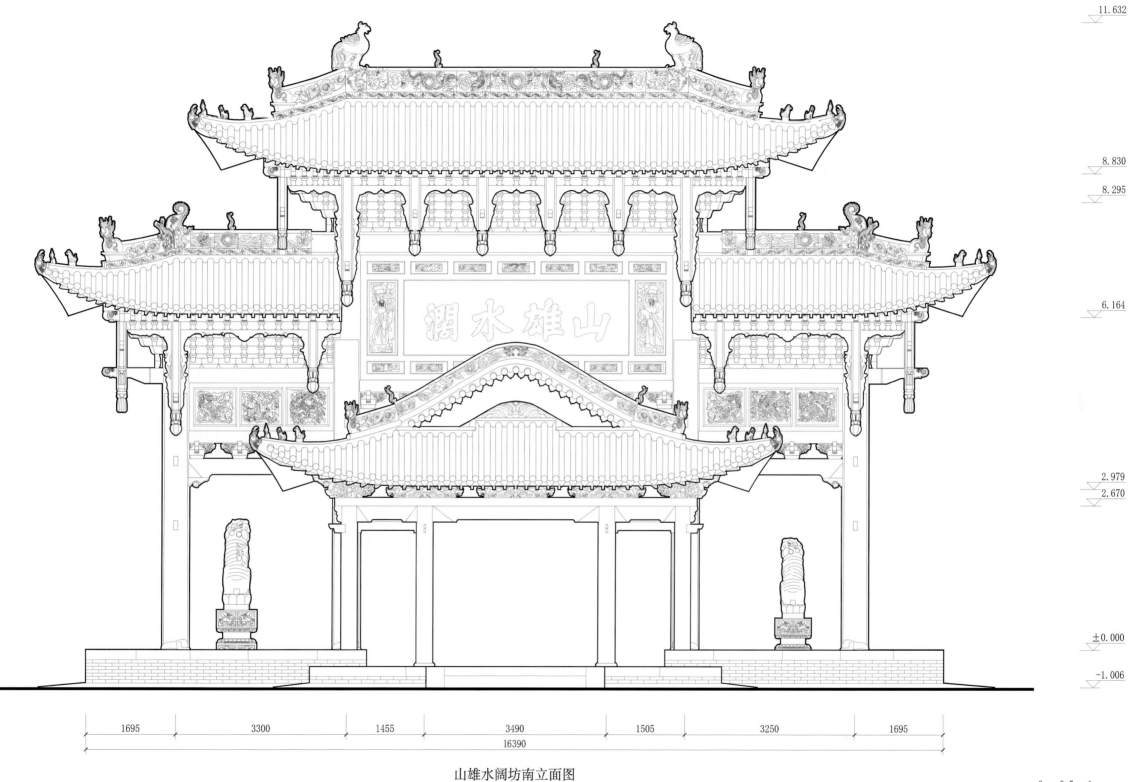

11.632

8.830

8.295

6.164

2.979

2.670

±0.000

-1.006

山水雄山

1695　　3300　　1455　　3490　　1505　　3250　　1695

16390

山雄水阔坊南立面图

South elevation of the Shanxiong shuikuo Memorial Archway

0　0.5　1m

11.632

8.830

7.746

5.745

2.979
2.670

±0.000

-0.695

| 3100 | 2655 | 1445 | 930 |
8290   160

| 3100 | 2465 | 1445 | 1090 |
190   8290

山雄水阔坊西立面图
West elevation of the Shanxiong shuikuo Memorial Archway

山雄水阔坊明间剖面图
Central bay section of the Shanxiong shuikuo Memorial Archway

0 0.5 1m

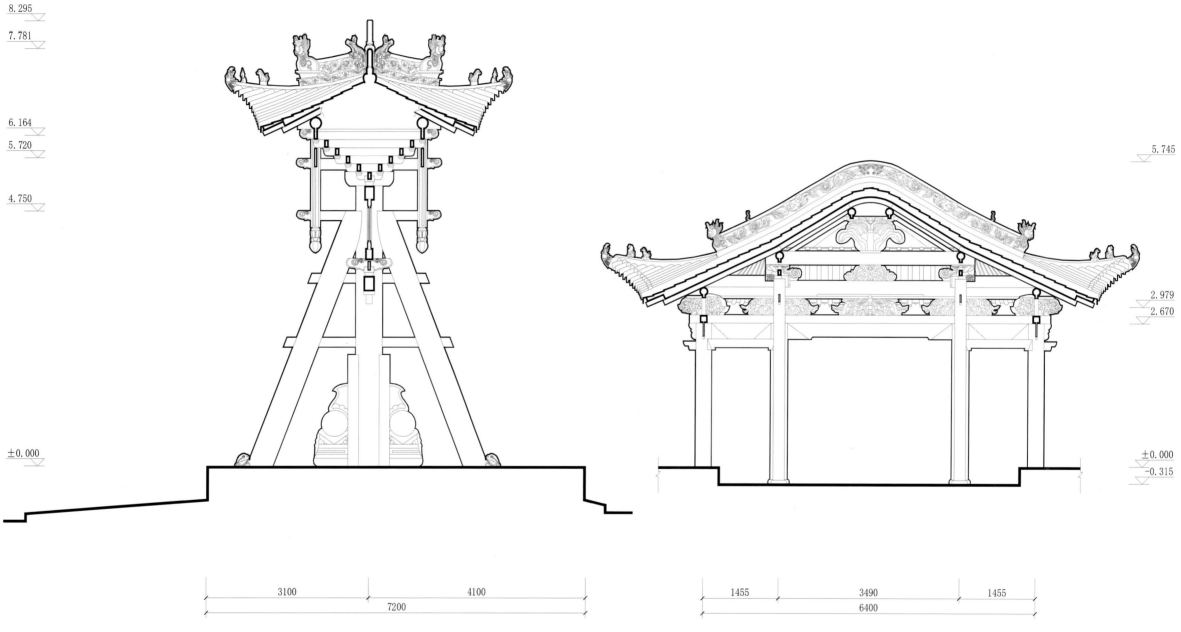

8.295

7.781

6.164

5.720

5.745

4.750

2.979

2.670

±0.000

±0.000

-0.315

3100   4100

7200

1455   3490   1455

6400

山雄水阔坊次间剖面图
Side bay section of the Shanxiong shuikuo Memorial Archway

山雄水阔坊抱厦纵剖面图
Longitudinal section of the portico of the Shanxiong shuikuo Memorial Archway

0   0.5   1m

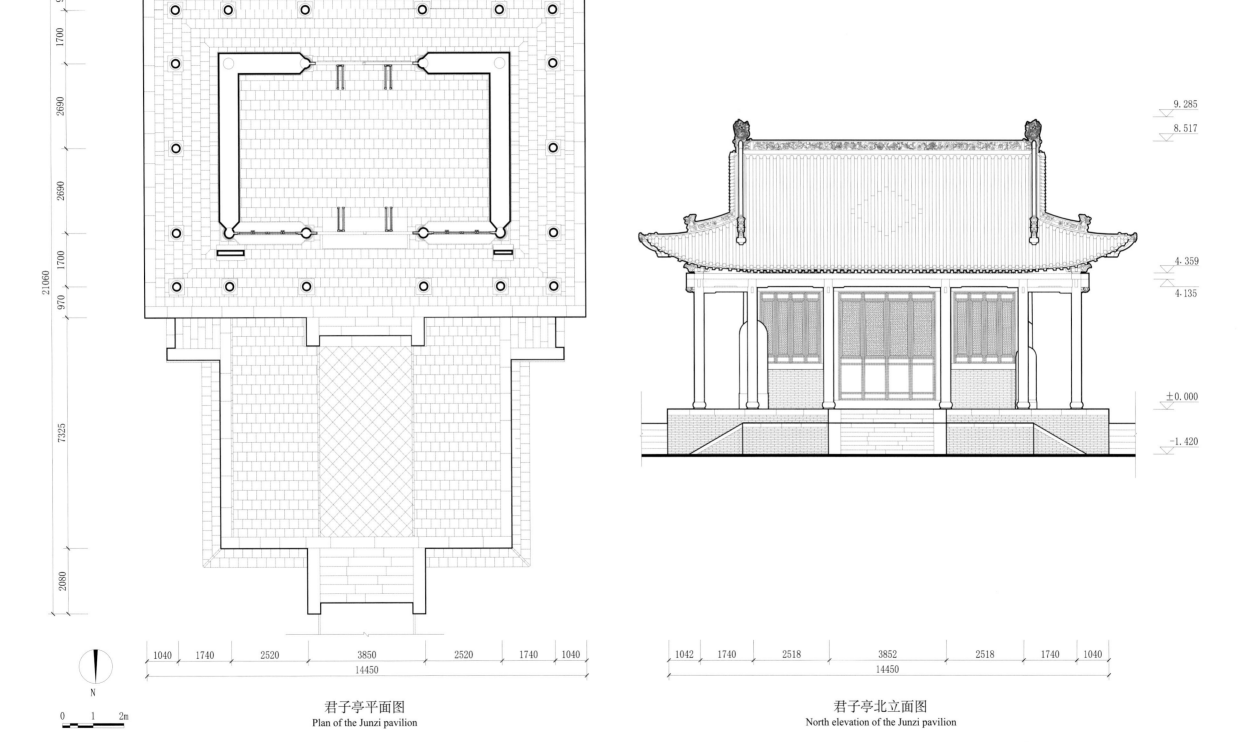

中国古建筑测绘大系·祠庙建筑——解州关帝庙

224

君子亭平面图
Plan of the Junzi pavilion

君子亭北立面图
North elevation of the Junzi pavilion

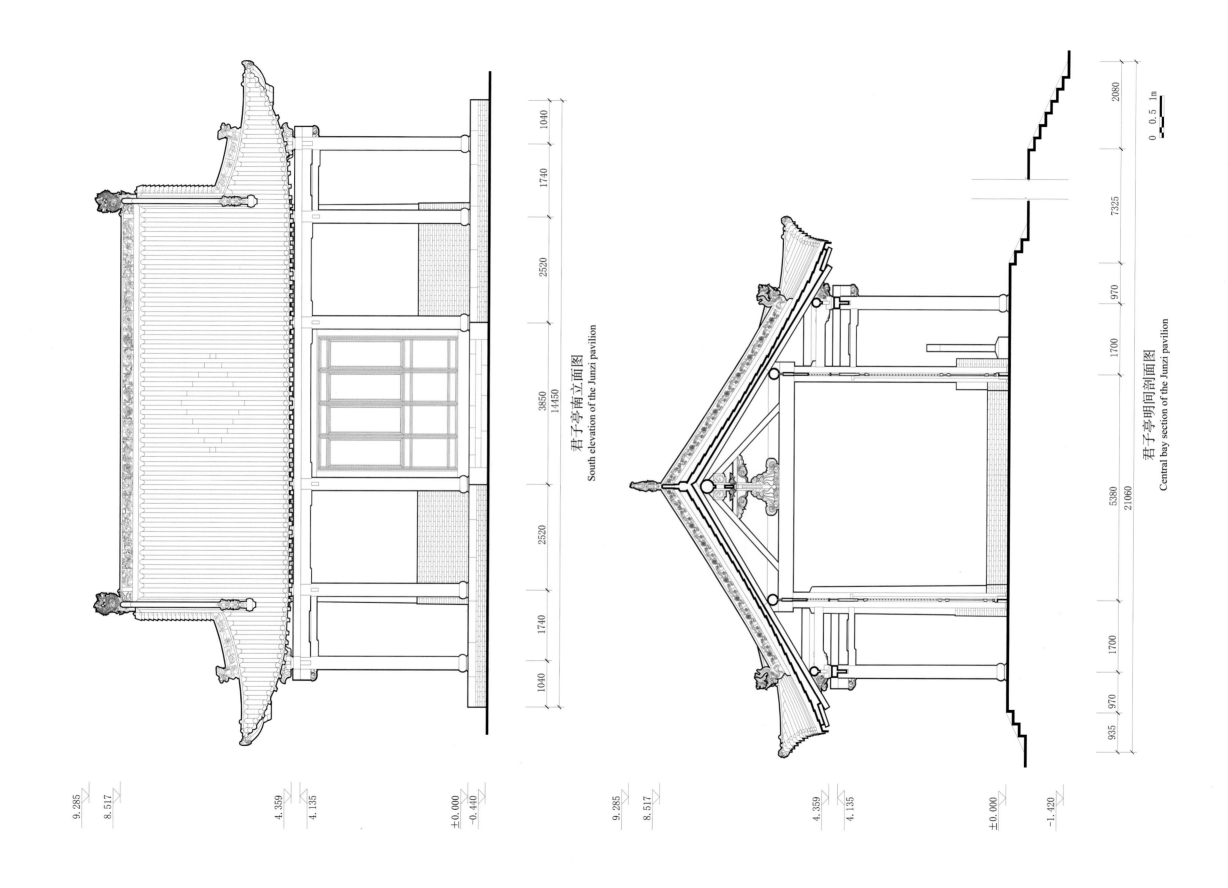

君子亭南立面图
South elevation of the Junzi pavilion

9.285
8.517
4.359
4.135
±0.000
-0.440

1040
1740
2520
3850
14450
2520
1740
1040

君子亭明间剖面图
Central bay section of the Junzi pavilion

9.285
8.517
4.359
4.135
±0.000
-1.420

2080
7325
970
1700
5380
21060
1700
970
935

0 0.5 1m

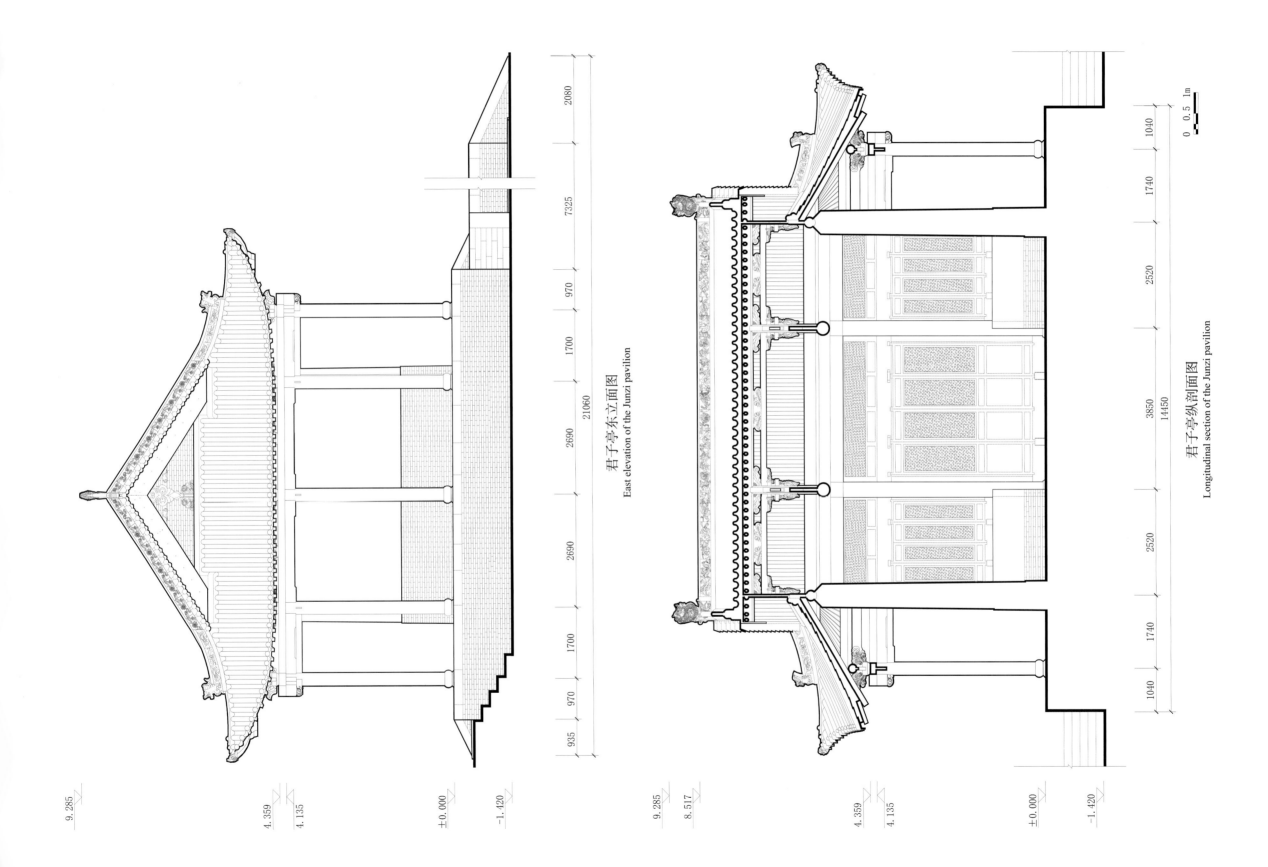

2080

7325

970

1700

2690

21060

2690

1700

970

935

君子亭东立面图
East elevation of the Junzi pavilion

9.285

4.359
4.135

±0.000

-1.420

1040

1740

2520

3850
14450

2520

1740

1040

0   0.5   1m

君子亭纵剖面图
Longitudinal section of the Junzi pavilion

9.285
8.517

4.359
4.135

±0.000

-1.420

结义亭
The Jieyi pavilion

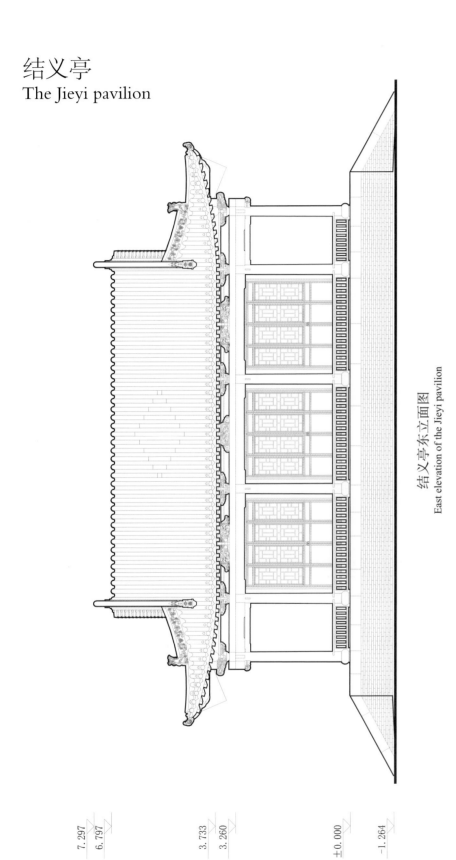

结义亭东立面图
East elevation of the Jieyi pavilion

7. 297
6. 797
3. 733
3. 260
±0. 000
-1. 264

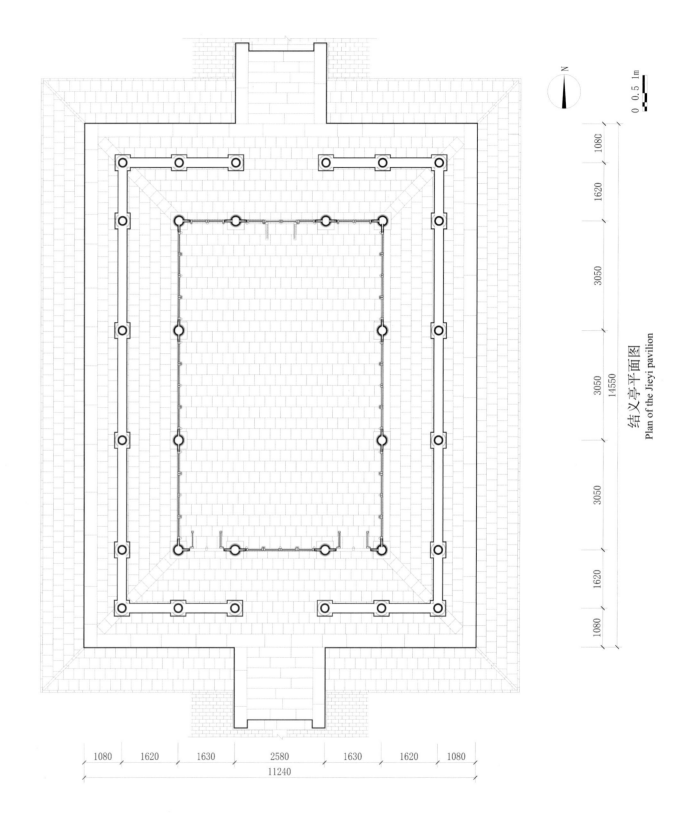

结义亭平面图
Plan of the Jieyi pavilion

0 0.5 1m

1080　1620　1630　2580　1630　1620　1080
11240

1080　1620　3050　3050　1620　1080
14550

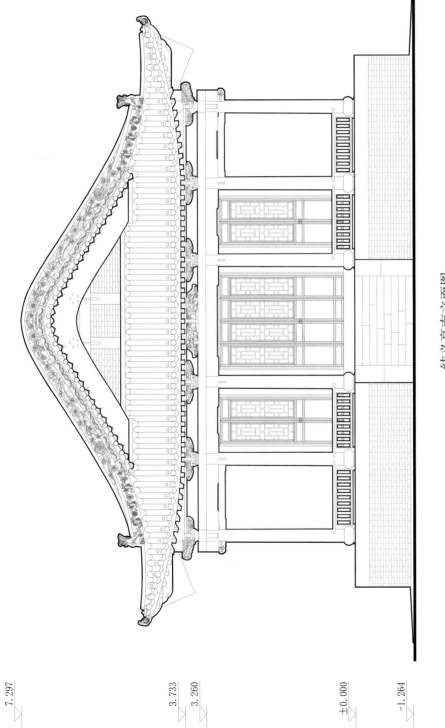

结义亭南立面图
South elevation of the Jieyi pavilion

7.297

3.733
3.260

±0.000

-1.264

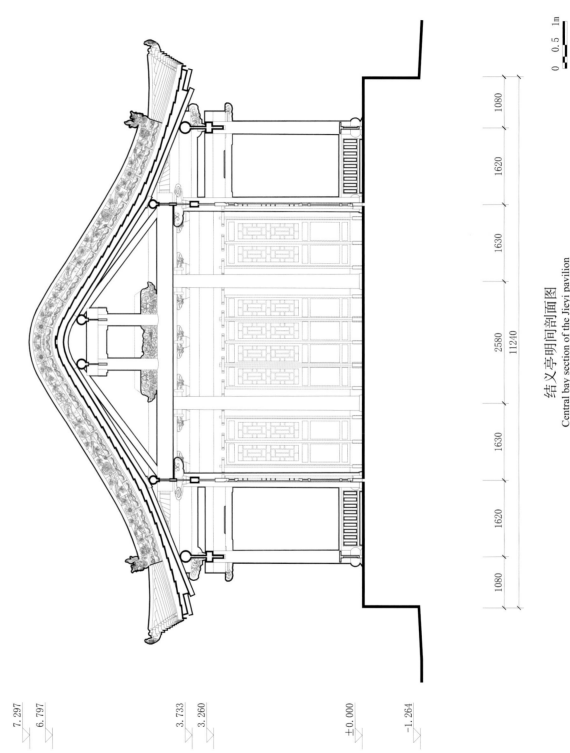

结义亭明间剖面图
Central bay section of the Jieyi pavilion

1080

1620

1630

2580
11240

1630

1620

1080

7.297
6.797

3.733
3.260

±0.000

-1.264

0　0.5　1m

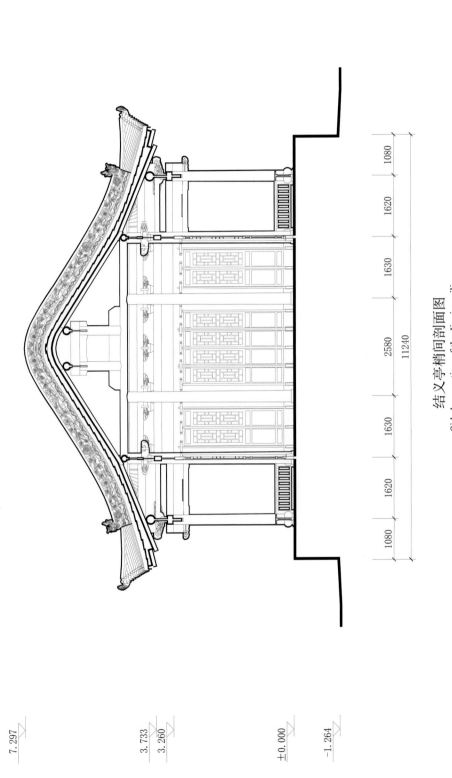

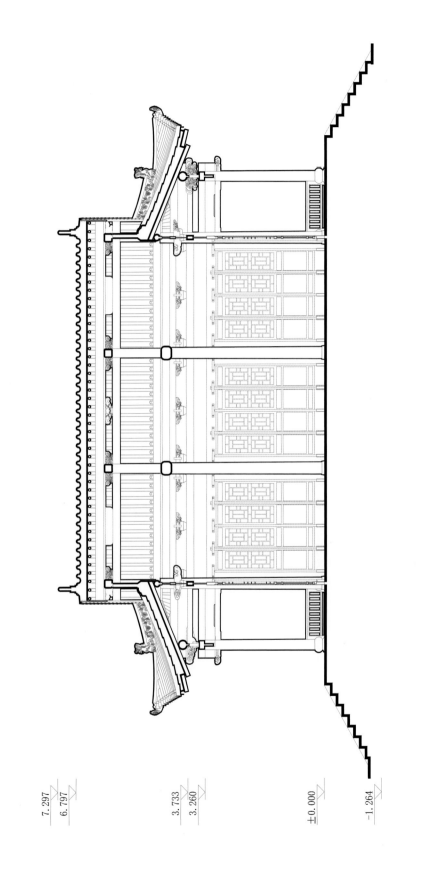

结义亭稍间剖面图
Side bay section of the Jieyi pavilion

结义亭纵剖面图
Longitudinal section of the Jieyi pavilion

0 0.5 1m

常平关帝庙

Changping Guandi
Temple

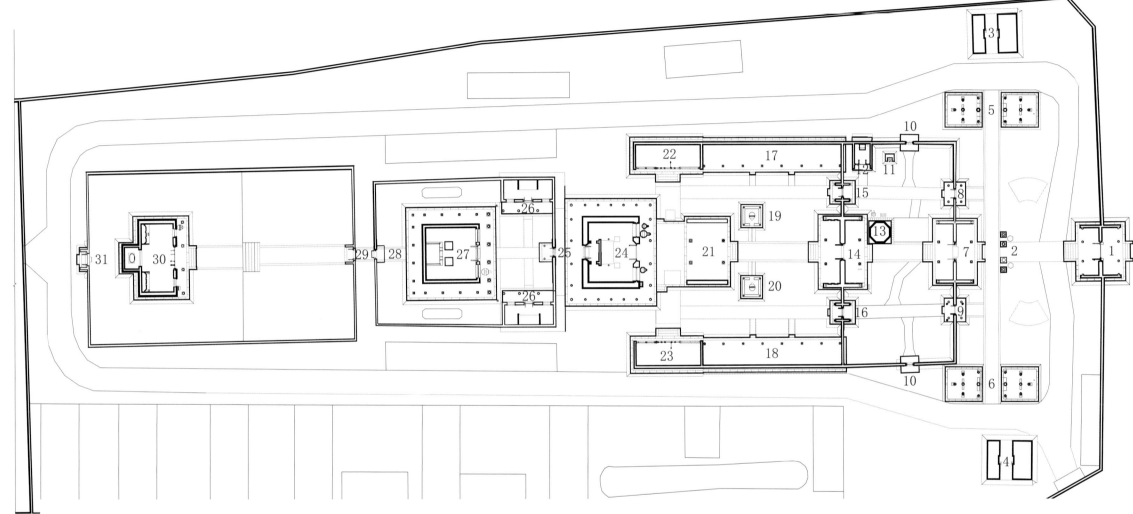

1　正门　The Main Gate
2　关王故里坊　The Guan's Hometown Memorial Stone Archway
3　钟楼　The Bell Tower
4　鼓楼　The Drum-Tower
5　秀毓条山坊　The Xiuyu tiaoshan Memorial Archway
6　灵钟嵯海坊　The Lingzhong cuohai Memorial Archway
7　山门　The Mountain Gate
8　山门东侧门　The East Gate of the Mountain Gate

9　山门西侧门　The West Gate of the Mountain Gate
10　仪门前院垂花门　The Floral Pendant Gate of the Frontyard of the Yi Gate
11　嘉庆碑碑亭　The Stele Pavilion of Jiaqing Monument
12　于宝庙　The Yubao Temple
13　祖宅塔　The Ancestral Pagoda
14　仪门　The Yi Gate
15　仪门东侧门　The East Gate of the Yi Gate

16　仪门西侧门　The West Gate of the Yi Gate
17　东廊　The East Corridor
18　西廊　The West Corridor
19　东碑亭　The East Stele Pavilion
20　西碑亭　The West Stele Pavilion
21　献殿　The Xian Hall
22　东官厅　The Dongguan Hall
23　西官厅　The Xiguan Hall

24　崇宁殿　The Chongning Hall
25　娘娘殿前院门　The Frontyard Gate of the Goddess Hall
26　太子殿　The Crown Prince Hall
27　娘娘殿　The Goddess Hall
28　娘娘殿后院门　The Backyard Gate of the Goddess Hall
29　圣祖殿院门　The Courtyard Gate of the Sage Hall
30　圣祖殿　The Sage Hall
31　圣祖殿后院门　The Backyard Gate of the Sage Hall

常平关帝庙组群平面图
Plan of the groups of the Changping Guandi Ancestral Temple

0　5　10m

关王故里坊
The Guan's Hometown Memorial Stone Archway

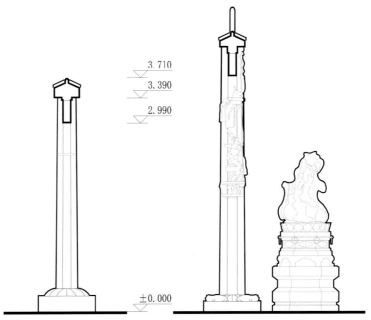

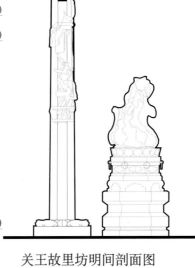

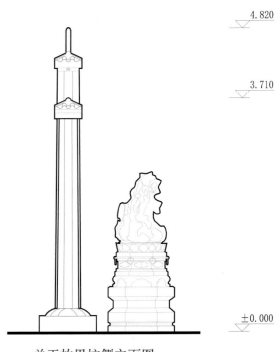

关王故里坊正立面图
Front elevation of the Guan's Hometown Memorial Stone Archway

关王故里坊次间剖面图
Side bay section of the Guan's Hometown Memorial Stone Archway

关王故里坊明间剖面图
Central bay section of the Guan's Hometown Memorial Stone Archway

关王故里坊侧立面图
Side elevation of the Guan's Hometown Memorial Stone Archway

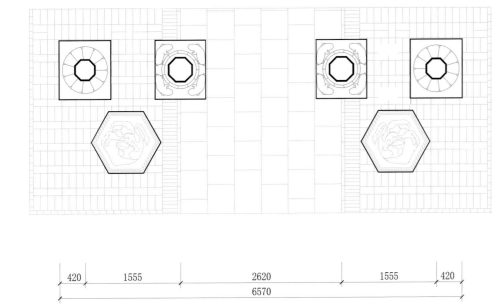

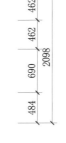

关王故里坊平面图
Plan of the Guan's Hometown Memorial Stone Archway

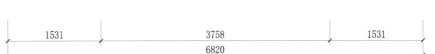

关王故里坊屋顶俯视图
Roof plan of the Guan's Hometown Memorial Stone Archway

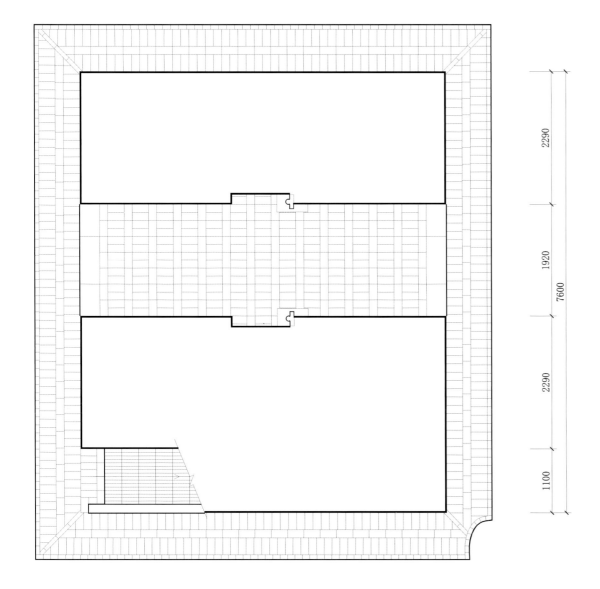

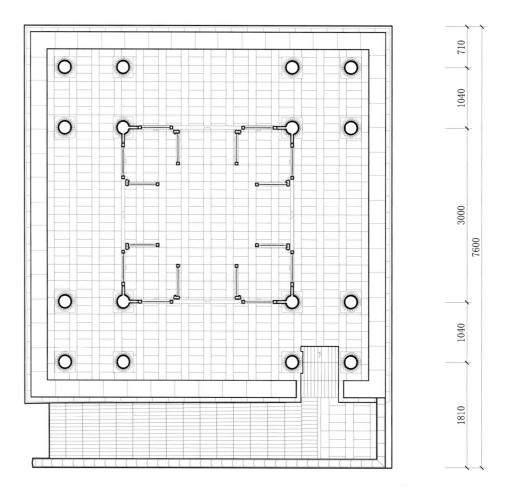

钟楼一层平面图
First Floor plan of the Bell Tower

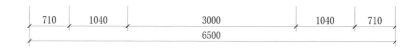

钟楼二层平面图
Second Floor plan of the Bell Tower

N

0   0.5   1m

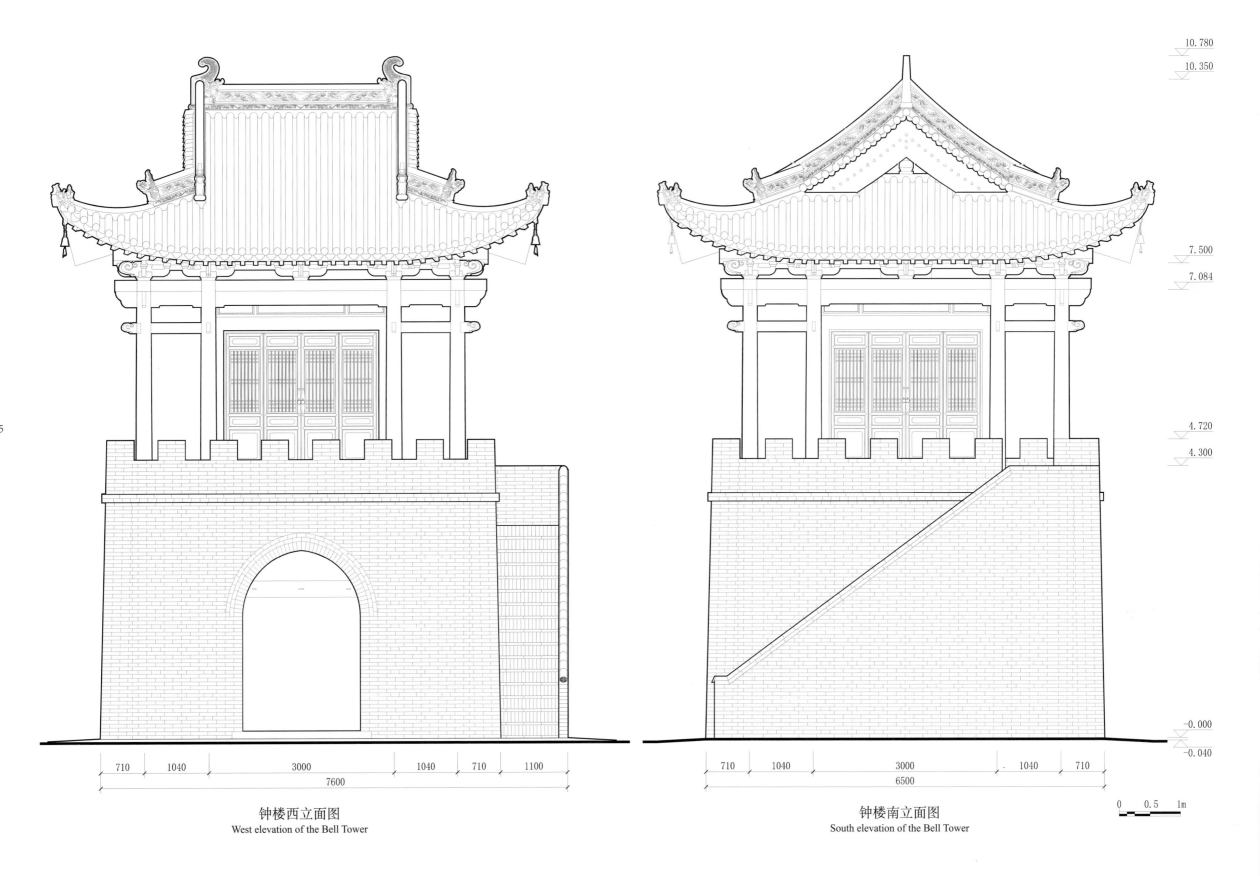

235

10.780
10.350

7.500
7.084

4.720
4.300

-0.000
-0.040

710 | 1040 | 3000 | 1040 | 710 | 1100
7600

710 | 1040 | 3000 | 1040 | 710
6500

钟楼西立面图
West elevation of the Bell Tower

钟楼南立面图
South elevation of the Bell Tower

0   0.5   1m

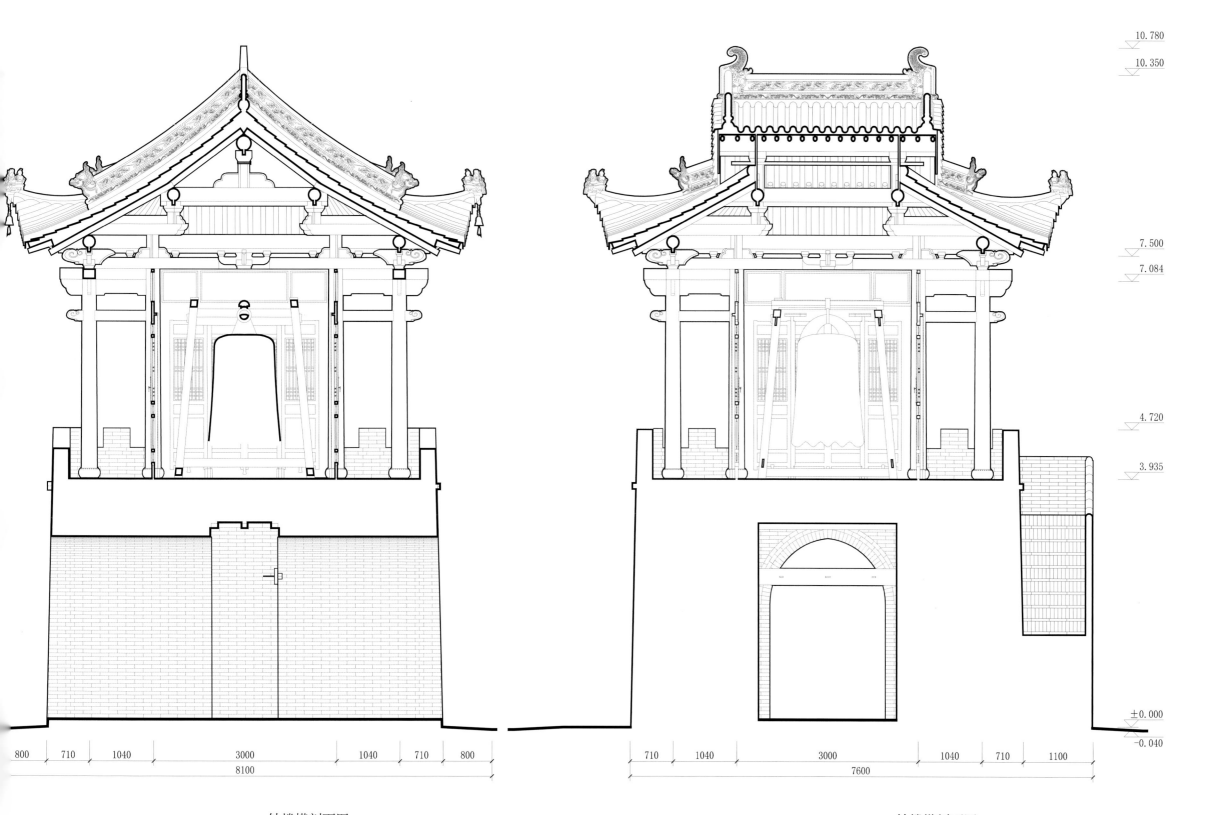

10.780

10.350

7.500

7.084

4.720

3.935

±0.000

-0.040

| 800 | 710 | 1040 | 3000 | 1040 | 710 | 800 |

8100

| 710 | 1040 | 3000 | 1040 | 710 | 1100 |

7600

钟楼横剖面图
Cross-section of the Bell Tower

钟楼纵剖面图
Longitudinal section of the Bell Tower

0   0.5   1m

中国古建筑测绘大系·祠庙建筑——解州关帝庙

237

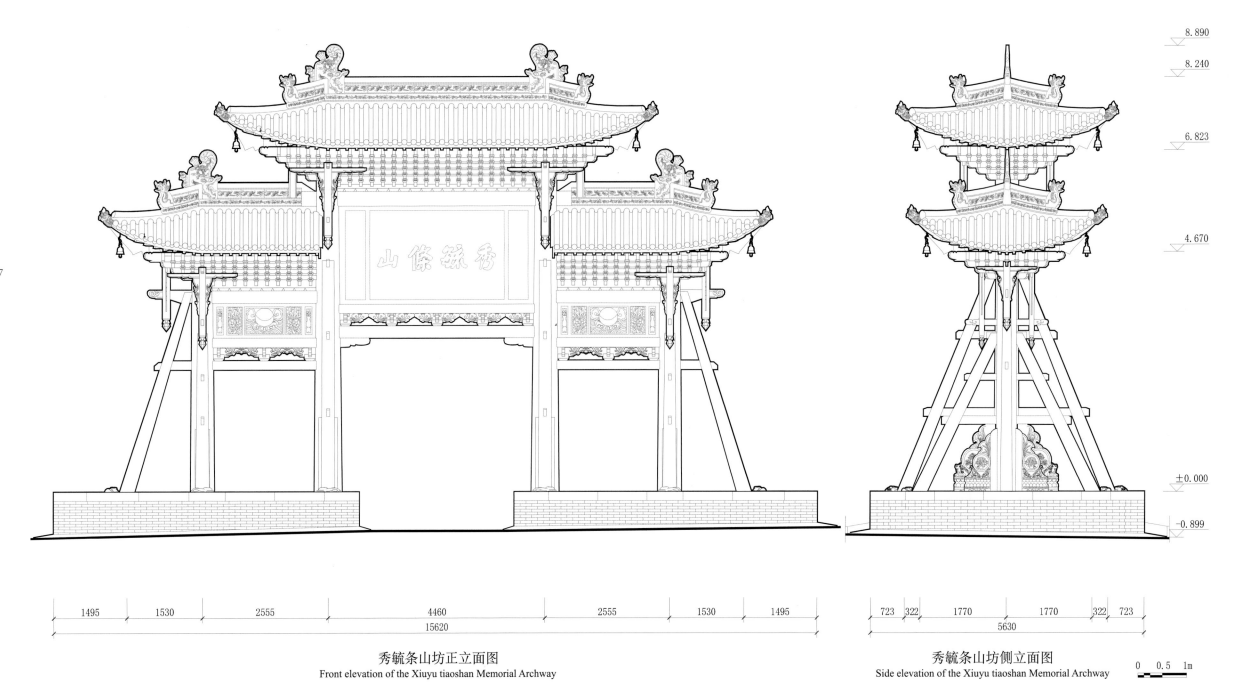

8.890

8.240

6.823

4.670

±0.000

-0.899

1495　1530　2555　4460　2555　1530　1495

15620

723　322　1770　1770　322　723

5630

秀毓条山坊正立面图
Front elevation of the Xiuyu tiaoshan Memorial Archway

秀毓条山坊侧立面图
Side elevation of the Xiuyu tiaoshan Memorial Archway

0　0.5　1m

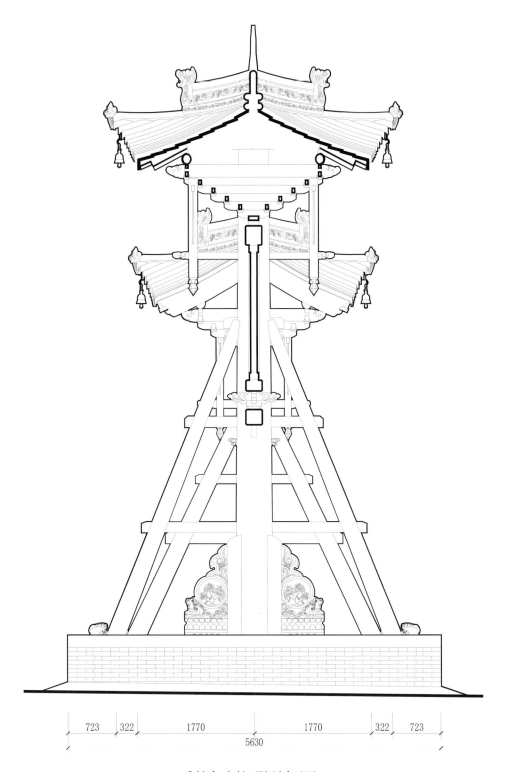

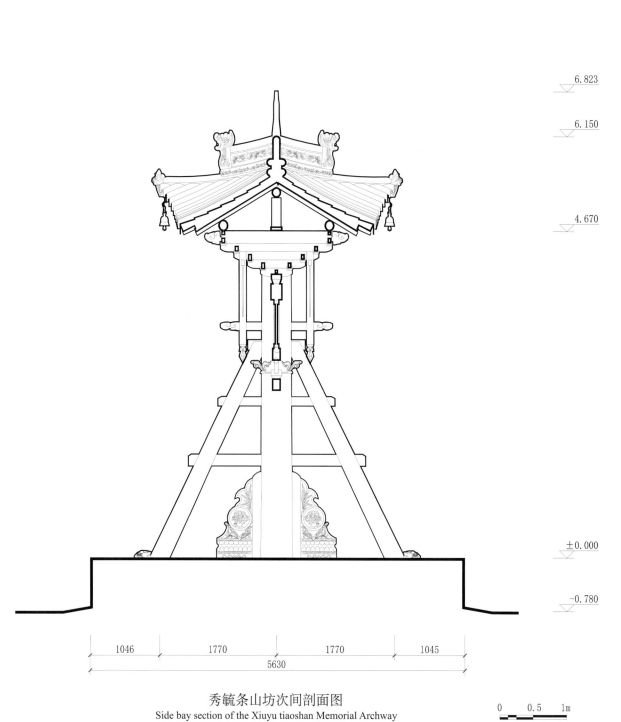

8.890

8.240

6.823

6.150

4.670

±0.000

-0.780

723 322 1770 1770 322 723

5630

1046 1770 1770 1045

5630

秀毓条山坊明间剖面图
Central bay section of the Xiuyu tiaoshan Memorial Archway

秀毓条山坊次间剖面图
Side bay section of the Xiuyu tiaoshan Memorial Archway

0   0.5   1m

灵钟醯海坊
The Lingzhong cuohai Memorial Archway

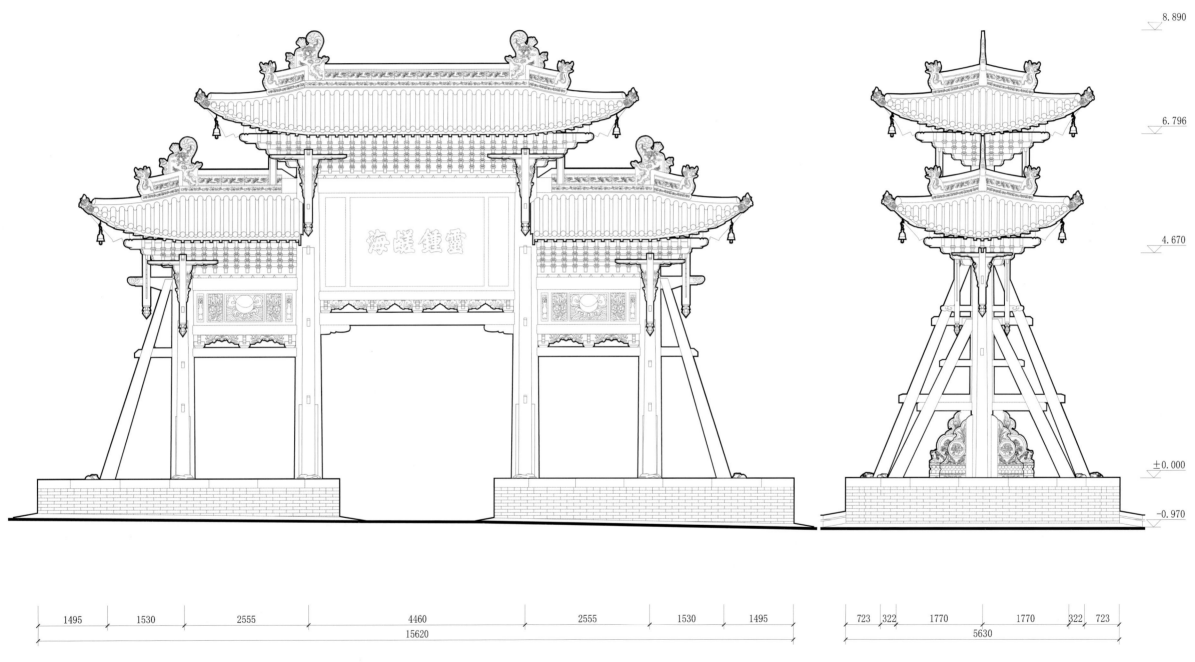

8.890

6.796

4.670

±0.000

-0.970

| 1495 | 1530 | 2555 | 4460 | 2555 | 1530 | 1495 |

15620

| 723 | 322 | 1770 | 1770 | 322 | 723 |

5630

灵钟醯海坊正立面图
Front elevation of the Lingzhong cuohai Memorial Archway

灵钟醯海坊侧立面图
Side elevation of the Lingzhong cuohai Memorial Archway

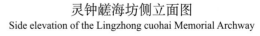

0   0.5   1m

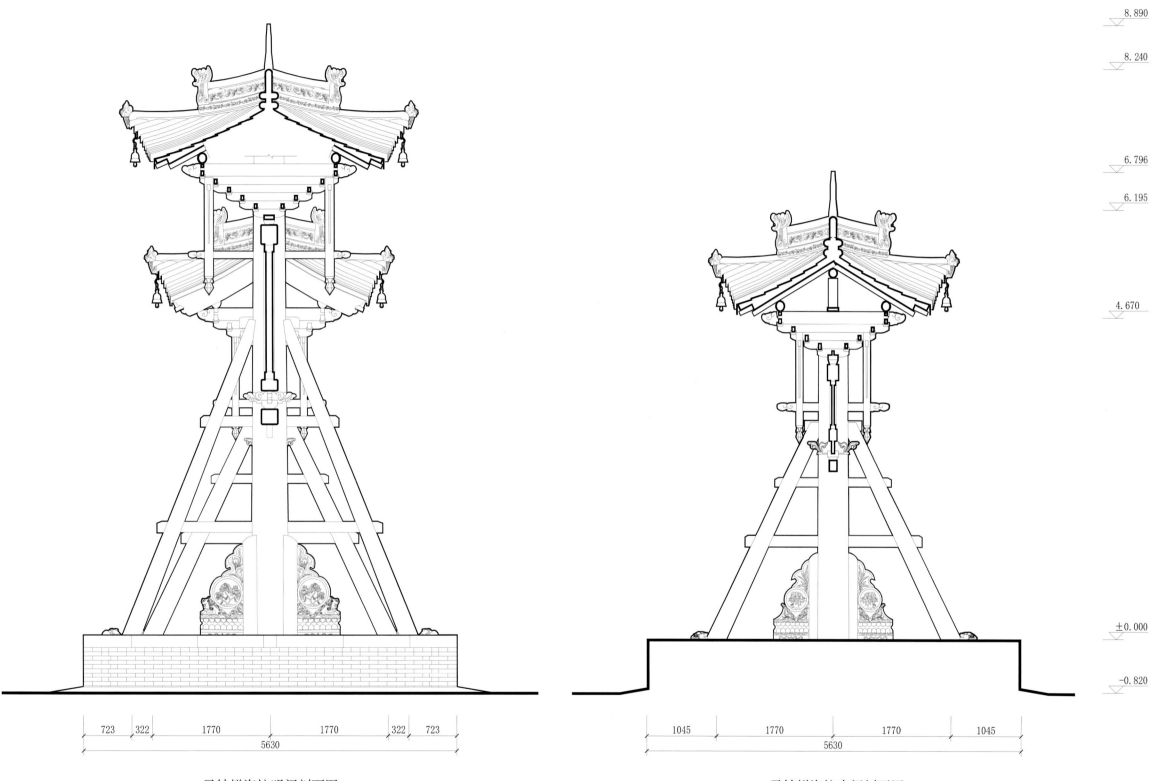

8.890

8.240

6.796

6.195

4.670

±0.000

-0.820

| 723 | 322 | 1770 | 1770 | 322 | 723 |

5630

| 1045 | 1770 | 1770 | 1045 |

5630

灵钟鹾海坊明间剖面图
Central bay section of the Lingzhong cuohai Memorial Archway

灵钟鹾海坊次间剖面图
Side bay section of the Lingzhong cuohai Memorial Archway

0   0.5   1m

山门
The Mountain Gate

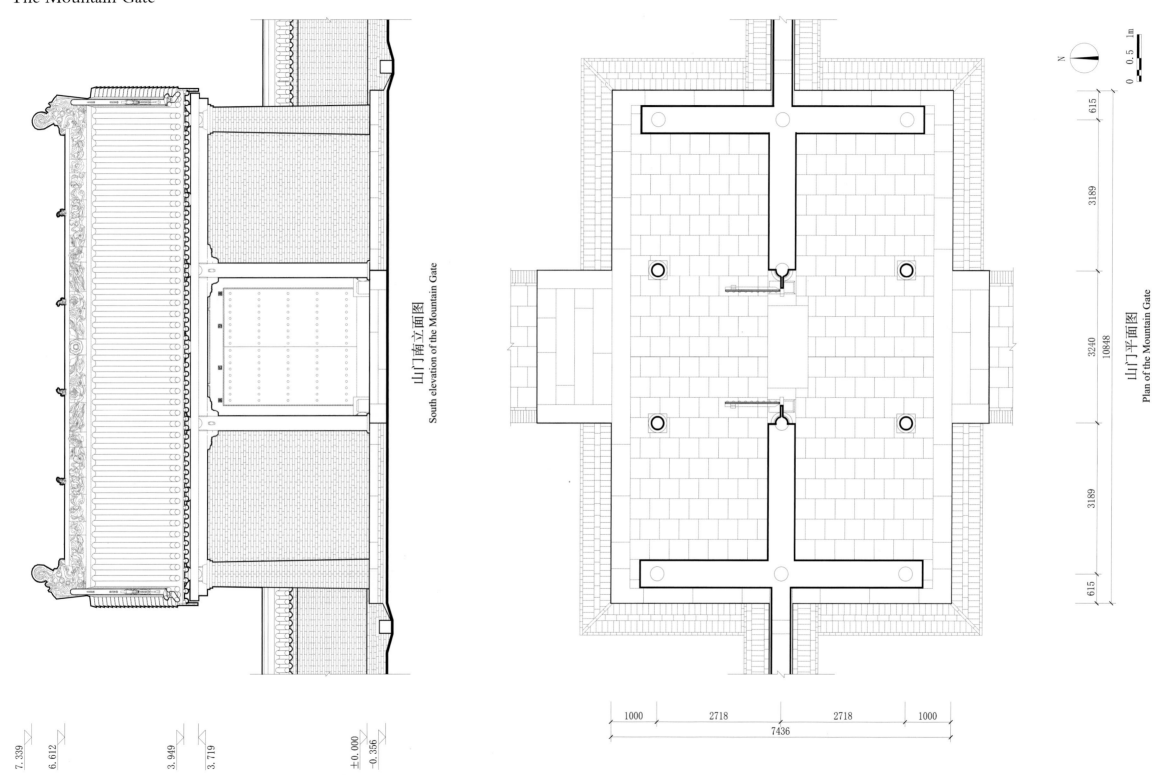

山门南立面图
South elevation of the Mountain Gate

山门平面图
Plan of the Mountain Gate

7.339

6.612

3.949

3.719

±0.000

-0.356

N

0 0.5 1m

615

3189

3240

10848

3189

615

1000   2718   2718   1000

7436

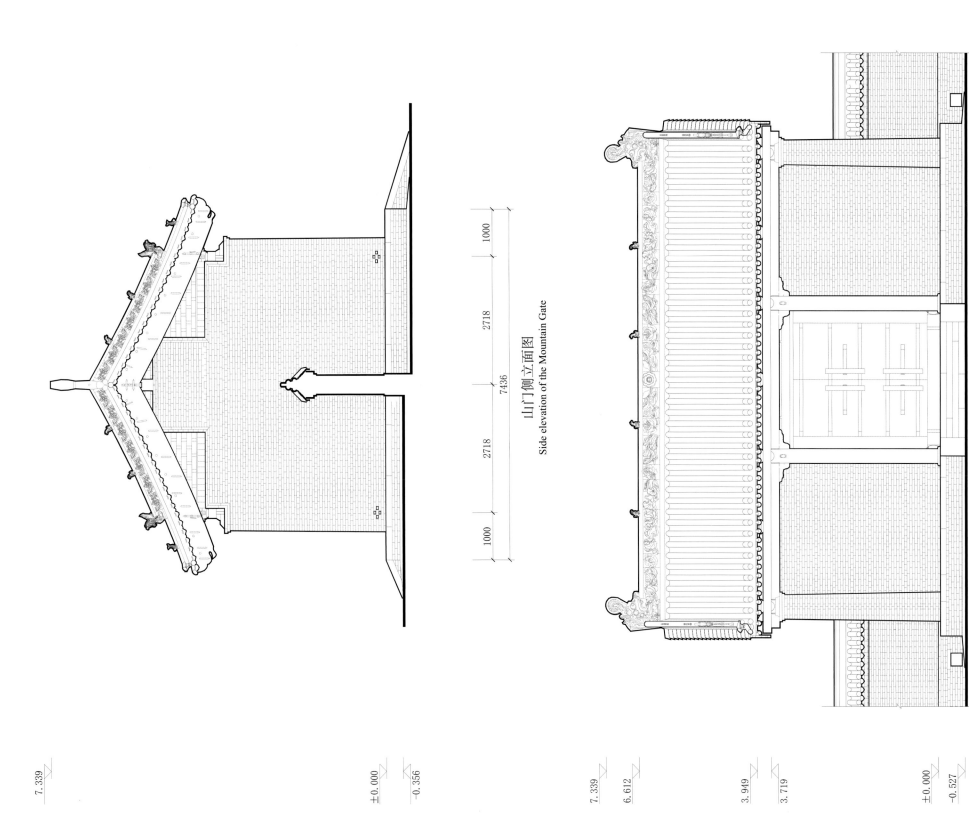

7.339

±0.000
-0.356

山门侧立面图
Side elevation of the Mountain Gate

1000

2718

7436

2718

1000

7.339

6.612

3.949

3.719

±0.000
-0.527

615

3189

3240

10848

3189

615

0 0.5 1m

山门北立面图
North elevation of the Mountain Gate

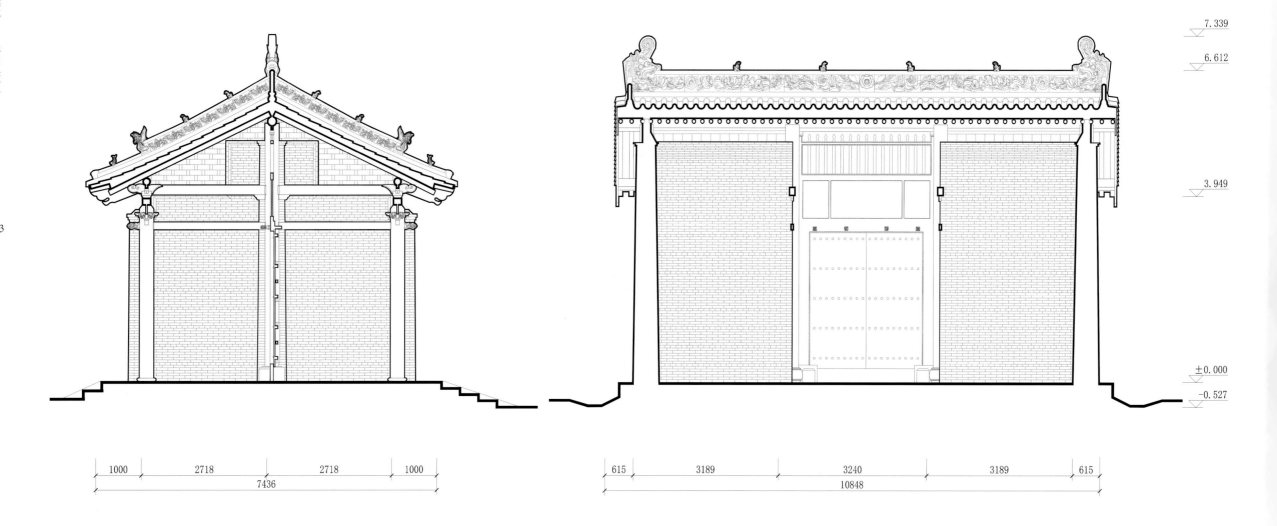

1000   2718   2718   1000
7436

山门明间剖面图
Central bay section of the Mountain Gate

7.339
6.612
3.949
±0.000
−0.527

615   3189   3240   3189   615
10848

山门纵剖面图
Longitudinal section of the Mountain Gate

0   0.5   1m

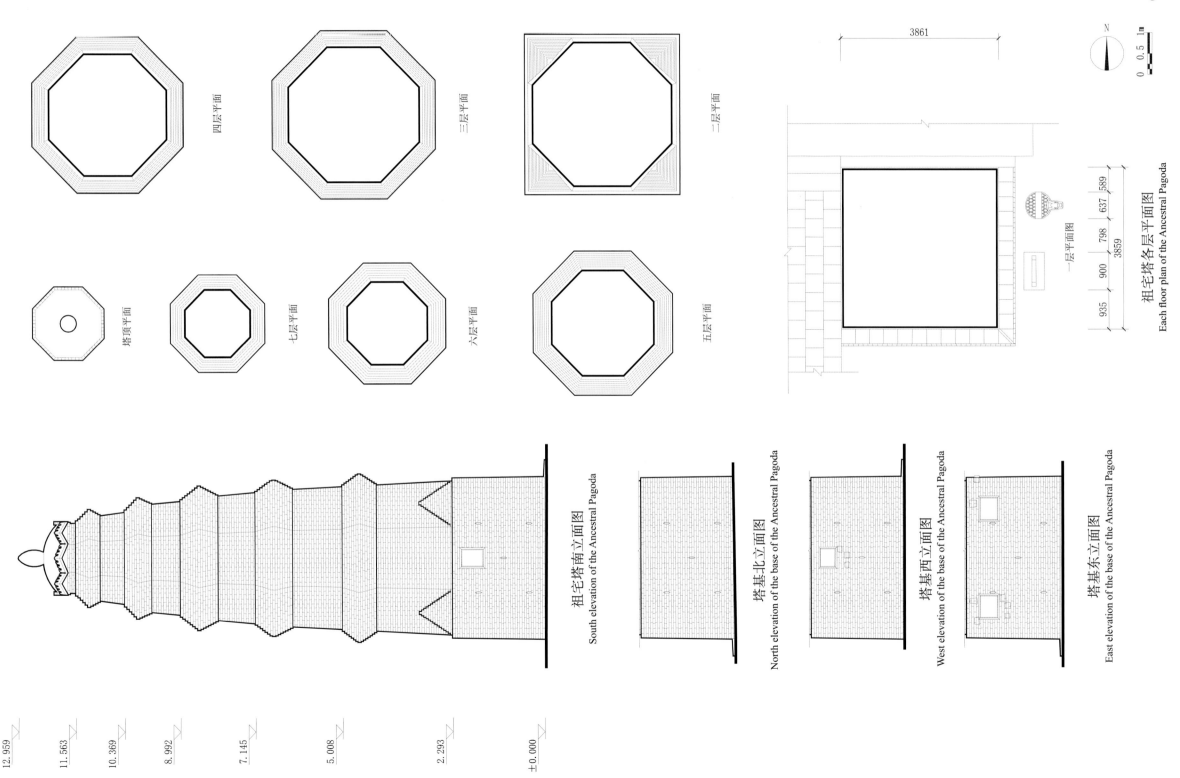

祖宅塔
The Ancestral Pagoda

四层平面

三层平面

二层平面

塔顶平面

七层平面

六层平面

五层平面

祖宅塔各层平面图
Each floor plan of the Ancestral Pagoda

一层平面图

3861

935 900 798 637 589
3859

祖宅塔南立面图
South elevation of the Ancestral Pagoda

塔基北立面图
North elevation of the base of the Ancestral Pagoda

塔基西立面图
West elevation of the base of the Ancestral Pagoda

塔基东立面图
East elevation of the base of the Ancestral Pagoda

12.959
11.563
10.369
8.992
7.145
5.008
2.293
±0.000

仪门
The Yi Gate

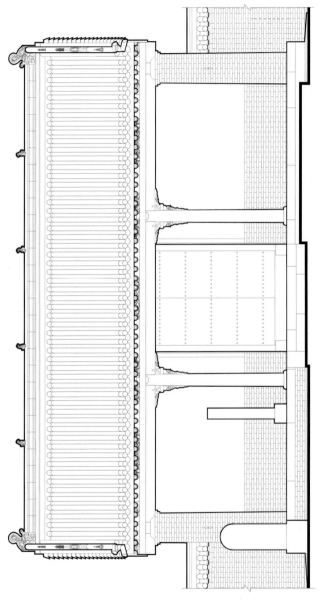

6.770

3.430

±0.000
-0.510

仪门正立面图
Front elevation of the Yi Gate

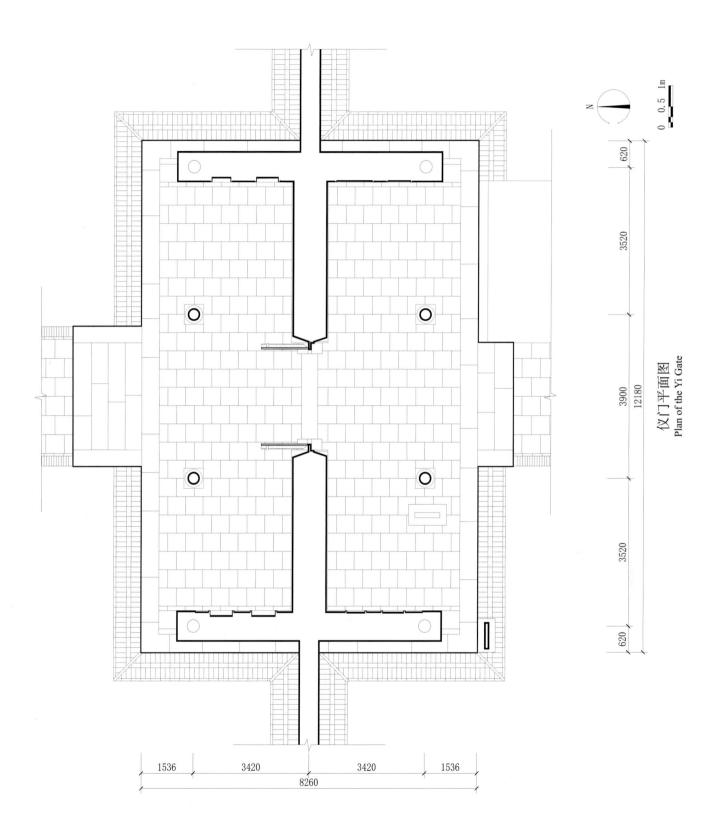

620

3520

3900
12180

3520

620

仪门平面图
Plan of the Yi Gate

1536　3420　3420　1536

8260

N

0　0.5　1m

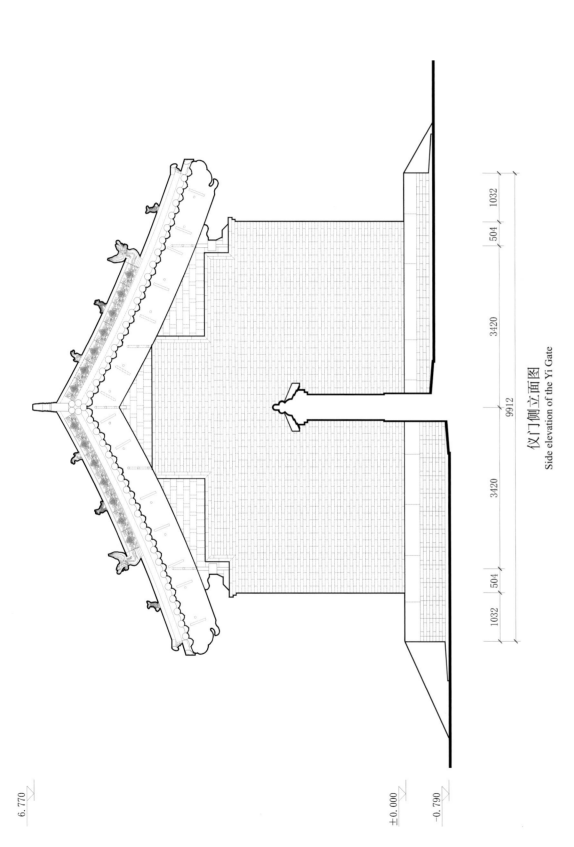

仪门侧立面图
Side elevation of the Yi Gate

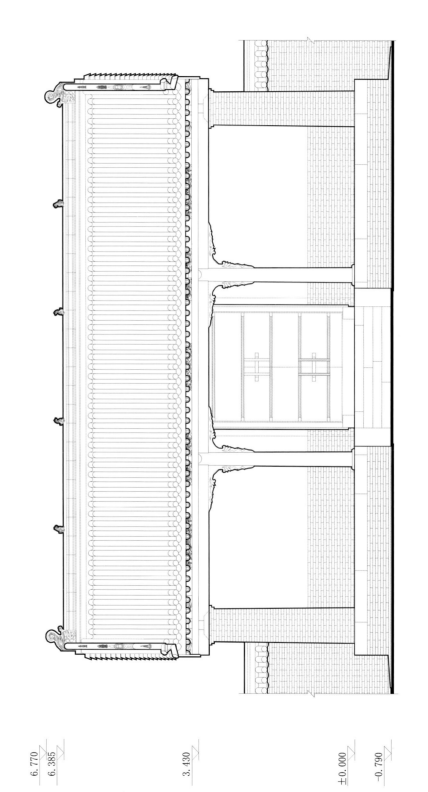

仪门背立面图
Back elevation of the Yi Gate

0 0.5 1m

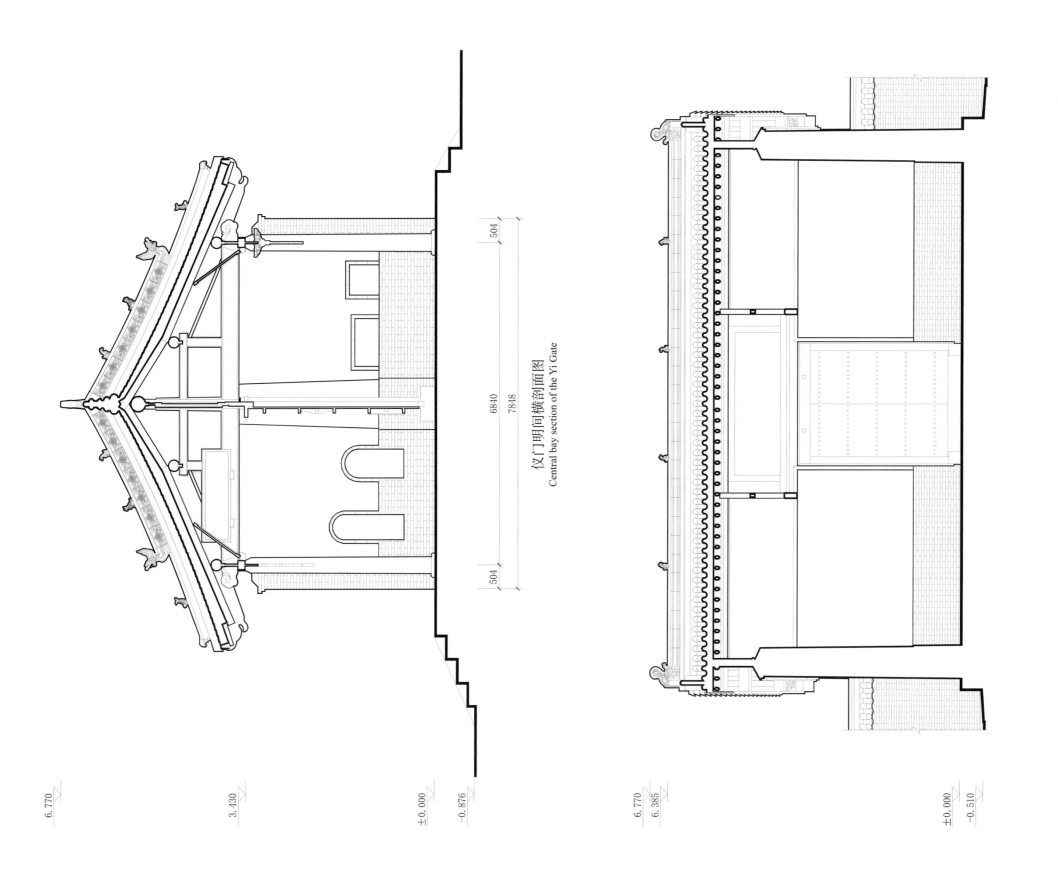

仪门明间横剖面图
Central bay section of the Yi Gate

6.770

3.430

±0.000

-0.876

504

6840

7848

504

仪门纵剖面图
Longitudinal section of the Yi Gate

620

3520

3900

12180

3520

620

6.770

6.385

±0.000

-0.510

0 0.5 1m

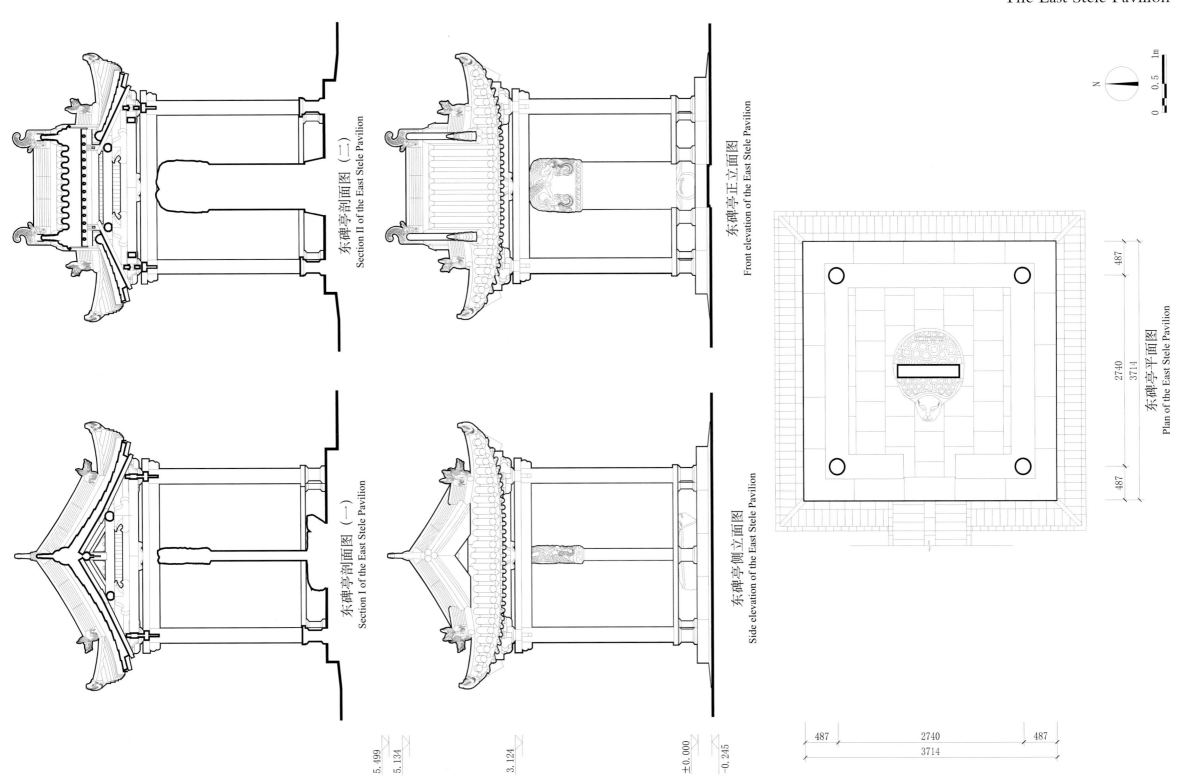

中国古建筑测绘大系·祠庙建筑——解州关帝庙

0 0.5 1m

N

东碑亭剖面图（二）
Section II of the East Stele Pavilion

东碑亭正立面图
Front elevation of the East Stele Pavilion

东碑亭剖面图（一）
Section I of the East Stele Pavilion

东碑亭侧立面图
Side elevation of the East Stele Pavilion

东碑亭平面图
Plan of the East Stele Pavilion

5.499
5.134

3.124

±0.000
−0.245

487
2740
3714
487

487
2740
487
3714

The West Stele Pavilion

249

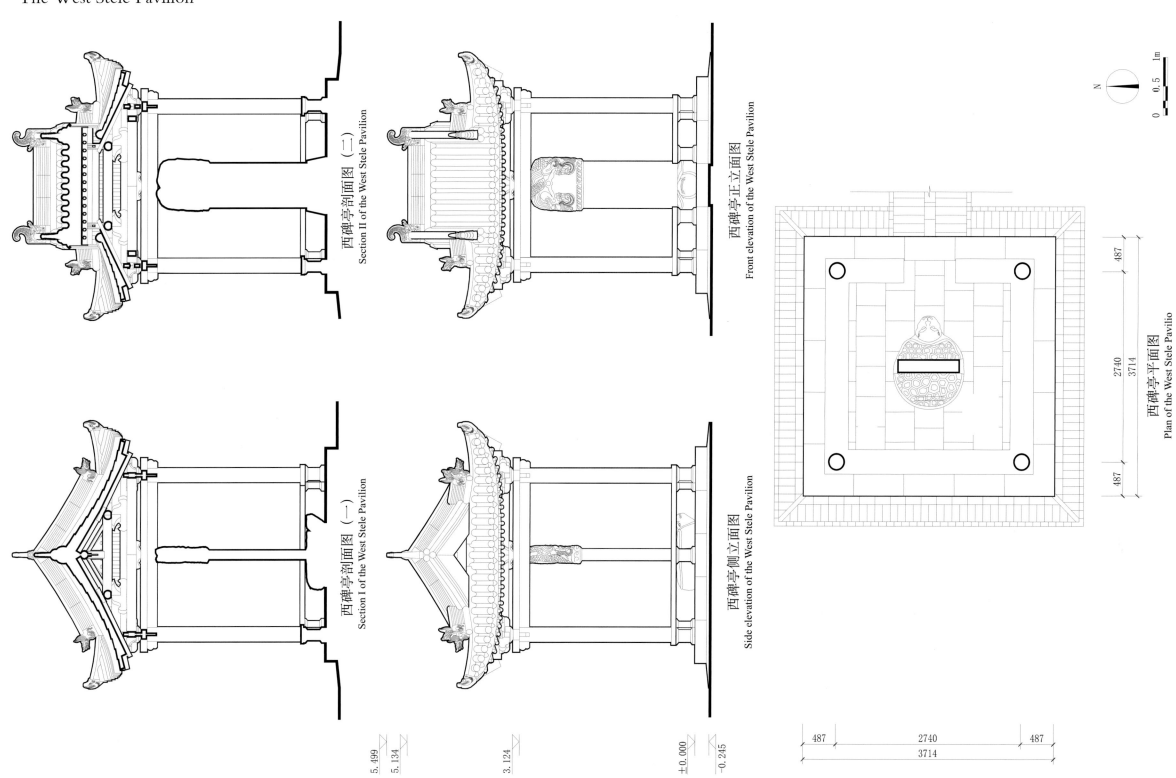

西碑亭剖面图（二）
Section II of the West Stele Pavilion

西碑亭正立面图
Front elevation of the West Stele Pavilion

西碑亭剖面图（一）
Section I of the West Stele Pavilion

西碑亭侧立面图
Side elevation of the West Stele Pavilion

西碑亭平面图
Plan of the West Stele Pavilio

5.499
5.134

3.124

±0.000
-0.245

487
2740
3714

487
2740
487
3714

0 0.5 1m

N

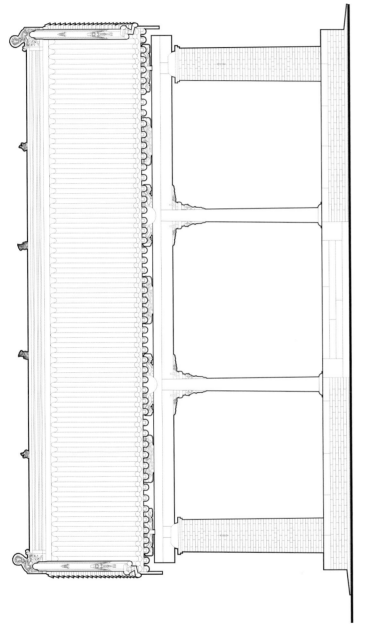

献殿立面图
Front elevation of the Xian Hall

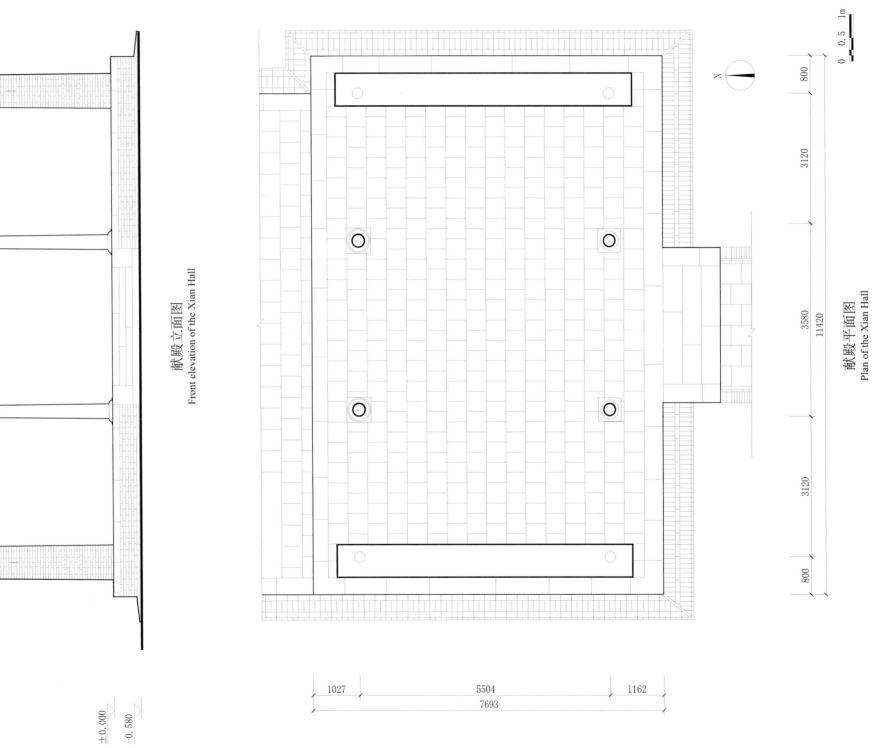

中国古建筑测绘大系 · 祠庙建筑 —— 解州关帝庙

献殿平面图
Plan of the Xian Hall

250

6.777
6.444
4.008
3.522
±0.000
-0.580

0　0.5　1m

N

800
3120
3580
11420
3120
800

1027
5504
1162
7693

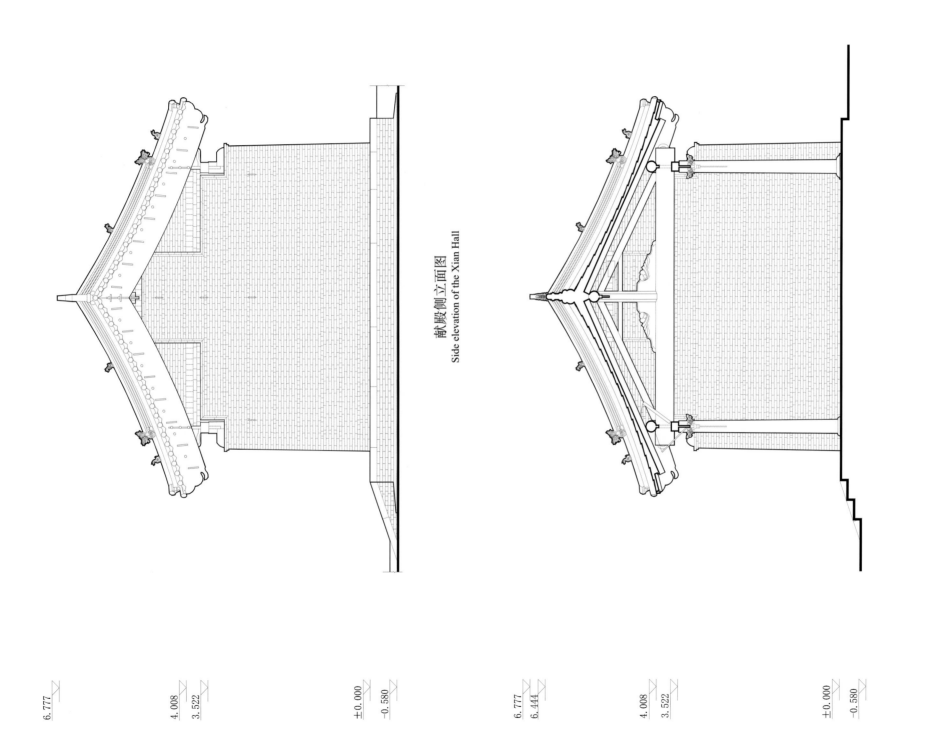

献殿侧立面图
Side elevation of the Xian Hall

献殿明间剖面图
Central bay section of the Xian Hall

6.777

4.008
3.522

±0.000
-0.580

6.777
6.444

4.008
3.522

±0.000
-0.580

1027

5504
7693

1162

0 0.5 1m

献殿次间剖面图
Side bay section of the Xian Hall

献殿纵剖面图
Longitudinal section of the Xian Hall

0 0.5   1m

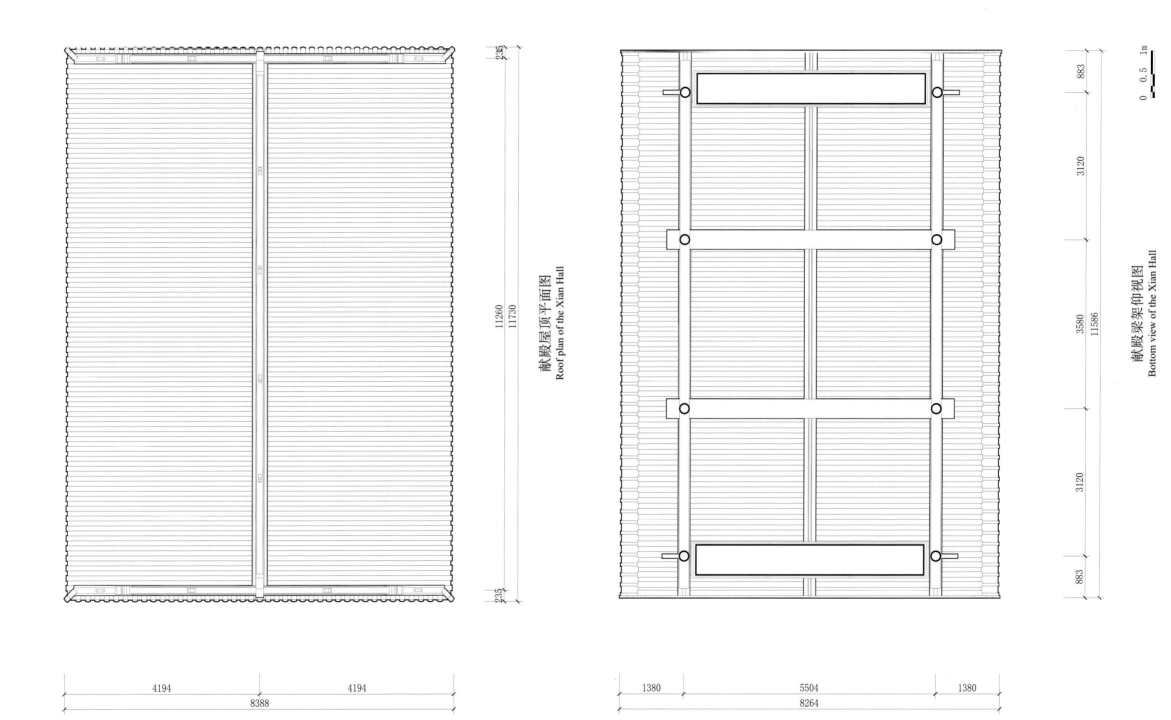

献殿屋顶平面图
Roof plan of the Xian Hall

献殿梁架仰视图
Bottom view of the Xian Hall

0 0.5 1m

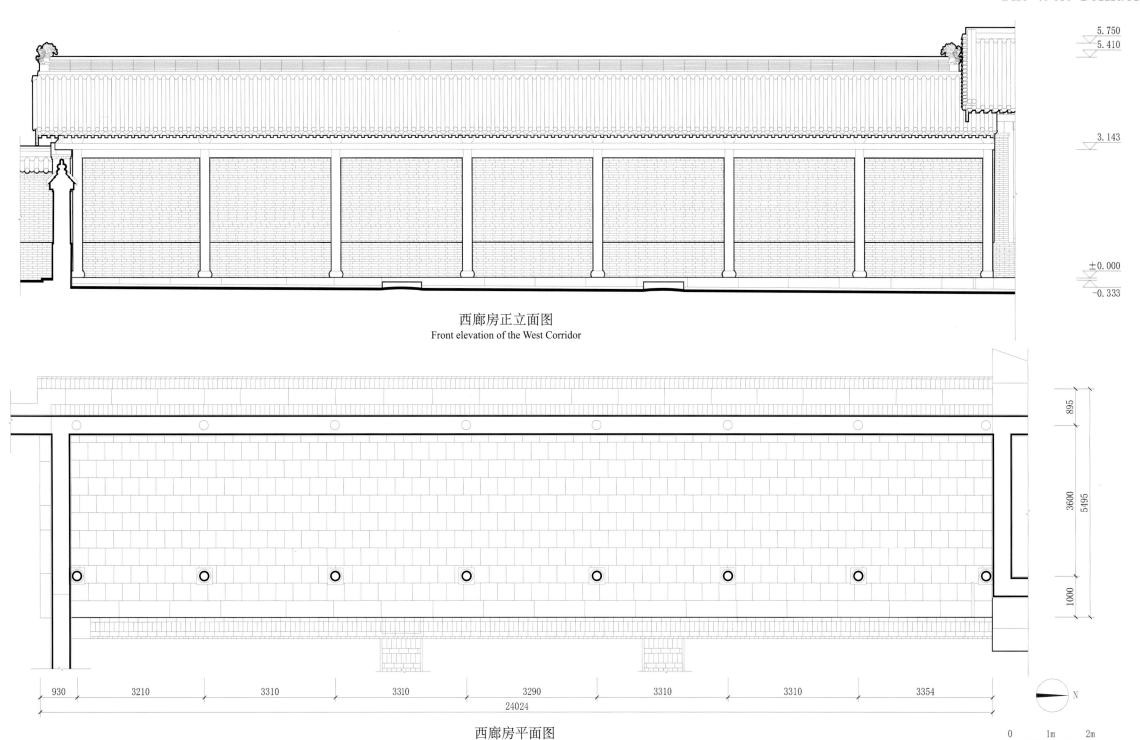

5. 750
5. 410

3. 143

±0. 000

-0. 333

西廊房正立面图
Front elevation of the West Corridor

895

3600

5495

1000

930   3210   3310   3310   3290   3310   3310   3354

24024

西廊房平面图
Plan of the West Corridor

N

0   1m   2m

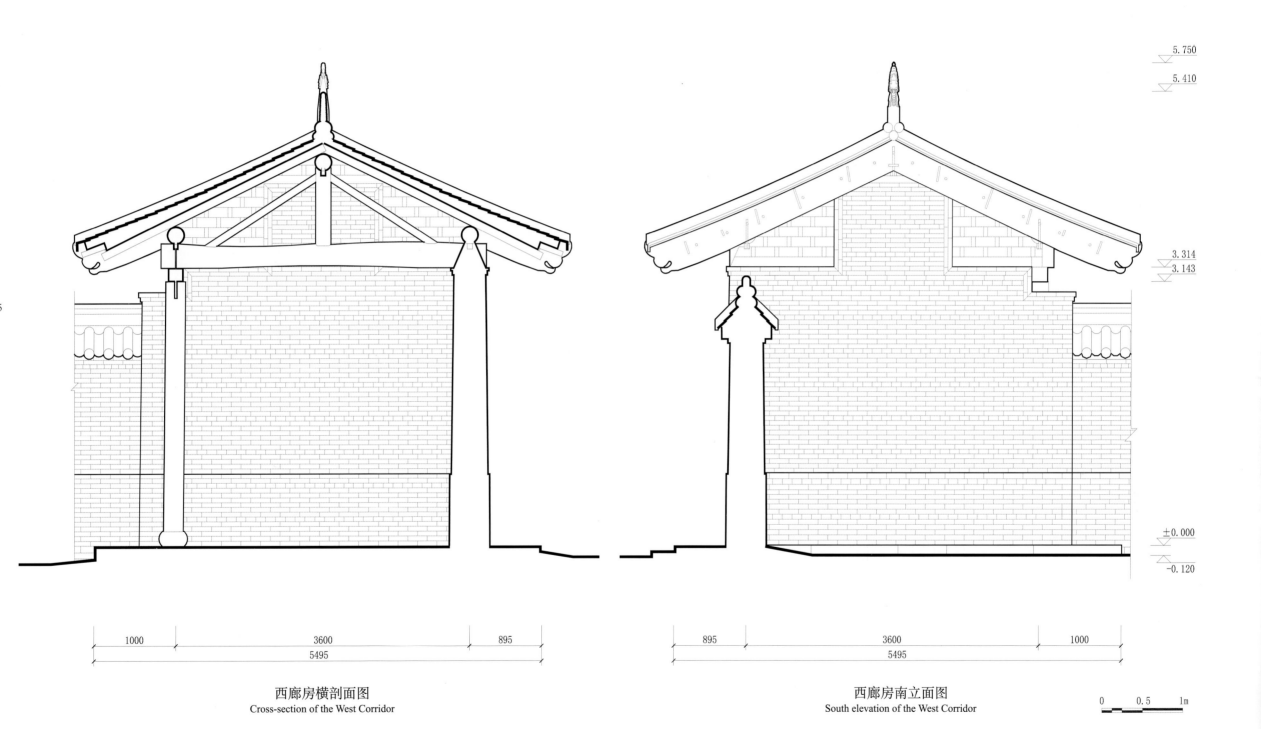

5.750

5.410

3.314
3.143

±0.000

-0.120

1000 3600 895
5495

895 3600 1000
5495

西廊房横剖面图
Cross-section of the West Corridor

西廊房南立面图
South elevation of the West Corridor

0   0.5   1m

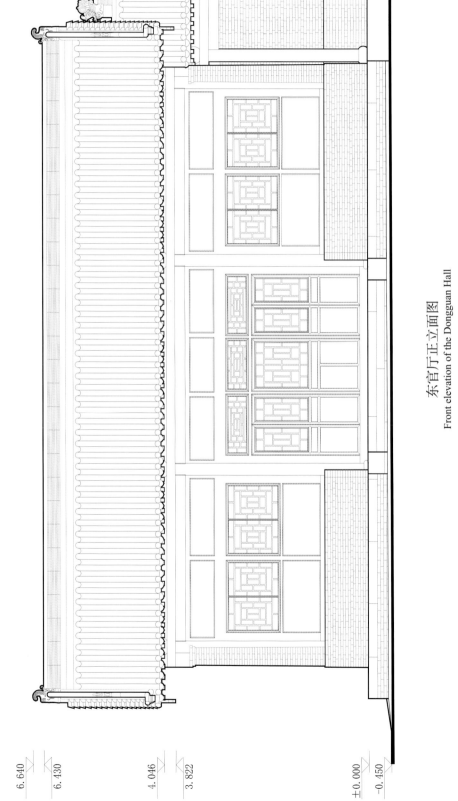

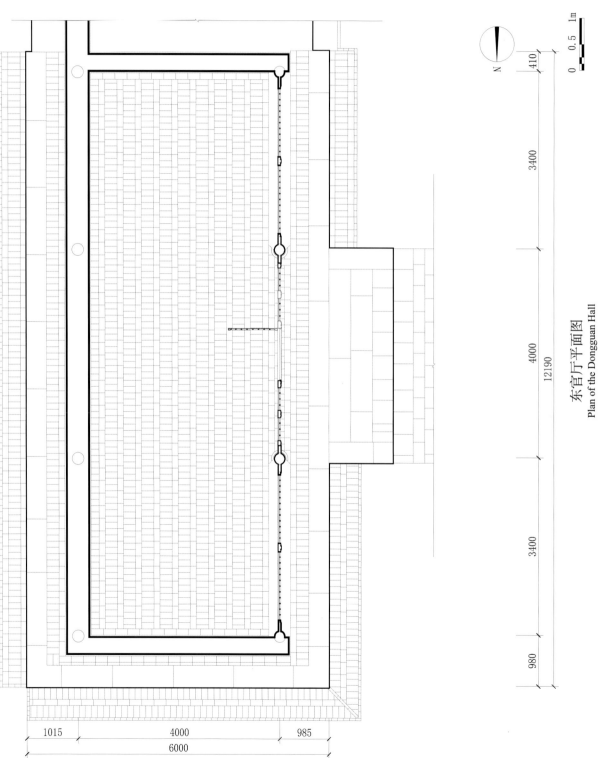

中国古建筑测绘大系·祠庙建筑——解州关帝庙

东官厅正立面图
Front elevation of the Dongguan Hall

东官厅平面图
Plan of the Dongguan Hall

西官厅
The Xiguan Hall

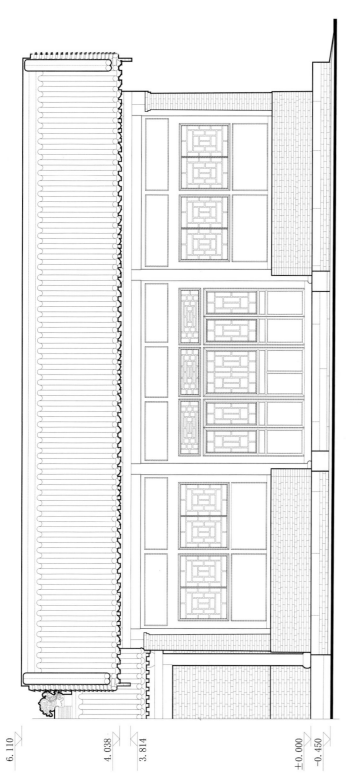

6.110
4.038
3.814
±0.000
-0.450

西官厅正立面图
Front elevation of the Xiguan Hall

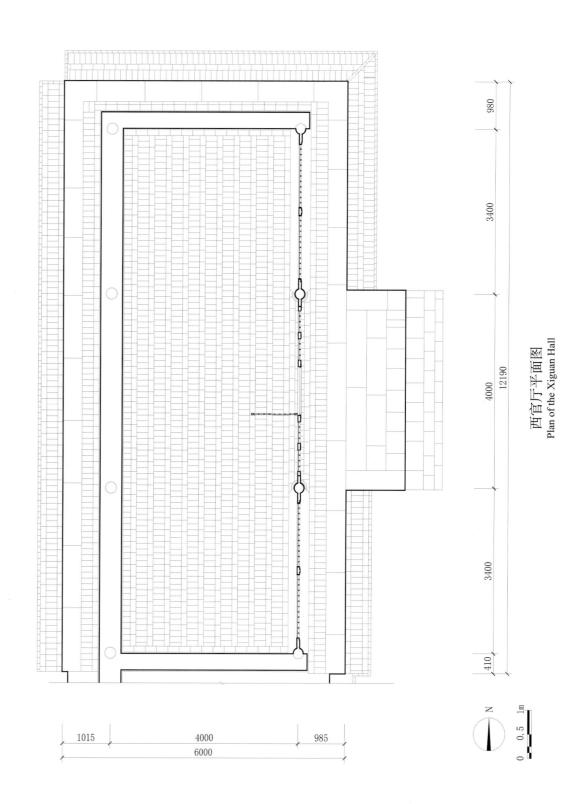

980
3400
4000
12190
3400
410

1015　4000　985
6000

N

0　0.5　1m

西官厅平面图
Plan of the Xiguan Hall

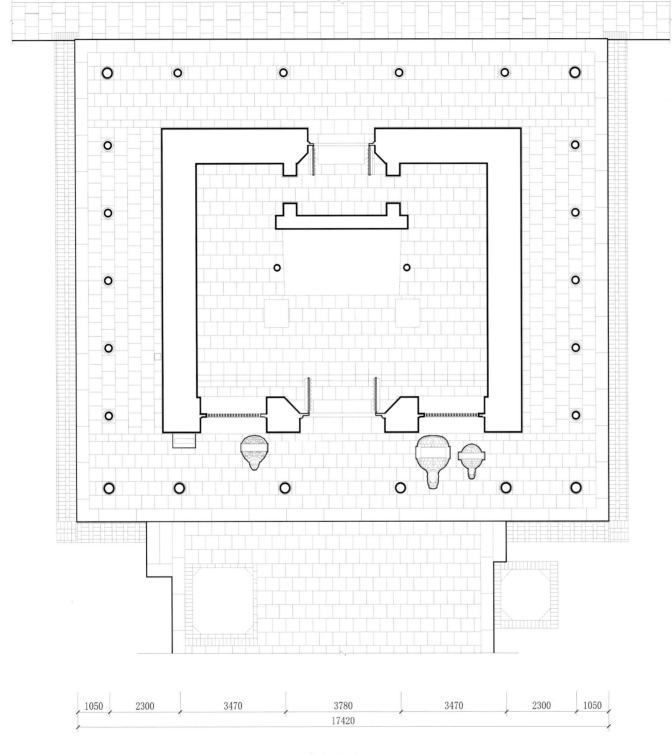

1050　2300　3470　3780　3470　2300　1050

17420

1050　2300　2150　2150　2150　2150　2300　1050

15300

N

0　0.5　1m

崇宁殿平面图
Plan of the Chongning Hall

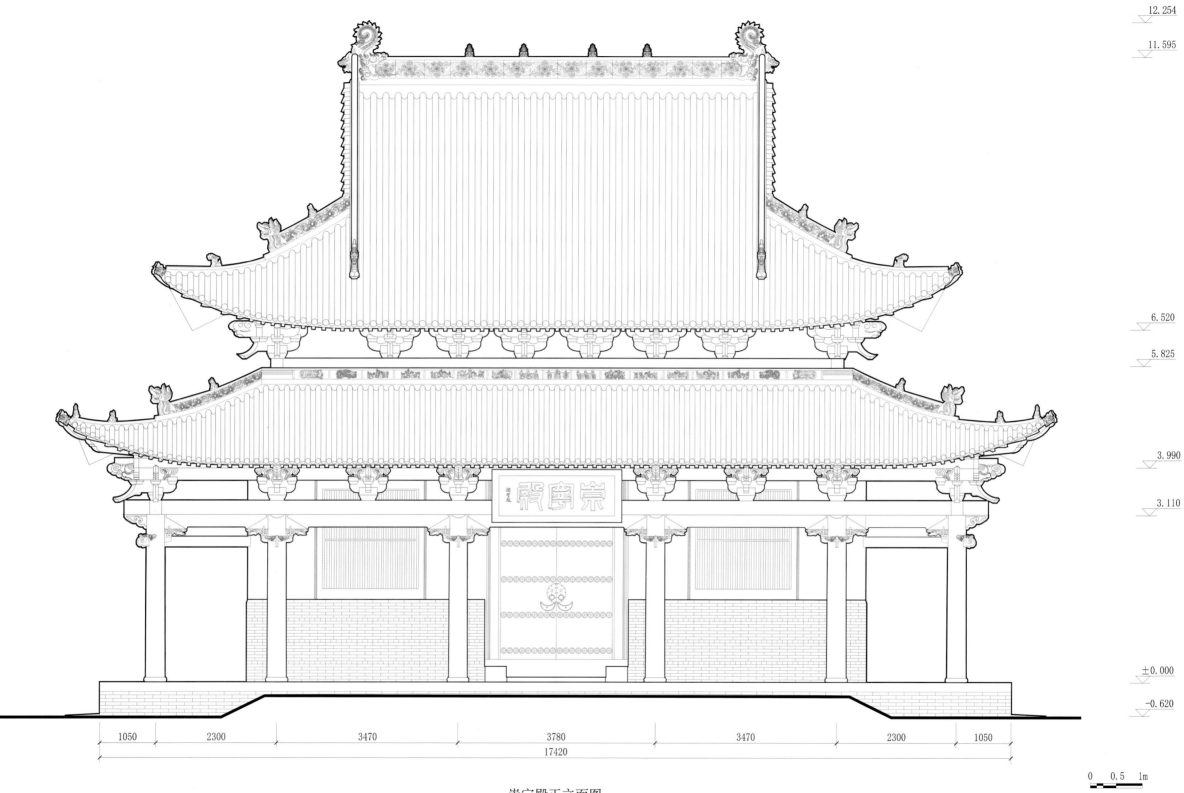

12.254

11.595

6.520

5.825

3.990

3.110

±0.000

-0.620

崇宁殿

1050   2300   3470   3780   3470   2300   1050

17420

0   0.5   1m

崇宁殿正立面图

Front elevation of the Chongning Hall

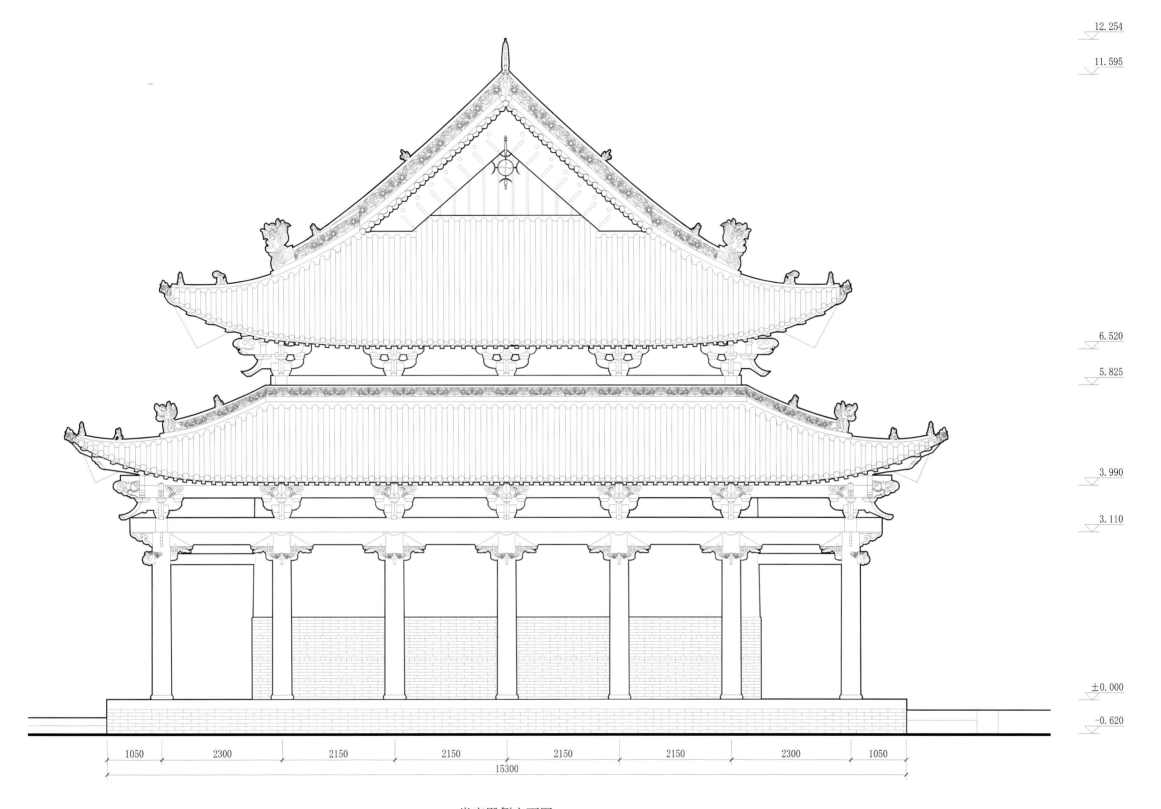

12.254

11.595

6.520

5.825

3.990

3.110

±0.000

-0.620

| 1050 | 2300 | 2150 | 2150 | 2150 | 2150 | 2300 | 1050 |

15300

崇宁殿侧立面图
Side elevation of the Chongning Hall

0 0.5 1m

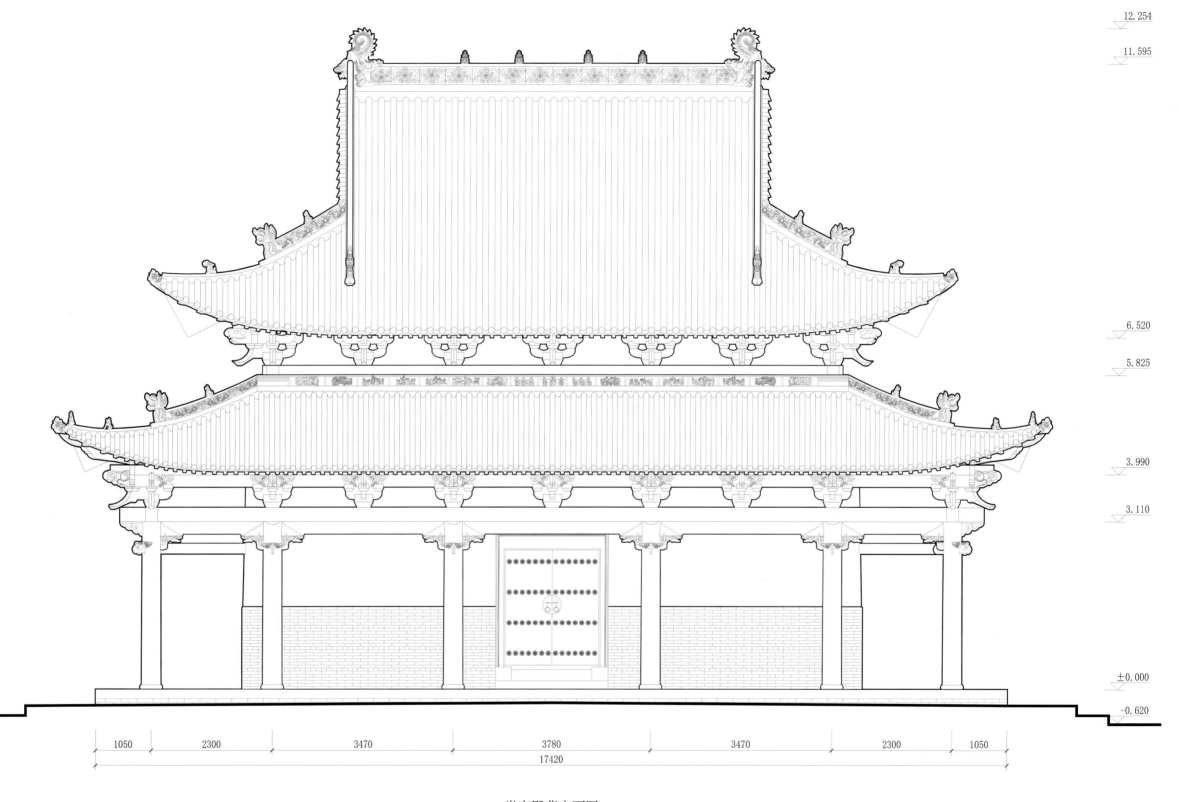

12.254

11.595

6.520

5.825

3.990

3.110

±0.000

-0.620

1050 2300 3470 3780 3470 2300 1050

17420

崇宁殿背立面图

Back elevation of the Chongning Hall

0 0.5 1m

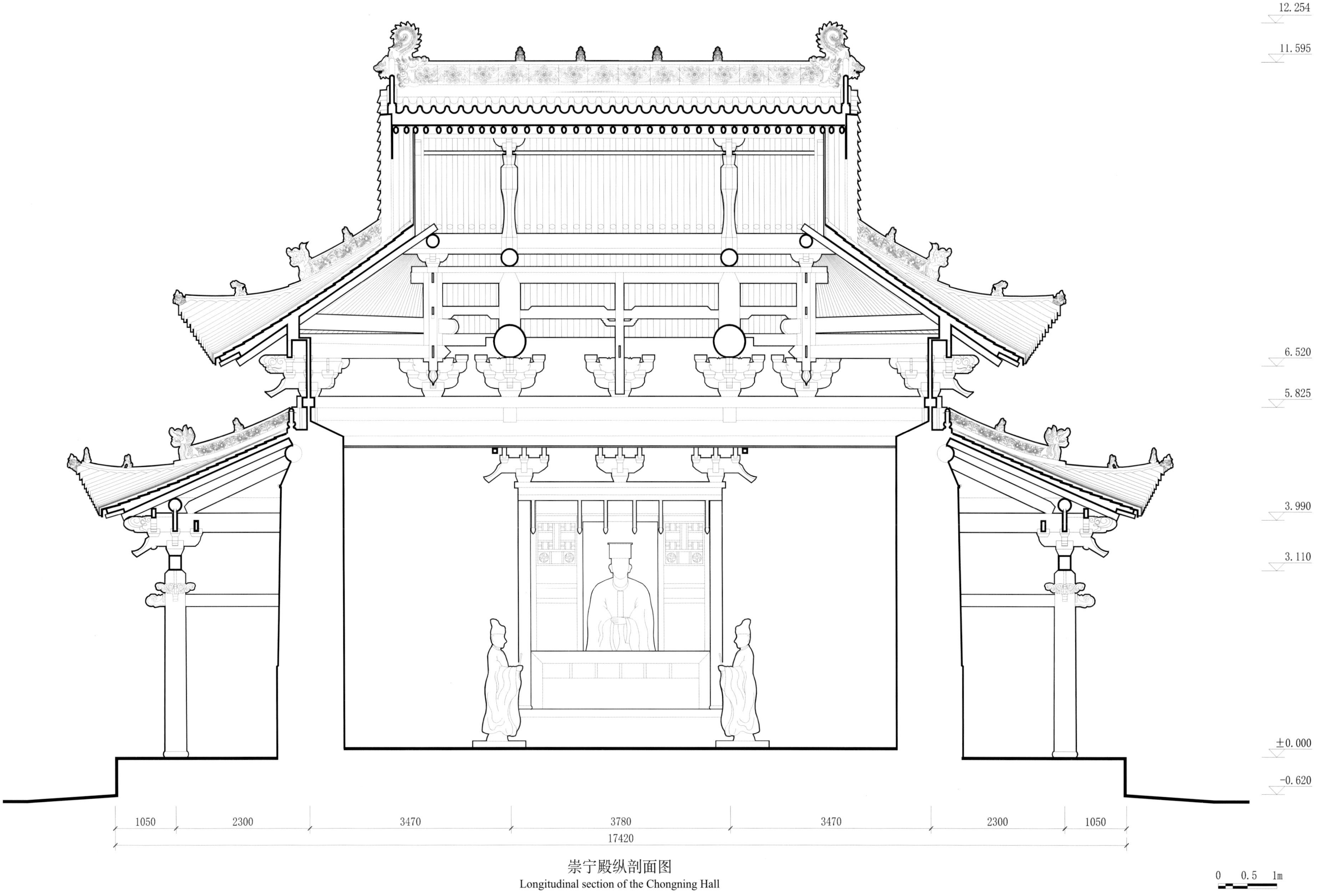

12.254
11.595
6.520
5.825
3.990
3.110
±0.000
-0.620

1050　2300　3470　3780　3470　2300　1050
17420

崇宁殿纵剖面图
Longitudinal section of the Chongning Hall

0　0.5　1m

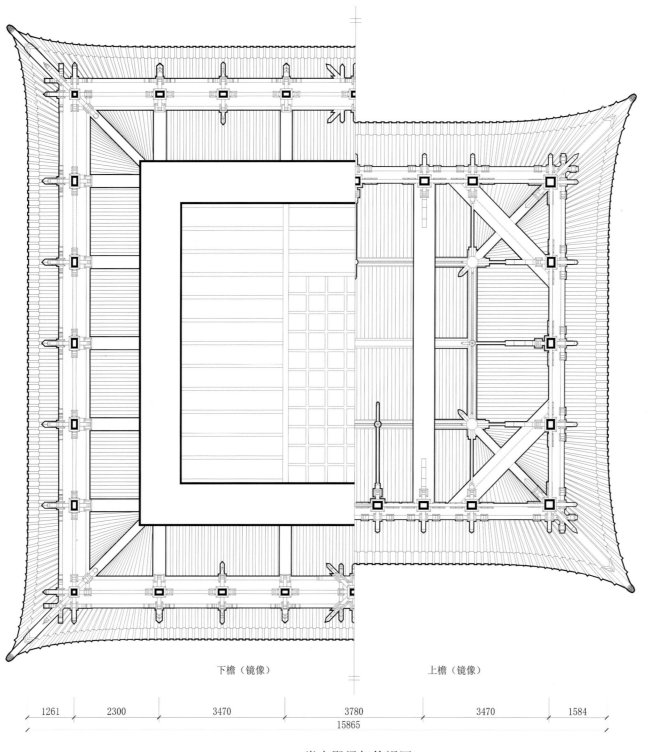

下檐（镜像）　　　　　上檐（镜像）

崇宁殿梁架仰视图
Bottom view of the lower and upper eaves and the beam frame of the Chongning Hall

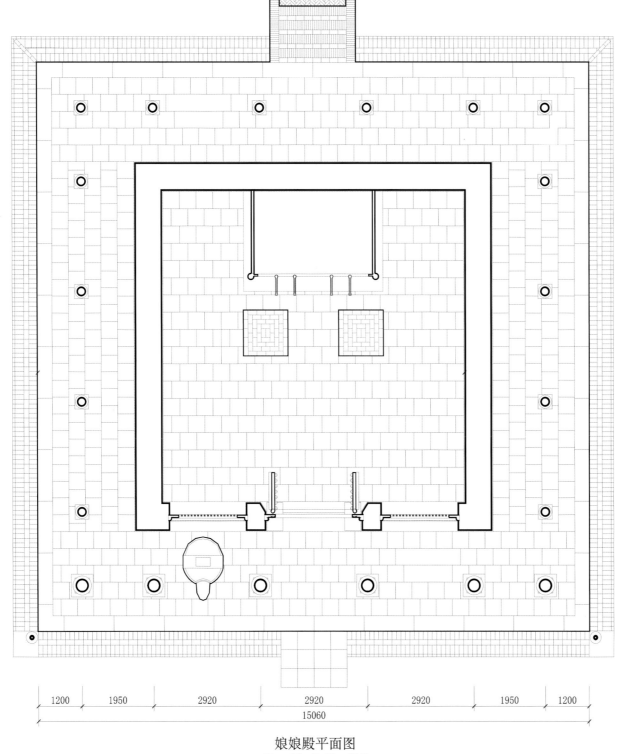

娘娘殿平面图
Plan of the Goddess Hall

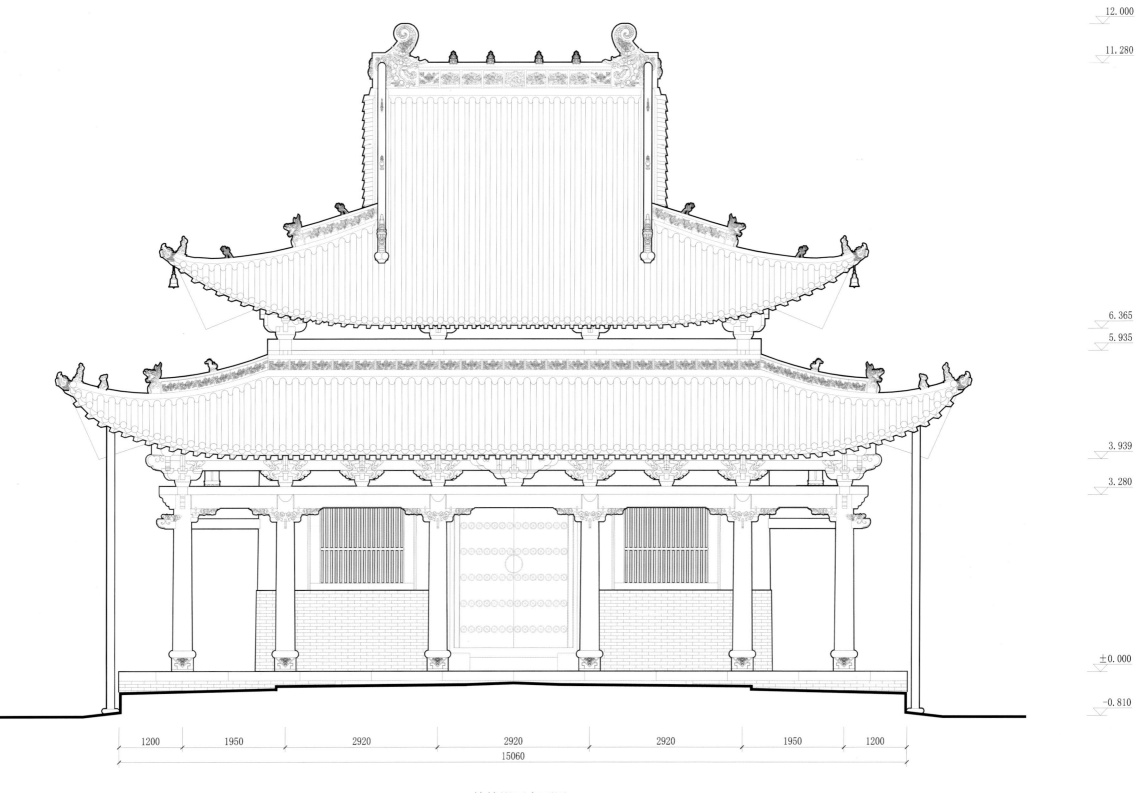

12.000

11.280

6.365

5.935

3.939

3.280

±0.000

-0.810

| 1200 | 1950 | 2920 | 2920 | 2920 | 1950 | 1200 |

15060

娘娘殿正立面图

Front elevation of the Goddess Hall

0   0.5   1m

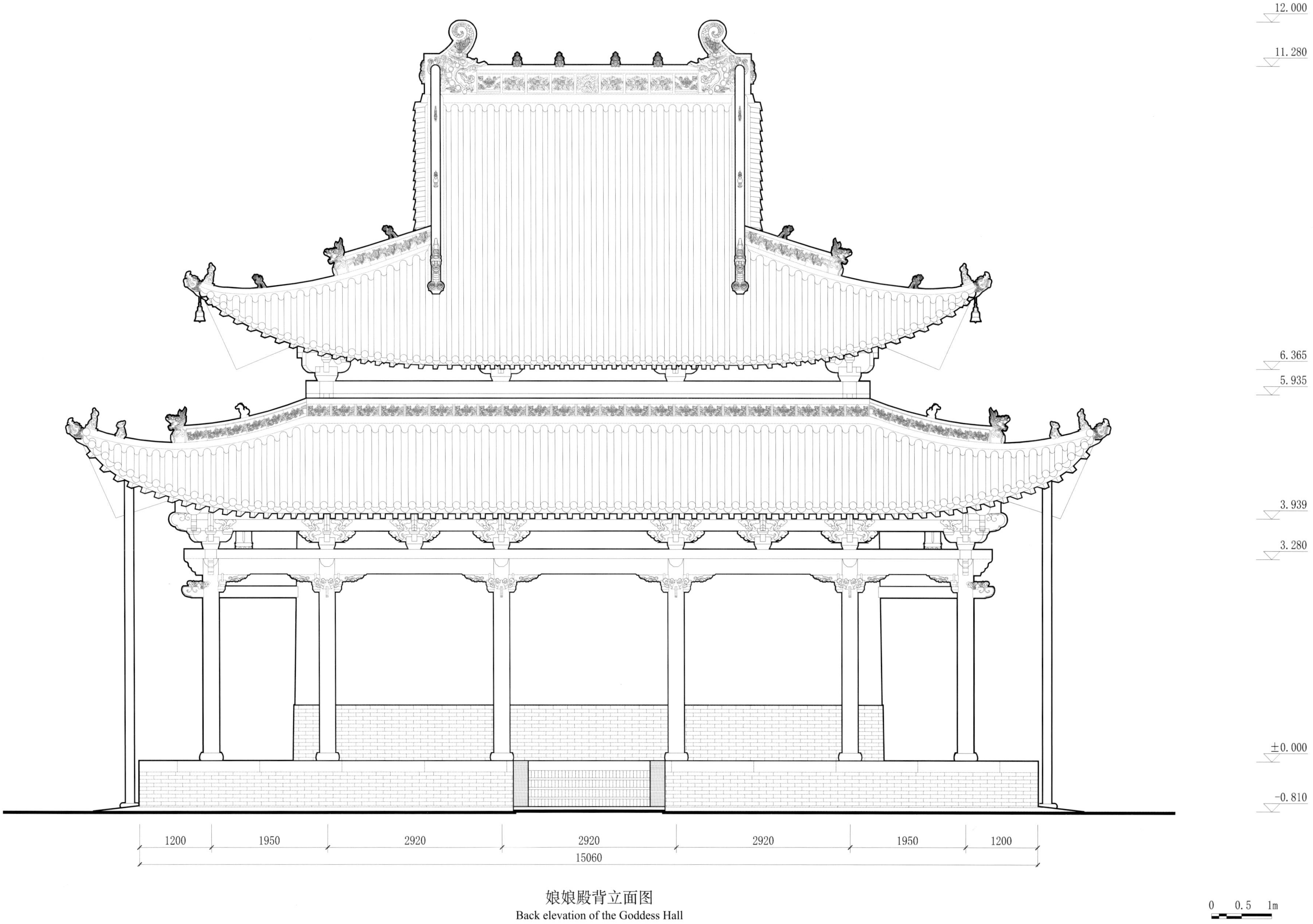

12. 000
11. 280
6. 365
5. 935
3. 939
3. 280
±0. 000
-0. 810

1200   1950   2920   2920   2920   1950   1200
15060

娘娘殿背立面图
Back elevation of the Goddess Hall

0   0.5   1m

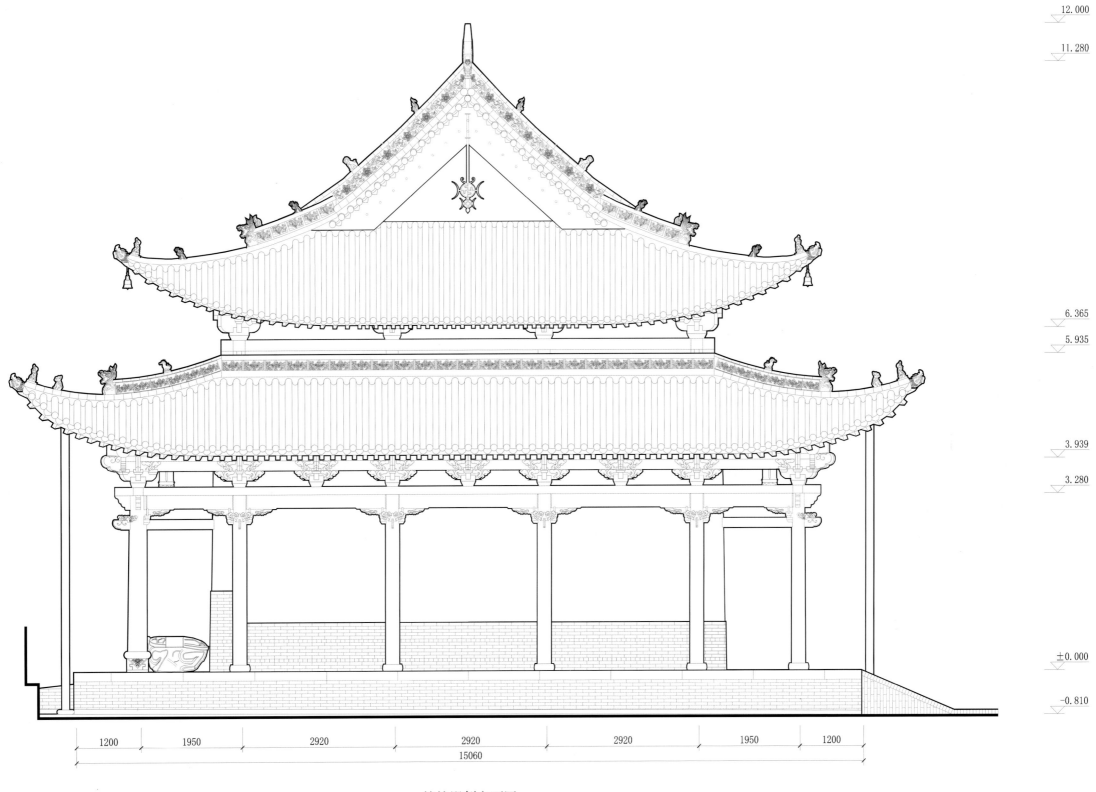

12.000

11.280

6.365

5.935

3.939

3.280

±0.000

-0.810

1200 1950 2920 2920 2920 1950 1200

15060

娘娘殿侧立面图
Side elevation of the Goddess Hall

0 0.5 1m

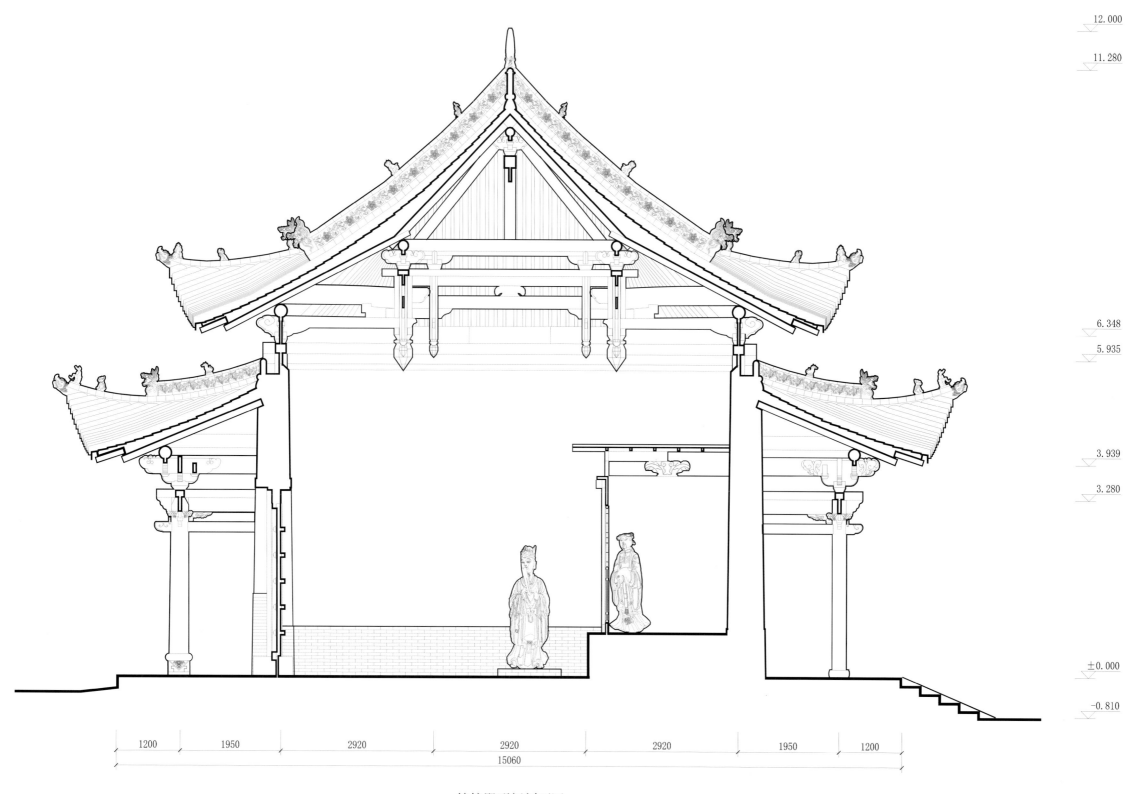

12.000

11.280

6.348

5.935

3.939

3.280

±0.000

-0.810

1200　1950　2920　2920　2920　1950　1200

15060

娘娘殿明间剖面图
Central bay section of the Goddess Hall

0　0.5　1m

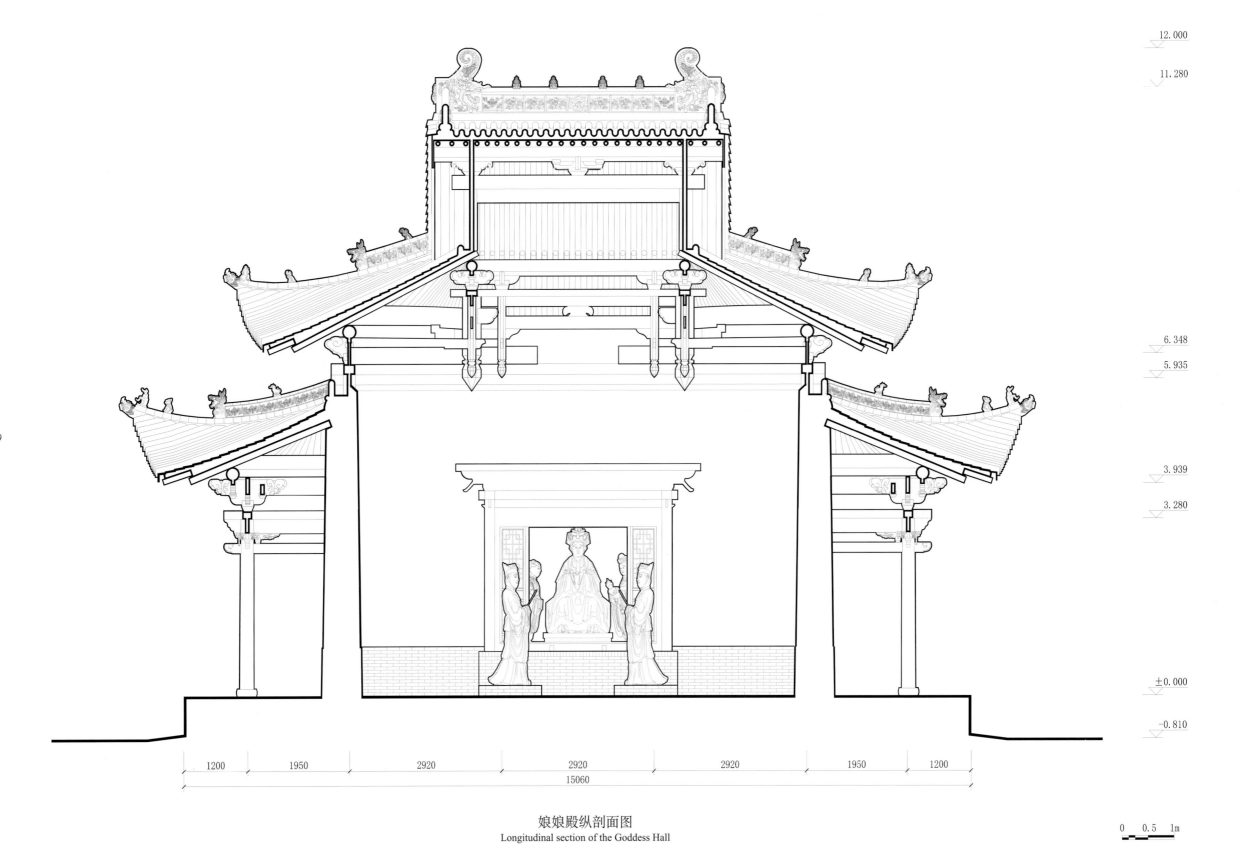

12.000

11.280

6.348

5.935

3.939

3.280

±0.000

-0.810

| 1200 | 1950 | 2920 | 2920 | 2920 | 1950 | 1200 |

15060

娘娘殿纵剖面图
Longitudinal section of the Goddess Hall

0 0.5 1m

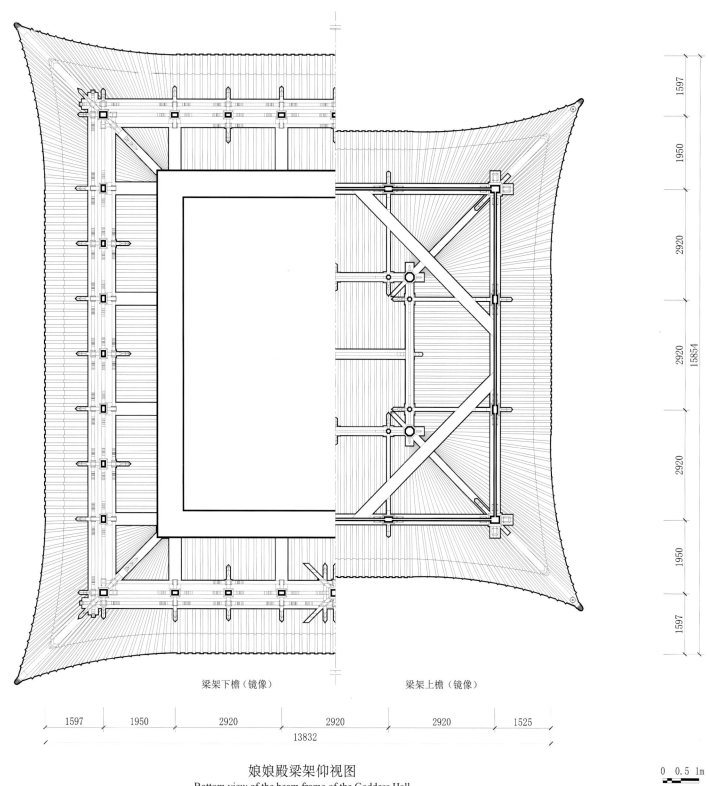

1597

1950

2920

15854

2920

2920

1950

1597

梁架下檐（镜像）　　　　梁架上檐（镜像）

1597　1950　2920　2920　2920　1525

13832

娘娘殿梁架仰视图
Bottom view of the beam frame of the Goddess Hall

0　0.5　1m

西太子殿
The West Crown Prince Hall

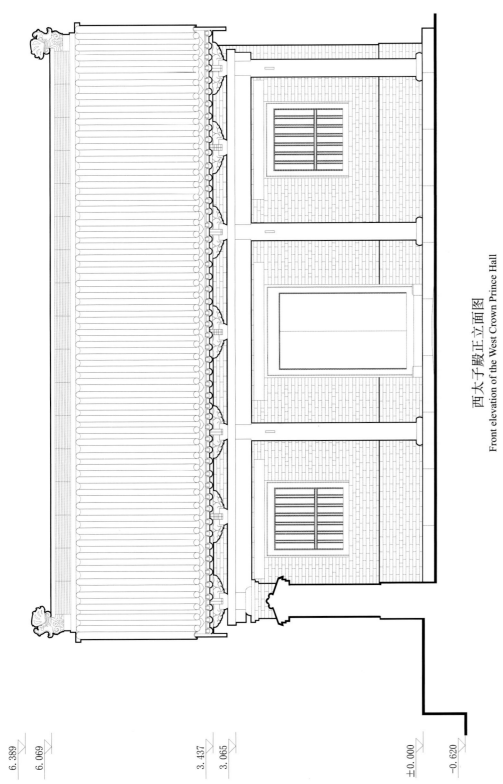

西太子殿正面图
Front elevation of the West Crown Prince Hall

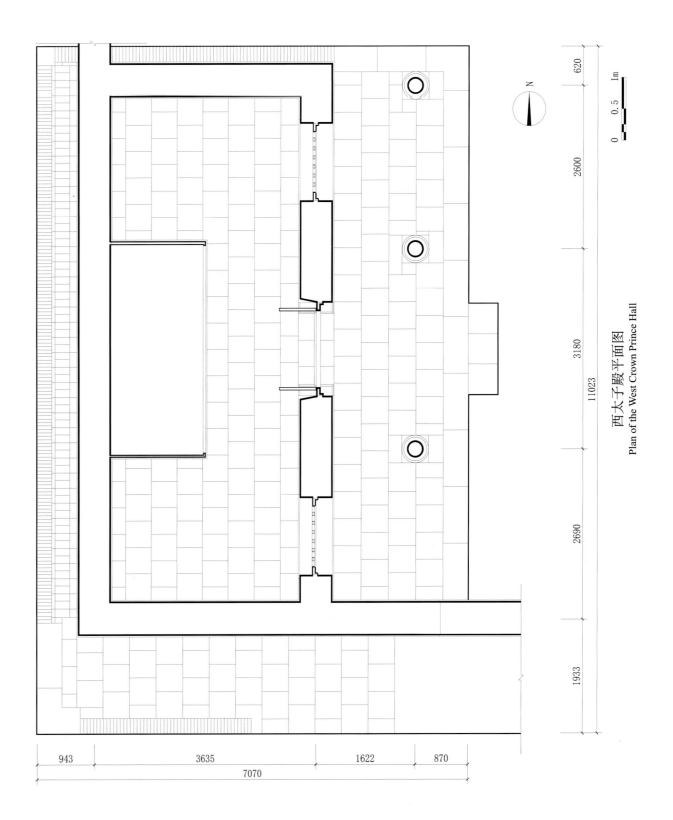

西太子殿平面图
Plan of the West Crown Prince Hall

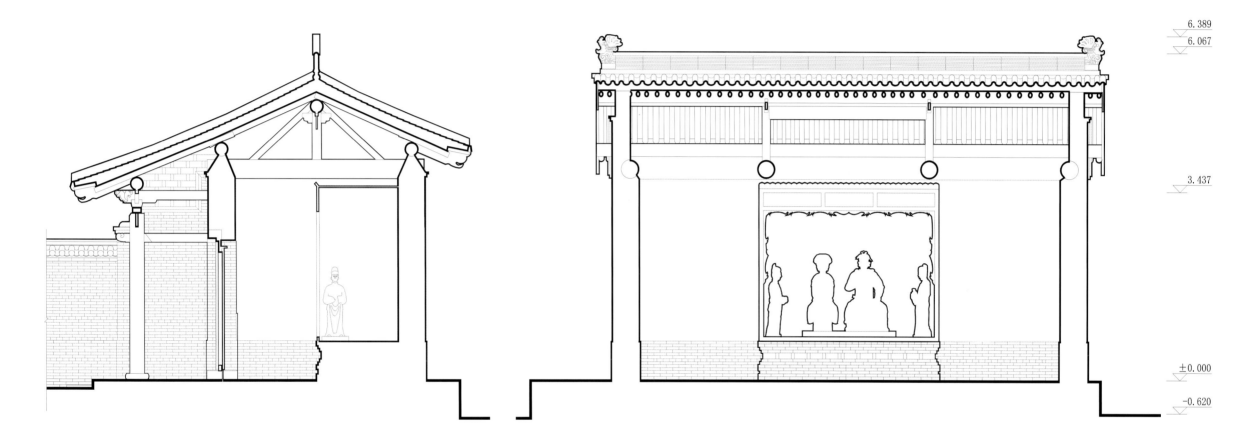

6.389
6.067
3.437
±0.000
-0.620

870 1622 3635 943
7070

1933 2690 3180 2600 620
11023

西太子殿横剖面图
Cross-section of the West Crown Prince Hall

西太子殿纵剖面图
Longitudinal section of the West Crown Prince Hall

0 0.5 1m

圣祖殿
The Sage Hall

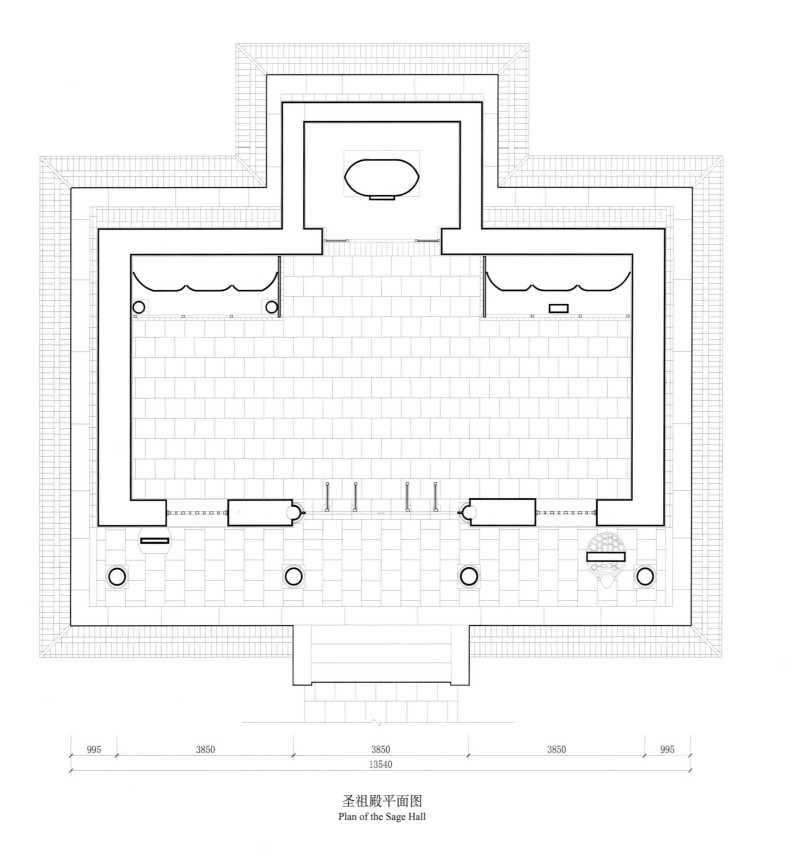

780

3091

1139

11662

5755

1350

1030

995 3850 3850 3850 995

13540

N

0 0.5 1m

圣祖殿平面图
Plan of the Sage Hall

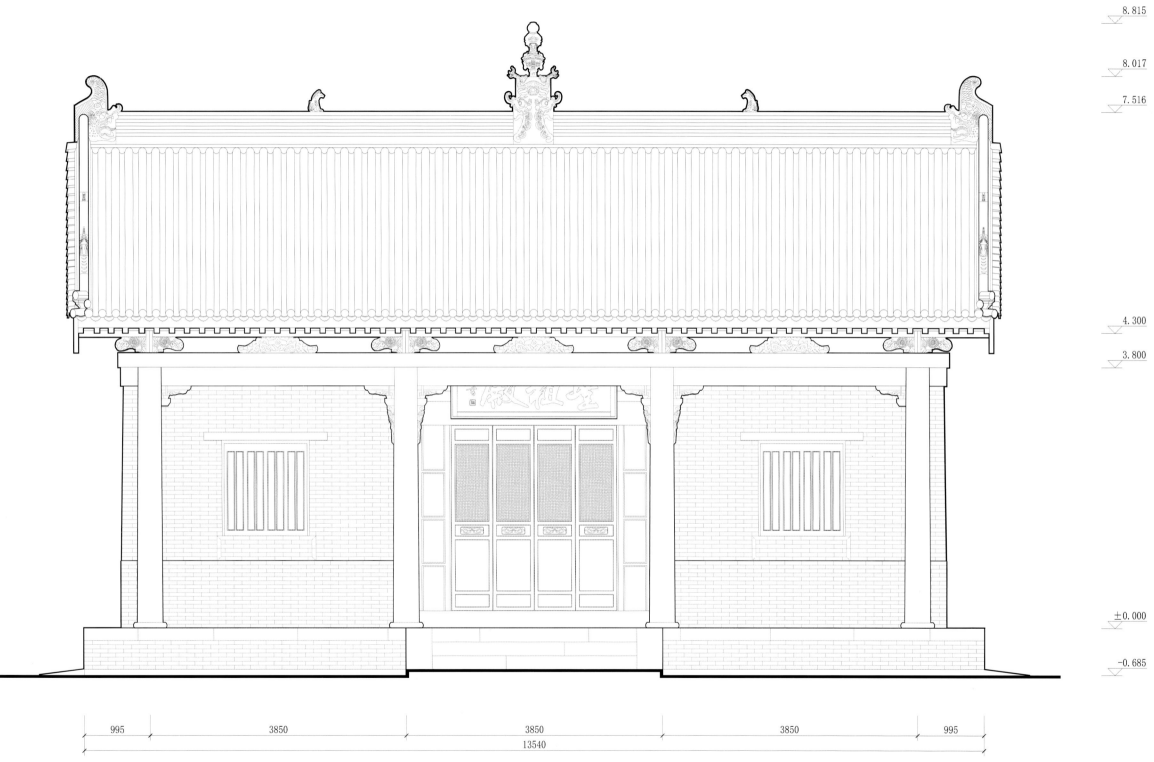

8.815

8.017

7.516

4.300

3.800

±0.000

-0.685

995　　3850　　3850　　3850　　995

13540

圣祖殿正立面图
Front elevation of the Sage Hall

0　0.5　1m

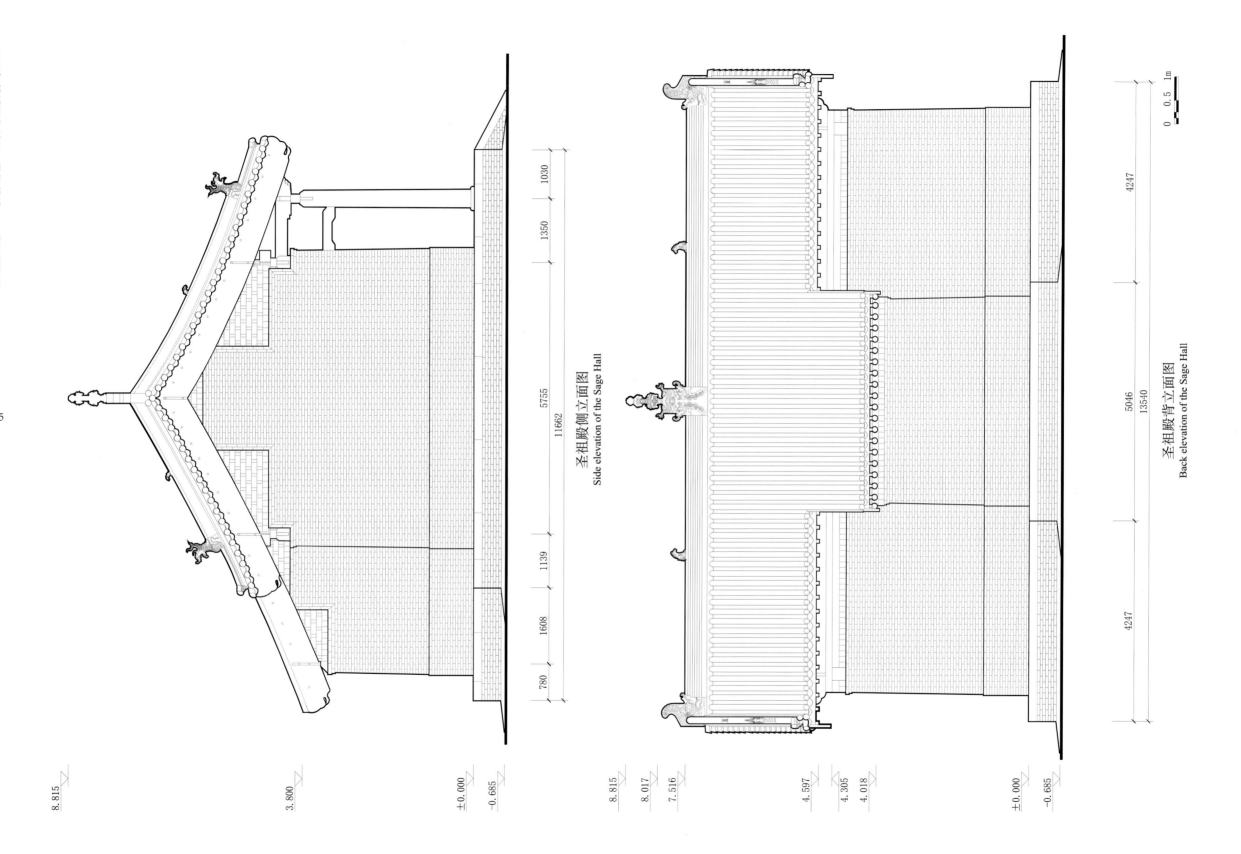

8.815

3.800

±0.000
-0.685

1030

1350

5755

11662

1139

1608

780

圣祖殿侧立面图
Side elevation of the Sage Hall

8.815

8.017

7.516

4.597

4.305

4.018

±0.000
-0.685

4247

5046

13540

4247

0 0.5 1m

圣祖殿背立面图
Back elevation of the Sage Hall

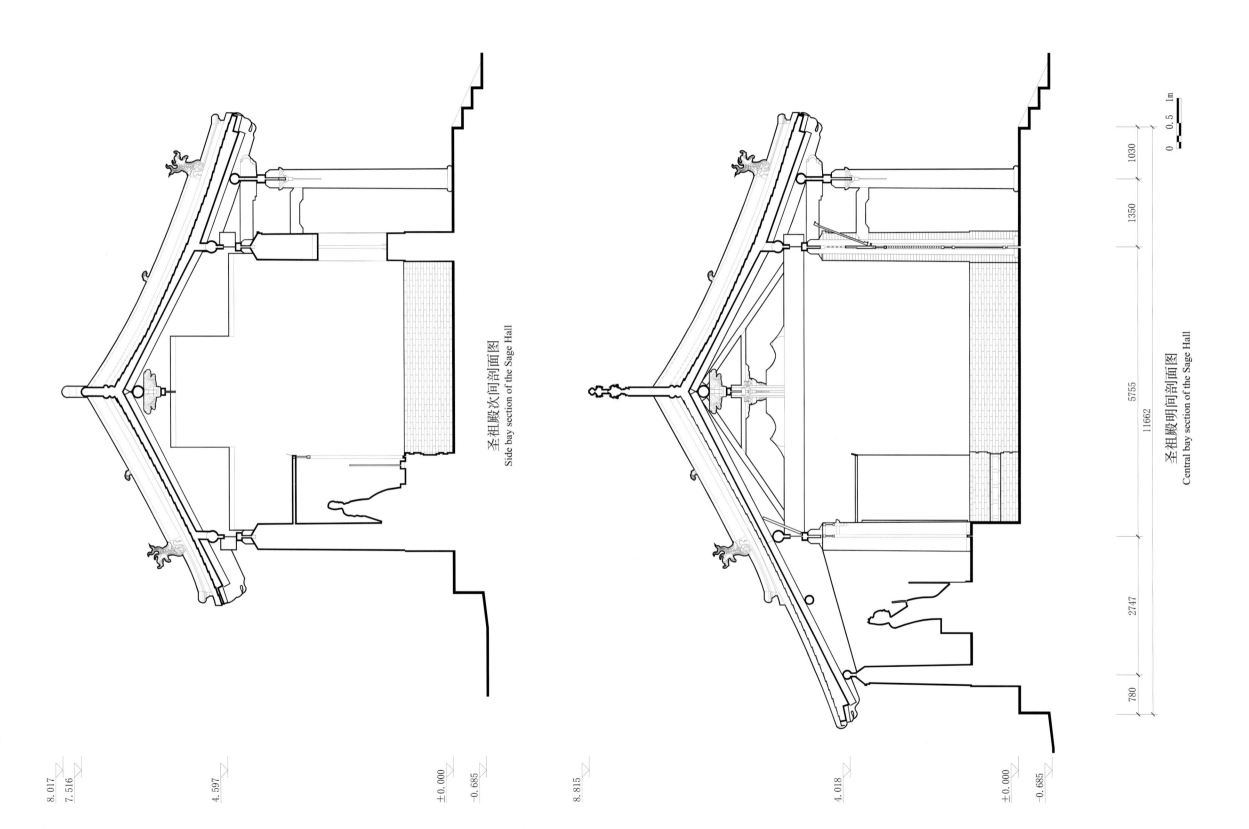

圣祖殿次间剖面图
Side bay section of the Sage Hall

圣祖殿明间剖面图
Central bay section of the Sage Hall

8.017
7.516
4.597
±0.000
-0.685

8.815
4.018
±0.000
-0.685

1030
1350
5755
11662
2747
780

0 0.5 1m

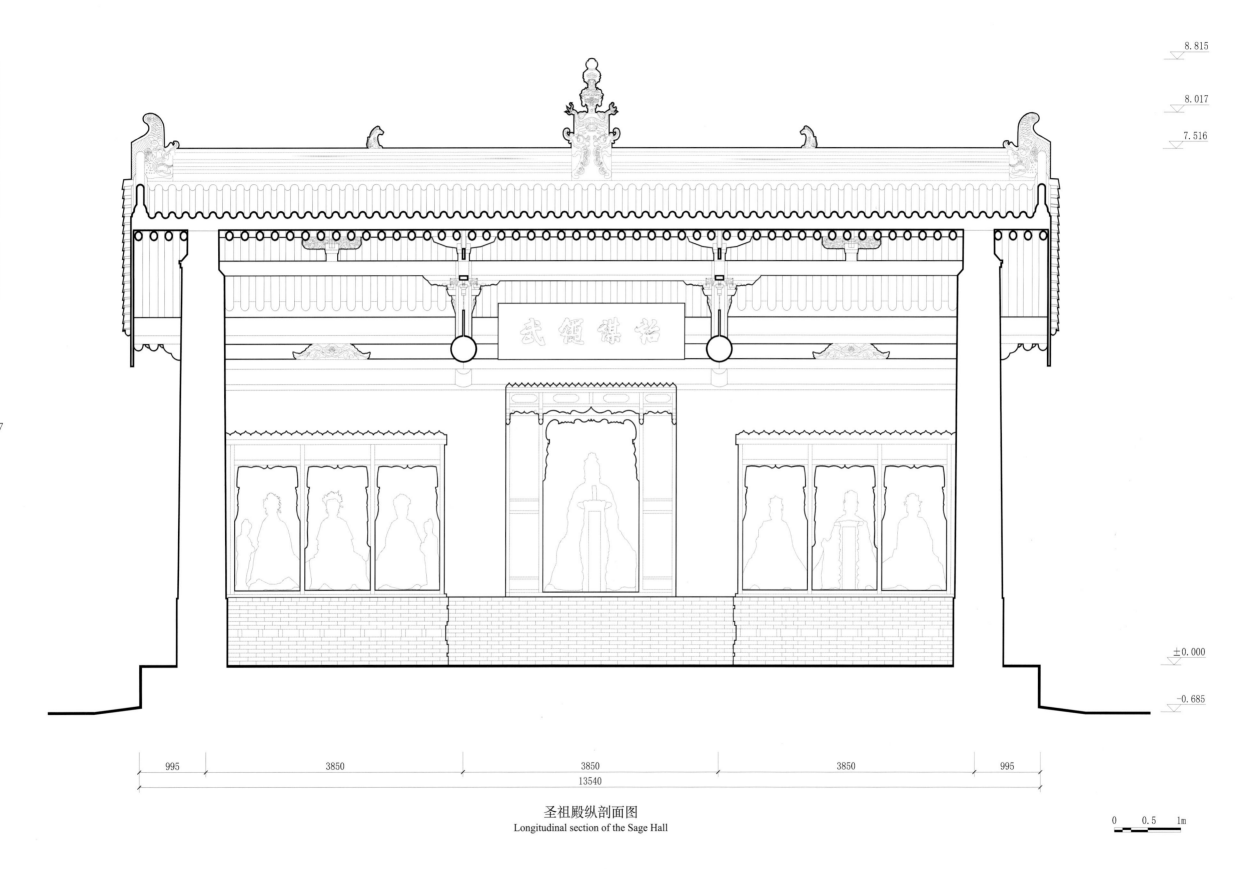

8.815

8.017

7.516

武襄揖裕

±0.000

−0.685

| 995 | 3850 | 3850 | 3850 | 995 |

13540

圣祖殿纵剖面图
Longitudinal section of the Sage Hall

0   0.5   1m

参考文献 References

[一] [明] 吕柟纂修，[清] 乔庭桂重修. 解州志（清康熙四年抄本）[M]. 山西省图书馆藏

[二] [清] 言如泗修. 解州全志（清乾隆二十七年）[M]. 解州关帝庙文管所藏复印件

[三] [清] 张镇辑，宋万忠，武建华 标点注释. 解梁关帝志 [M]. 清乾隆二十一年. 太原：山西人民出版社，1992

[四] [清] 马丕瑶，魏象干修，张承熊纂. 解州志（清光绪七年）[M]. 国家图书馆藏

[五] [清] 卢湛辑，[清] 汪潮绘，[清] 王尔臣刻. 关帝圣迹图 [M]. 上海：上海书店出版社，2006

[六] [清] 徐嘉清修，曲乃锐纂. 解县志（民国九年石印本）[M]. 国家图书馆藏

[七] 山西旅游景区志丛书编委会编. 关公文化旅游志 [M]. 太原：山西人民出版社，2006

[八] 刘致平. 内蒙古、山西等处古建筑调查纪略（下），建筑历史研究（第二辑）. 北京：中国建筑科学研究院建筑情报研究所，建筑工程情报资料第 8285 号，1982. 24-26

[九] 刘致平. 中国建筑类型及结构（新一版）[M]. 北京：中国建筑工业出版社，1987

[十] 柴泽俊. 中国古代建筑·解州关帝庙 [M]. 北京：文物出版社，2002

[十一] [美] 杜赞奇. 刻划标志：中国战神关帝的神话，载 [美] 韦斯谛编，陈仲丹译. 中国大众宗教 [M]. 南京：江苏人民出版社，2006. 07. p93-114

[十二] 卫龙，杨明珠主编. 山西解州关帝祖庙楹联牌匾 [M]. 北京：文物出版社，2006

[十三] 郑土有. 关公信仰 [M]. 北京：学苑出版社，1994

[十四] 蔡东洲，文廷海. 关羽崇拜研究 [M]. 成都：巴蜀书社，2001

[十五] 胡小伟. 关公崇拜溯源（上、下）[M]. 太原：北岳文艺出版社，2009

[十六] 马书田，马书侠. 全像关公 [M]. 南昌：江西美术出版社，2008

[十七] 刘海燕. 关羽形象与关羽崇拜的演变史论 [D]. 博士学位论文，福建师范大学，2002

[十八] 郭华瞻. 民俗学视野下的祠庙建筑研究——以明清山西为中心 [D]. 博士学位论文，天津大学，2011

参与解州关帝庙、常平关帝庙古建筑测绘与保护规划的人员名单

前期调查：
王其亨　丁垚

测绘指导：：（共 19 人）
王其亨　吴葱　曹鹏　王蔚（教师，共 4 人）
郭华瞻　林佳　王茹茹　沙黛诺　李哲　陈燕丽
傅东雁　陈学　曾引　王刚　曹苏　刘婷婷
刘翔宇　王蕊佳（研究生，共 14 人）
闫金强（2004 级本科生，共 1 人）

实习学生：（共 96 人）
蔡一鸣　薛文博　朱少华　陈俤俤　冯香梅　董苏　李丛　孙琦
朱文林　高晓雪　王云鹏　郭君君　丁旭　左龙　李良树　杨蒙
吴涵儒　柯隽　张晓旭　朱妙　曾波　吴晓旭　张梦恬　马庆阳
马欣　孙熹　李诗娴　常威　孙铭泽　王睿　李鹏　陈萌
刘威　任大任　陈骁　程功　李慧　宁之滔　息悦　朱一
张帆　赵琛　李睿　张一览　李鑫　钱辰元　齐飞宇　王珏蓉
郭璐瑛　张磊　蒋力克　于炜光　赵婧怡　丁寿颐　刘冠男　邓文博
刘婧　郭万里　陈越　牟文昊　蔡美玲　宋谷笙　陈宇冰　胡晓雨
郑阳洁　李驰宇

（以上解州关帝庙，2006 级本科生，共 66 人）

朱晓娟　庞博　李然然　王中钰　刘雪娇　李璐　张又天　孙众
林经宇　杨昕　郭浩　王彬　白继明　钟婷　陈志洋　高敏
王思源　段卓伟　陈阳　杨舒明　李源媛　梁颖　张谐　赵竹君
刘锦鑫　席慧敏　刘思朔　何涧　苏金

（以上常平关帝庙，2005 级本科生，共 29 人）

孙丽娜（2005 级本科生，共 1 人）

解州关帝庙文管所：（共4人）

卫 龙 郝平生 韩秋霞 郭 波

解州关帝庙测绘图整理：（共3人）

王其亨 郭华瞻 王蕊佳

解州关帝庙及常平关帝庙保护规划参编人员：（共21人）

王其亨 郭华瞻 朱 阳 王蕊佳 刘婷婷 王茹茹 叶 青 江 蓓
任 远（以上天津大学）
卫 龙 郝平生 何秀兰 王兴中 卫建斌 付文元 贾德彰 李甲寅
韩秋霞 陈春荣 王 锐 吴天福（以上解州关帝庙文物保管所）

解州关帝庙测绘图出版整理：（共14人）

郭华瞻 敖欣然 王 雨 高怡洁 唐元登 田 文 陈 颖 程煊桐
丁 哲 吕 茵 孟凡瑜 苏晓婉 郑宇航（以上北京交通大学）
卫 龙 何晋阳（以上解州关帝庙文物保管所）

List of Participants in the Surveying and Mapping Project and the Preservation Planning of Haizhou Guandi Temple and Changping Guandi Temple

Participants of Preliminary Research: WANG Qiheng, DING Yao

Instructors of the Surveying and Mapping Project: (19 persons in total)

WANG Qiheng, WU Cong, CAO Peng, WANG Wei (4 university faculties in total)

GUO Huazhan, LIN Jia, WANG Ruru, SHA Dainuo, LI Zhe, CHEN Yanli, FU Dongyan, CHEN Xue, ZENG Yin, WANG Gang, CAO Su, LIU Tingting, LIU Xiangyu, WANG Ruijia (14 postgraduates in total)

YAN Jinqiang (1 undergraduate of grade 2004 in total)

Intern Students: (96 persons in total)

CAI Yiming, XUE Wenbo, ZHU Shaohua, CHEN Didi, FENG Xiangmei, DONG Su, LI Cong, SUN Qi, ZHU Wenlin, GAO Xiaoxue, WANG Yunpeng, GUO Junjun, DING Xu, ZUO Long, LI Liangshu, YANG Meng, WU Hanru, KE Jun, ZHANG Xiaoxu, ZHU Miao, ZENG Bo, WU Xiaoxu, ZHANG Mengtian, MA Qingyang, MA Xin, SUN Xi, LI Shixian, CHANG Wei, SUN Mingze, Wang Rui, LI Peng, CHEN Meng, LIU Wei, REN Daren, CHEN Xiao, CHENG Gong, LI Hui, NING Zhitao, XI Yue, ZHU Yi, ZHANG Fan, ZHAO Chen, LI Rui, ZHANG Yilan, LI Xin, QIAN Chenyuan, QI Feiyu, WANG Yurong, GUO Luying, ZHANG Lei, JIANG Like, YU Weiguang, ZHAO Jingyi, DING Shouyi, LIU Guannan, DENG Wenbo, LIU Jing, GUO Wanli, CHEN Yue, MOU Wenhao, CAI Meiling, SONG Gusheng, CHEN Yubing, HU Xiaoyu, ZHENG Yangjie, LI Chiyu (responsible for Haizhou Guandi Temple, 66 undergraduates of grade 2006 in total)

ZHU Xiaojuan, PANG Bo, LI Ranran, WANG Zhongyu, LIU Xuejiao, LI Lu, ZHANG Youtian, SUN Zhong, LIN Jingyu, YANG Xin, GUO Hao, WANG Bin, BAI Jiming, ZHONG Ting, CHEN Zhiyang, GAO Min, WANG Siyuan, DUAN Zhuowei, CHEN Yang, YANG Shuming, LI Yuanyuan, LIANG Ying, ZHANG Xie, ZHAO Zhujun, LIU Jinxin, XI Huimin, LIU Sishuo, HE Jian, SU Jin (responsible for Changping Guandi Temple, 29 undergraduates of grade 2006 in total)

SUN Lina (1 undergraduate of grade 2005 in total)

Heritage Management Institute of Haizhou Guandi Temple: (4 persons in total)

WEI Long, HAO Pingsheng, HAN Qiuxia, GUO Bo

Participants of the Arrangement of Haizhou Guandi Temple Surveying and Mapping Drawings: (3 persons in total)

WANG Qiheng, GUO Huazhan, WANG Ruijia

Participants in the Preservation Planning of Haizhou Guandi Temple and Changping Guandi Temple: (21 persons in total)

WANG Qiheng, GUO Huazhan, ZHU Yang, WANG Ruijia, LIU Tingting, WANG Ruru, YE Qing, JIANG Bei, REN Yuan (from Tianjing University) WEI Long, HAO Pingsheng, HE Xiulan, WANG Xingzhong, WEI Jianbin, FU Wenyuan, JIA Dezhang, LI Jiayin, HAN Qiuxia, CHEN Chunrong, WANG Rui, WU Tianfu (from the Heritage Management Institute of Haizhou Guandi Temple)

Participants of the Arrangement and Publish of Surveying and Mapping Drawings of Haizhou Guandi Temple: (14 persons in total)

GUO Huazhan, AO Xinran, WANG Yu, GAO Yijie, TANG Yuandeng, TIAN Wen, CHEN Ying, CHENG Xuantong, DING Zhe, LYU Yin, MENG Fanyu, SU Xiaowan, ZHENG Yuhang (from Beijing Jiaotong University)

WEI Long, HE Jinyang (from the Heritage Management Institute of Haizhou Guandi Temple)

图书在版编目（CIP）数据

解州关帝庙 = HAI ZHOU GUANDI TEMPLE：汉英对照 /
王其亨主编；郭华瞻，王其亨编著 . — 北京：中国建
筑工业出版社，2019.12
（中国古建筑测绘大系，祠庙建筑）
ISBN 978-7-112-24545-1

Ⅰ . ①解… Ⅱ . ①王… ②郭… Ⅲ . ①寺庙－宗教建
筑－建筑艺术－运城－图集 Ⅳ. ① TU-885

中国版本图书馆CIP数据核字（2019）第284380号

丛书策划 / 王莉慧
责任编辑 / 李 鸽 刘 川
英文翻译 / 刘仁皓 郭 涵
书籍设计 / 付金红
责任校对 / 王 烨

中国古建筑测绘大系·祠庙建筑

解州关帝庙

天津大学建筑学院
山西解州关帝庙文物保管所　　合作编写

王其亨 主编 郭华瞻 王其亨 编著

Traditional Chinese Architecture Surveying and Mapping Series:
Shrines and Temples Architecture
HAIZHOU GUANDI TEMPLE
Compiled by School of Architecture, Tianjin University &
Guandi Temple Cultural Relics Preservation Bureau in Haizhou, Shanxi
Chief Edited by WANG Qiheng
Edited by GUO Huazhan, WANG Qiheng

*

中国建筑工业出版社出版、发行（北京海淀三里河路9号）
各地新华书店、建筑书店经销
北京海视强森文化传媒有限公司制版
北京雅昌艺术印刷有限公司印刷

*

开本：787毫米×1092毫米 横1/8 印张：40 字数：1060千字
2022年9月第一版 2022年9月第一次印刷
定价：**328.00**元
ISBN 978-7-112-24545-1
（35215）